GW00889092

Photovoltaic Power Generation

Solar Energy Development - Third Programme

Series Editor: W. Palz

Volume 3

Photovoltaic Power Generation

Volume 1: Photovoltaic Power Generation
Proceedings of the First Contractors' Meeting, held in Brussels, 18 April 1986

Volume 2: Solar Energy Applications to Buildings and Solar Radiation Data

Photovoltaic Power Generation

Proceedings of the
Second Contractors' Meeting held in
Hamburg, 16-18 September 1987

edited by

R. van Overstraeten

and

G. Caratti

Commission of the European Communities,
Brussels, Belgium

KLUWER ACADEMIC PUBLISHERS

DORDRECHT / BOSTON / LONDON

for the Commission of the European Communities

Library of Congress Cataloging in Publication Data

Photovoltaic power generation : proceedings of the second contractors' meeting held in Hamburg, 16-18 September 1987 / edited by R. van Overstraeten and G. Caratti.
p. cm. -- (Solar energy development--third programme : v. 3)
"Commission of the European Communities, Brussels, Belgium."
Includes index.
ISBN 9027726914
1. Photovoltaic power generation--Congresses. I. Overstraeten, R. van. II. Caratti, G., 1956- III. Commission of the European Communities. IV. Series.
TK2960.P467 1988
621.31'244--dc19 87-36102
CIP

ISBN 90-277-2691-4

Publication arrangements by
Commission of the European Communities
Directorate-General Telecommunications, Information Industries and Innovation, Luxembourg

EUR 11368

Published by Kluwer Academic Publishers,
P.O. Box 17, 3300 AA Dordrecht, The Netherlands.

Kluwer Academic Publishers incorporates the publishing programmes of
D. Reidel, Martinus Nijhoff, Dr W. Junk and MTP Press.

Sold and distributed in the U.S.A. and Canada
by Kluwer Academic Publishers,
101 Philip Drive, Norwell, MA 02061, U.S.A.

In all other countries, sold and distributed
by Kluwer Academic Publishers Group,
P.O. Box 322, 3300 AH Dordrecht, The Netherlands.

Printed in The Netherlands

C O N T E N T S

Photovoltaic Power Generation R&D Programme ix

AMORPHOUS SILICON SOLAR CELLS (AMOR)

1. A-Si Solar cells prepared by the glow discharge technique

. Development of the scientific and technical basis for integrated amorphous silicon modules

- Deposition techniques for a-Si
  J.P.M. SCHMITT, Solems S.A., France 1

- Physics of the pin device
  G. WINTERLING, MBB, F.R. Germany 13

- Improvement of the device structure
  G. WILLEKE, IMEC, Belgium 21

- Description of the research work carried out at SCK/CEN
  P. NAGELS, SCK/CEN, Belgium 33

- AES analysis of the SnO_2/a-Si:H inferface
  H.H. BRONGERSMA, A.S. VERLINDE, H.J. VAN DAAL, Eindhoven University of Technology, The Netherlands 38

- The French contribution to the project
  G. de ROSNY, CNRS, France 44

- Light trapping in a-Si:H solar cells by texturization of the glass substrate
  O. MARIS, P. VILATO, Saint-Gobain Recherche, France 48

- Photochemical process induced by a low pressure Hg lamp and a pulsed ArF excimer laser for depositing hydrogenated silicon films
  E. FOGARASSY, P. SIFFERT, C.R.N. Lab. Phase, France 53

- Vacuum UV photo CVD for amorphous silicon carbon alloys
  G.H. BAUER, University of Stuttgart, F.R. Germany 60

- Preparation of improved amorphous silicon cells with p-i-n structure based on new volatile hydrides and fluorides of silicon and germanium
  H. SCHMIDBAUR, Technical Univ. Munich, F.R. Germany 65

- Development of the scientific and technical basis for integrated amorphous silicon modules
  L. DELGADO, M.T. GUTIERREZ, P. ROMAN, CIEMAT-IER, Spain 70

. Low-gap alloys for amorphous silicon based solar cells

- Narrow band-gap alloys for the improvement of efficiency of amorphous silicon-based solar cells
J. BULLOT, Université Paris-Sud, France 76

- The Italian contribution to the project
R. GISLON, ENEA/FARE, Italy 90

- Characterization of amorphous silicon alloys through thermal H effusion
J.L. MORENZA, G. SARDIN, C. ROCH, University of Barcelona, Spain 105

- Characterization of hydrogenated amorphous silicon by photo deflection spectroscopy
J.W. METSELAAR, M. KLEEFSTRA, Delft Univ. of Technology The Netherlands 110

. Kinetic study of the deposition process of glow discharge a-Si:H for high efficiency solar cells

- The Greek contribution to the project
D. MATARAS, D. RAPAKOULIAS, S. KAVADIAS, Univ. Campus-Patras, Greece 114

- Characterization of $SiH_4:H_2$ plasmas for a-Si:H film deposition
G. BRUNO, P. CAPEZZUTO, F. CRAMAROSSA, Univ.of Bari, Italy 125

. Electro-Optical properties of a-Si;H/μC-Si;H and a-Si:C:H/μC-Si:C:H undoped and doped films produced by a TCDDC system
L.J.M. GUIMARAES, Universidade Nova de Lisboa, Portugal 131

. Evaluation of promising alternative a-Si deposition methods

. Development of a high deposition rate technique for device quality thin films of hydrogenated amorphous silicon
A. CHRISTOU, Research Centre of Crete, Greece 137

. Preparation of solar grade amorphous silicon for PV cells by means of an electrolytic process
M.V. GINATTA, Elettrochimica Marco Ginatta, Italy 140

. Hydrogenated amorphous silicon by disilane LPCVD
A. SHAMSI, TEAM srl, Italy 149

THIN FILM SOLAR CELLS FROM ALTERNATIVE MATERIALS (ALTERNA)

1. High efficiency crystalline silicon thin-film solar cells

. High efficiency thin-film solar cells on upgraded metallurgical grade silicon substrates
M. CAYMAX, IMEC, Belgium 157

. The French contribution to the project
M. RODOT, G. REVEL, CNRS, France 174

. Modelling of epitaxial cells on UMG silicon and feasibility of a high throughput epitaxial reactor
A. LUQUE, G. SALA, I.E.S.-E.T.S.I., Spain 180

. Growth and characterization of poly-Si ingots
D. MARGADONNA, R. PERUZZI, R. SPOSITO, Italsolar, Italy 184

. Growth and characterization of poly-Si ingots
S. PIZZINI, A. SANDRINELLI, D. NARDUCCI, M. BEGHI
University of Milano, Italy 191

2. Thin film solar cells based on II-VI and ternary chalcopyrite semiconductor materials

. Thin film solar cells based on the chalcopyrite semiconductor $CU(GA,IN)SE_2$
H.W. SCHOCK, University of Stuttgart, F.R. Germany 199

. Research and development on sprayed $CdS-CuInSe_2$ thin film solar cells
M. SAVELLI, J. BOUGNOT, S. DUCHEMIN, V. CHEN, J.C. YOYOTTE,
Université des Sciences et Techniques du Languedoc, France 205

. $CuInSe_2/CdS$ thin film solar cells by R.F. sputtering
N. ROMEO, University of Parma, Italy 210

. Electrochemical study of chalcopyrite semiconductors
J. VEDEL, CNRS, France 214

. Thin film of copper indium diselenide for PV devices
R. HILL, Newcastle Polytechnic, United Kindgom 220

. Development of CdSe thin film solar cells with regard to the requirements for the use in tandem structures
H. RICHTER, Battelle-Institut, F.R. Germany 225

. The fabrication of merocyanine dyes for use as PV materials
R.W. BUCKLEY, Twyford Church of England High School,
United Kindgom 230

3. III-V compound semiconductors for use in thin film cells or in monolytic multilayer cells

. Optimization of high efficiency multilayer solar cells based on III-V compounds

- MOCVD and characterization
  M.H.J.M. DE CROON, L.J. GILING, University of Nijmegen, The Netherlands 234

- MBE and physical analysis
  M. MEURIS, G. BORGHS, W. VANDERVORST, IMEC, Belgium 243

- Inhomogeneities in LEC GaAs substrates for solar cell applications
  L. ZANOTTI, C. FRIGERI, G. FRIGERIO, Istituto Maspec, Italy 249

. High efficiency multispectral cells based on III-V compound semiconductors
  C. VERIE, CNRS, France 254

. GaAs thin film solar cells obtained by vapour phase homoepitaxy (VPE) on reusable substrates

- The contribution by ENEA
  M. GAROZZO, ENEA/FARE, Italy 261

- The contribution by CISE
  C. FLORES, B. BOLLANI, D. PASSONI, CISE, Italy 274

. Study on the applicability of the ionized cluster beam deposition technology for GaAs thin film solar cells
  D. BONNET, Battelle Institute, F.R. Germany 279

. High efficiency III-V solar cells for use with fluorescent concentrators

- The German contribution to the project
  K. HEIDLER, Fraunhofer Gesellschaft, F.R. Germany 285

- The Belgian contribution to the project
  G. BORGHS, M. MAUK, P. DE MEESTER, S. Xu, R. BAETS, IMEC, Belgium 290

. Hydrogen effect on the electrical properties of amorphous gallium arsenide
  H. CARCHANO, University of Aix-Marseille, France 297

List of authors 303

PHOTOVOLTAIC POWER GENERATION R&D PROGRAMME

The Programme covers the period 1985-1988. The overall budget is approximately 20 million ECU. The programme was adopted as part of the sub-programme Solar Energy which in turn is part of the Non-nuclear Energy R&D programme.

The Commission of the European Communities is responsible for the implementation of the programme.

Contracts have mostly been concluded via calls for proposals published in the Official Journals C 69/3 of 16 March 1985 and C 146/2 of 13 June 1986. These calls for proposals comprised :

- Action C1: Amorphous silicon solar cells
- Action C2: Thin film solar cells from alternative materials
- Action C3: Photovoltaic pilot programme

Action C1: Amorphous silicon solar cells (AMOR)

The main thrust of this action is based on projects concerning a-Si solar cells prepared mainly by the glow discharge technique, complemented by a small number of projects on promising alternative a-Si deposition methods.

a) a-Si solar cells mainly prepared by the glow discharge technique

The silane glow discharge and related techniques have been identified as today's most reliable and promising preparation methods. In this field, two large projects are being supported which include two leading European a-Si solar cell manufacturers (Solems of France and MBB of West Germany) and major research institutions. These projects, which focus on the development of a European scientific and technical basis for integrated single and multi-junction a-Si modules, rely on strong German, French and Italian collaborative initiatives. In addition, support is given to a more fundamental and detailed study of the glow discharge deposition process.

b) Evaluation of promising alternative a-Si deposition methods

In this field a small number of experimental studies are being carried out, which are devoted to promising alternative deposition methods. These techniques include the ionized cluster beam (ICB) method, disilane LPCVD as well as a novel electrolytic process.

Action C2: Thin film solar cells from alternative materials (ALTERNA)

This action can be subdivided according to the different materials involved, covering topics on group IV semiconductors as well as III-V and II-VI compounds.

a) High efficiency crystalline silicon thin film solar cells

In the field of high efficiency thin film solar cells on upgraded metallurgical grade (UMG) silicon substrates, a new approach has been initiated by five research groups including IMEC (Belgium) as project coordinator. This project aims at the realization of 12% efficient cells.

b) Thin film solar cells based on II-VI and ternary chalcopyrite semiconductor materials

A large multinational project headed by the University of Stuttgart (FRG) aims at the evaluation of different deposition methods for $Cu(Ga, In)Se_2$ solar cells. In another project CdSe thin film structures are developed for use in tandem cell configurations. It is interesting to note a small effort on the fabrication and performance of merocyanine CdS solar cells in a school laboratory.

c) III-V compound semiconductors for use in thin film cells or in monolytic multilayer cells

Considerable attention is paid to the area of III-V compound semiconductors. Several different preparation methods are being investigated including MBE, MOCVD, IMLPE and ICB among others. Cell technologies range from monolytic multilayer cells to VPE deposited thin film cells on reusable substrates as well as structures based on cr-Si substrates. The different activities are coordinated by a Euro GaAs PV working group headed by the LPSES group of the French CNRS.

Action C3: Photovoltaic pilot programme

The aim of this action is the development of innovative photovoltaic power systems, with emphasis both on the single components and the overall system.

During its 1979-1984 solar energy R&D programme the Commission has financed the erection of sixteen pilot plants distributed over eight Member States.

Further research is now needed to upgrade their performance as well as to resolve specific problems occurring in some parts of these plants.

This programme includes in particular concerted actions on advanced subsystems, such as battery control, power conditioning, modules etc.

DEVELOPMENT OF THE SCIENTIFIC AND TECHNICAL BASIS FOR INTEGRATED AMORPHOUS SILICON MODULES

DEPOSITION TECHNIQUES FOR AMORPHOUS Si

Contract number : EN3-S-0053-F

Duration : 36 months 1 September 1986-31 August 1989

Total budget : 36 MFF CEE Contribution : 8.9 MFF

Head of project : Pr. J.P.M. SCHMITT

Contractor : SOLEMS S.A.

Address : SOLEMS
3, rue Léon Blum - Z.I. Les Glaises
91120 PALAISEAU
FRANCE

Summary

We report here the progress on the design and optimization of the plasma deposition technique for the manufacture of large amorphous silicon PV panels. A large experimental deposition chamber capable of handling 1.2 x 0.6 m^2 substrate was built and tested. Finite wave length effect was found to become a limiting factor for deposition uniformity. This effect becomes effective at 13.56 MHz driving frequency when the machine size is above 1 m. Satisfactory uniformity was achieved by lowering the driving frequency. Parallely the relation between gas flow configuration and gas composition uniformity is analyzed. Solems has designed and tested a single chamber machine with no load lock with many technical features providing low cost solutions to the process problems usually solved by a multichamber design.

1. INTRODUCTION

Amorphous silicon PV panel mass production will require to master plasma chemical deposition in terms of large sizes, cost, maintenance and all other problems related to industrialization. Since plasma deposition is a novel technique, the development of all this production related know how involves a considerable technical research effort.
The major problems related to the design of a production deposition machine are the following :

- deposition should be uniform on very large area substrate (typical dimension 1 meter) ;
- the deposited amorphous silicon should have good electronic properties (density of state of the order or less than 10^{16} cm^3/eV) and very low impurities concentrations (for example oxygen atomic concentration should idealy be less than 0.01 %) ;
- the film stress should be limited, the density of ponctual defects (particulates) should remain reasonable (less than 1 per 100 cm^2) ;
- dopant level control should be stable and efficient ;
- silane consumption should remain reasonably efficient ;
- financial cost being important the machine productivity should be high hence deposition rate optimized ;
- downtime due to maintenance should be reduced to a minimum.

We present here some results on the R & D effort addressed to the above mentioned problems. An original single chamber was designed. This machine will be made available on the market for R & D purposes by a process machine company.
Finally the maintenance problem is considered. Plasma cleaning based on a fluorine containing etchant gases is studied and evaluated.

2. LARGE AREA DEPOSITION

2.1. Large test machine in Solems

A large deposition machine was built in Solems in order to study the problems related to large area deposition. The reactor is shown in Fig. I. This reactor is placed in a large vacuum vessel. The silane plasma is confined in the reactor which is a gas tight box with a flat rectangular shape. The two sides consist of two 1.2 x 0.6 m^2 substrates extending all the way to the reactor box edges. The plasma is driven by a very transparent mesh grid in the middle of the inter substrate gap. The reactive gas is flowing laminarly from the top to the bottom of the box then expand through throttling apertures in the main vacuum vessel and is finally pumped by a root blower. Two grounded counter electrodes are backing the substrates. In this experimental system devoted to plasma study no heating was provided, nor any doping possibility.

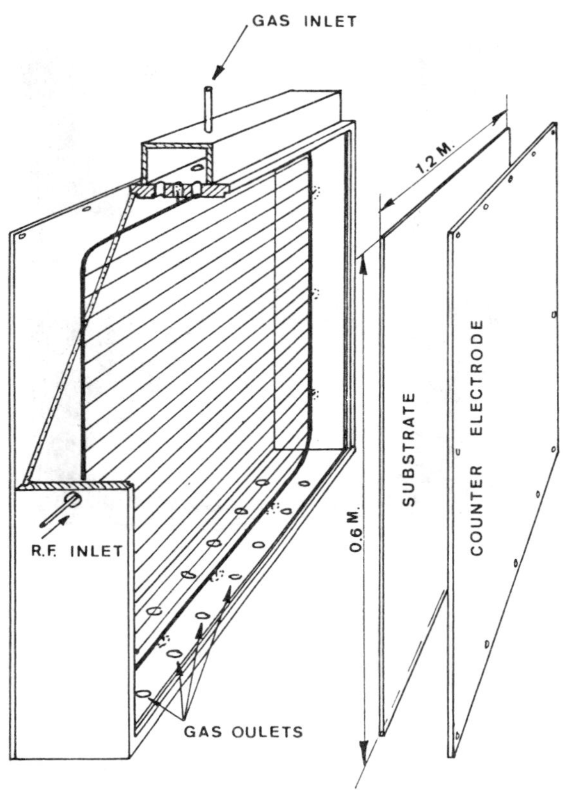

Fig. I. Schematic diagram of the large area plasma reactor. This box like reactor is lying in a larger vacuum vessel. It contains two 1.2 x 0.6 m^2 substrates.

2.2. <u>Finite wave length effect</u>
The machine was initially operated with a 13.56 MHz generator. The deposition rate was monitored in real time by optical transmission interference though both substrates. When the deposited film thickness in the central region was of the order of 500 nm the test run was stopped and the substrates unloaded for uniformity measurement. For this a photograph of the substrate was taken when illuminated by diffuse light from a low pressure sodium lamp. Very well defined fringes are observed, they map the contours of the amorphous silicon film optical thickness. Two photographs are shown in Fig. II. As observed in the top photograph of Fig. II, when deposited with a 13.56 MHz driving frequency the film was showing a systematic thickness deficiency on the RF inlet side. A tentative explanation is based on the finite wave length effect as following. Note that :

- the plasma impedance is only a perturbation with respect to the unloaded setting of the antenna matching box ensemble ;

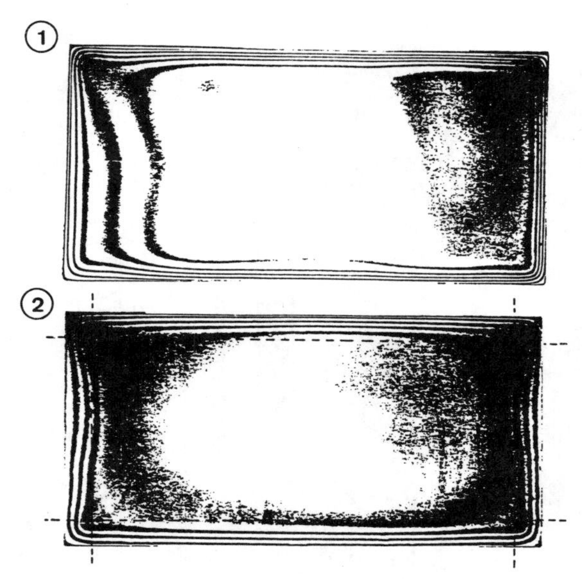

Fig. II. Interferometric mapping of amorphous silicon film thickness on 1.2 x 0.6 m^2 substrate.
1) Run at 13.56 MHz ; 2) Same run but at half the frequency. The dotted line correspond to a cut off at 7 cm from the plasma box side walls.

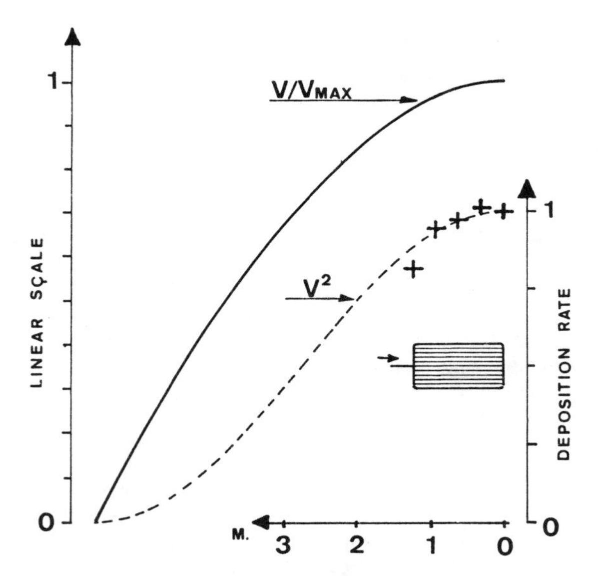

Fig. III. Standing wave pattern for a 13.56 MHz wave on an open end coaxial line superposed to the plasma antenna. Locally injected power (dotted line) is compared on the same relative scale to the deposition rate as measured from uper case of Fig. II (crosses).

- the power deposited in a section of the plasma is proportional to the square of the local peak voltage of the radiofrequency ;
- the deposition rate is rougthly proportional to the local power injected in the plasma.

The plasma antenna in our machine geometry can be identified to a coaxial line ending by an open circuit. When driven by the machine box a standing wave shall be present along the antenna with a voltage maximum at the open end. For the 13.56 MHz driving frequency the corresponding quarter of a wave length is around 5.5 m. The variation of locally injected power is then

$$\cos^2 (\pi\ 1.2/11) \simeq 0.89,$$

hence, although the antenna is typically 20 times shorter than the free space wave length a variations of 10 % of the deposition rate is to be expected. The experimental results of Fig. II are reported in Fig. III where the trend in non uniformity is very well correlated with a prediction based on a very simple standing wave model.

This interpretation is further confirmed by the second experiment seen in the lower part of Fig. II where the substrate was deposited in similar conditions except for the driving frequency which was divided by 2. The corresponding non uniformity is expected to be reduced by a factor 4 and indeed Fig. II exhibits a strongly improved uniformity. The example shown in the lower part of Fig. II has a satisfactory uniformity except for the margins which extend 7 cm from the side walls of the rectangular plasma box. This should be regarded as a very satisfactory edge effect when compared to the intersubstrate gap of 10 cm in the plasma box.

2.3. Gas compositional uniformity

When decomposed by the plasma, the silane follows grossly this path :

$$SiH_4 \longrightarrow Si + 2H_2$$

Some minor fraction of the hydrogen remains in the solid where it plays a very important role, but for the overall gas phase balance as silane is consumed it is substituted by approximately twice the amount of hydrogen. Moreover higher order silanes can be important gaseous intermediates in the decomposition process in particular for large electrode substrate gap ℓ and large process pressure P, typically for

$$\ell P \gtrsim 1 \text{ Torr cm}$$

Gas composition should vary in the reactor, however as one expects to keep the deposition rate and the thin film composition and structure as uniform as possible, the gas phase composition should be kept as uniform as possible. The main distribution schemes are shown in Fig. IV, laminar flow is the concept used in Solems design, many laboratory scale deposition systems are relying on diffusion to feed the central region of the reactive discharge, finally many large machine use a porous electrode for the gas distribution.

In both laminar flow and porous electrode concepts, one easy solution to the gas compositional uniformity problem is to waste silane.

Indeed if the silane relative concentration is n_o at the gas inlet and $n_o - \Delta n_o$ at the pumping aperture, assuming a simple volumetric pumping the maximum silane compositional relative variation is $\Delta n_o/n_o$, this ratio is also the reactor silane efficiency. If, for a given deposition rate the silane flow is strongly increased, the gas composition uniformity increases accordingly but this will be at the expense of large silane waste. Fortunately diffusion can further reduce the actual gas compositional variations $\Delta n/n_o$ by intermixing along the reactor.

For Solems' geometry a simplified solution of the diffusion equation brings

$$(\Delta n/n_o) = (\Delta n_o/n_o) \quad (e^{-x} - 1 + x)/x,$$

with $x = L/\lambda$, $\lambda = D/v$,

where L is the reactor length along the gas flow and D and v are respectively the gas diffusion coefficient and average flow velocity. Note that, for a given molar flow, the inter-diffusion length λ does not depend on pressure. The gas intermixing factor is evaluated in Fig. V. The range of parameters covered by the machines in Solems corresponds to the box on the curve, hence it appears that in typical conditions, for a laminar flow regime, diffusion reduces the extreme gas compositional variation estimate by a factor 2 to 3. Altogether for a 60 cm wide machine this corresponds to composition variations between 5 and 10 %, a value which is marginally tolerable. This effect appears to be particularly concerning for larger machines or machines where deposition rate is increased. With respect to gas compositional uniformity, without going through the mathematics the results are the following : for the diffusion scheme if the uniformity is extremely good for small size it will drastically deteriorate for large machine and/or fast deposition rate ; most promissing is the porous electrode concept where, in theory, all variations can be compensated by the uniform distribution of fresh gas injection, however the pressure drop due to finite impedance along the reactor will remain and provide potential inhomogeneities.

The analysis above was concentrating on silane partial pressure variations and the behaviour of higher silanes and potential powder precursors were not considered. For lack of knowledge on the kinetic chemistry associated to these species the problem remains untractable. For powder precursor we can however estimate the following trends. The laminar flow scheme will limit powder formation by "flushing" of the heavy gases. The diffusion concept is probably the worst case with respect to gas phase nucleation, the porous electrode being a somewhat intermediate situation.

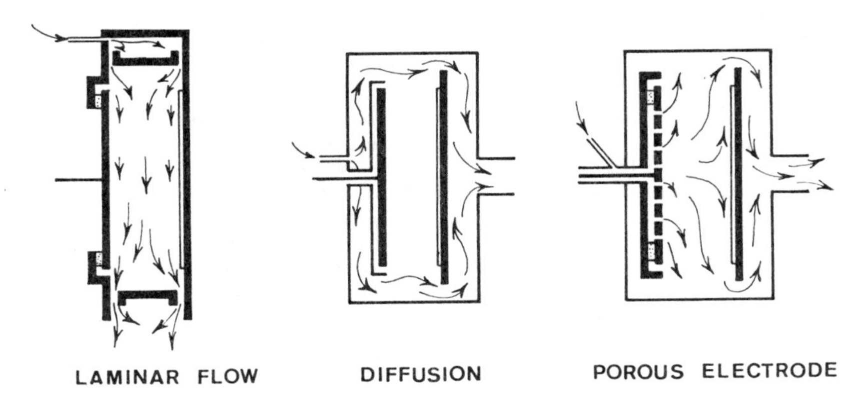

Fig. IV. Schematic of various gas flow patterns in typical silane plasma reactors.

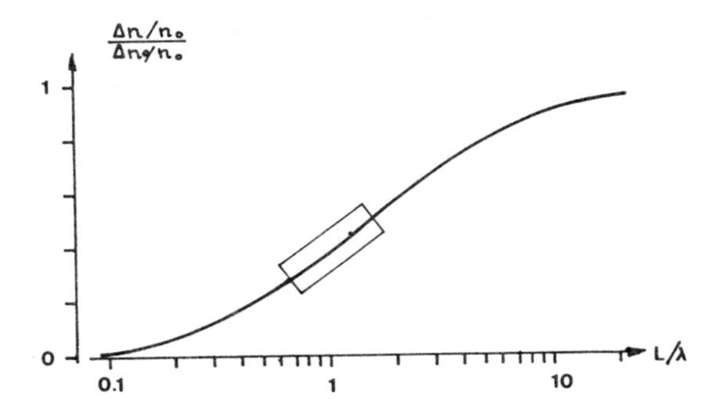

Fig. V. Effect of gas diffusion for reducing the gas compositional maximum non uniformity in a laminar flow reactor. The ratio L/λ is defined in the text. The central box corresponds to the standard machine range of parameters.

2.4. Silane efficiency

Due to the very specific geometry of the plasma box in the 1.2 x 0.6 m^2 machine, assuming 7 cm for the substrate frame width more than 54 % of the wall surface exposed to the plasma is bare substrate. The RF antenna cross section being very small, typically half of the deposited silane is actually used to build up a solar cell. In a porous electrode geometry or a planar electrode geometry, the situation with respect to silane consumption efficiency is systematically less favourable since one side of the planar plasma gap is a metallic electrode where the deposited silicon is wasted and will have to be removed later. From our estimation, cumulating the constraint on gas compositional uniformity and geometrical effects the best silane efficiency which can be achieved in the various geometry is 5 % in a diffusion scheme, 10 % in a porous electrode geometry and about 15 % in Solems' laminar flow geometry. Better silane overall consumption efficiency can probably be achieved but in rather extreme conditions such as slow deposition rate and low flow rate, both options were rejected because they are prone to bring film contamination. In any case for a 10 % silane efficiency and assuming a conservative price for silane(about 250 $ per kg) the incidence for the silane cost on a 2 $ per watt cell is only 0.5 %. Further improvement is not a key issue.

3. MACHINE AND PROCESS DESIGN

3.1. Machine maintenance and plasma cleaning

One very important problem to be solved when running for production a plasma deposition machine is the chamber maintenance. Amorphous silicon deposits on the surfaces permanently exposed to the plasma where it could accumulate and build exceedingly thick layers. Then stress build up leads to chipping and flaking, hence particulate contamination of the film. When a load lock is present in a machine the typical thickness which can be accumulated on the wall without flaking is larger than in a batch system because regular opening to air seems to reduce the cohesion of the successive layers. But in any case, at a given repetition rate, the amorphous silicon, including alloys and dopants have to be removed from the walls. Three options are to be considered :

- Disposable walls : the exposed parts are covered by a foil which is renewed during maintenance.
- Wet cleaning : the chamber is open, dismantled and the parts are cleaned in a reactive bath, rinced, dried and reassembled.
- Dry cleaning : a plasma of fluorine rich gas mixture is run in the chamber, silicon is etched to form gazeous silicon tetrafluoride which is subsequently pumped.

Solems' technology was so far based on wet cleaning but dry cleaning appears in principle more attractive because of its straight forward compatibility with the deposition process itself and because it can be done without even opening the reactor chamber. Development work was therefore initiated in order to define and evaluate a dry cleaning process.

Three gases were considered, CF_4-O_2 mixture, NF_3 and SF_6. Sulfure hexafluoride was ruled out rapidly because sulfur is known to be very poisonous to amorphous silicon. A systematic analysis of the etching rate was performed on a Solems 30 x 30 cm^2 machine using a laser reflecto-meter to monitor in real time the rate of silicon removal. Typical results are illustrated in Fig. VI.

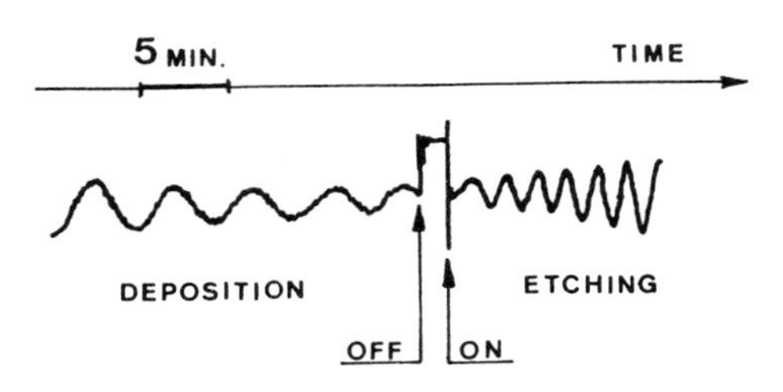

Fig. VI. Helium-Neon laser reflectronic measurement as observed in real time first while depositing amorphous silicon by a silane plasma, then removing the film by nitrogen trifluoride plasma etching.

Comparison between CF_4-O_2 and NF_3 demonstrated a very strong advantage for NF_3 (typically 5 times faster etching rate in equivalent conditions). As a consequence more emphasis was put on the NF_3 process despite of drawbacks such as corrosion activity or toxicity. Systematic scanning of the parameters led to an etching process where etching rate is 4 times faster than deposition. The limitation is not intrinsic but rather due to voluntary restriction in RF power (to limit electrode heating) and gas flow rate (due to pump capacity). Inspection of the machine has revealed that only a limited amount of materials (metallic alloys and insulating materials) are compatible with the sereve environment of a reactor regularly switching from a the strongly reducing atmosphere of a silane plasma to the exceedingly corrosive and oxidizing fluorine plasma. Obvious problems were also identified at the rotary pump level. The solution explored today relies on two separated pumping stands for hydride and fluoride respectively. Etching of Si-C alloys with pure trifluoride is found to be slow and unreliable, one solution is based on the addition of a small amount of oxygen to the etching gas.
Finally the analysis of possible contamination of the amorphous silicon photodiodes by various chemical left over was started. Indeed some slow degasing material is observed in mass spectroscopy after a dry cleaning process (see Fig. VII). Preliminary results show that in Solems single chamber batch machines, if only one hour of pumping is allowed between dry cleaning and the pin deposition process, some decay is observed on the first deposited diode (mostly affecting fill factor).

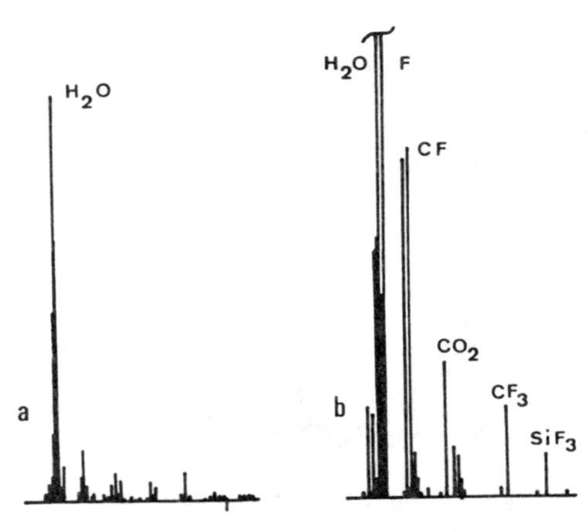

Fig. VII. Residual gas mass spectrum before dry etching (a) and 2 days of pumping after a NF_3/O_2 plasma cleaning (b).

If one night of pumping is allowed no significant effect is observed. This program will continue at the pilot level for qualification of the dry cleaning process (long term evolution, pump problems), at the research level to identify the mechanism of diode degradation (fluorine, oxygen, others) and at the technical level to conceive methods for speeding up degasing of dry cleaning left over (baking, plasma).

3.2. Design of an advanced single chamber machine

One of the attractive feature of Solems plasma box concept is related to the fact that the reactive zone is fully enclosed by substrates and walls except for the gas outlet. This geometry already ensures that no reactive species can escape from the plasma and either deposit a loosely bounded material in a cold hidden corner of the vacuum vessel or chemically extract some impurity which is then likely to come back in the chamber.
In the machine presently used in Solems the reactive gases after flowing in the plasma box expand in the main vacuum chamber and then are pumped. There is a moderate pressure drop between the plasma chamber and the surroudings vacuum vessel, therefore plasma contamination due to degasing of the outer vessel is reduced by the pressure differential. An improved geometry uses two different pumps, one directly connected via a throttle valve to the plasma box and one large pump operating at full capacity on the main chamber. If the plasma box is of a relatively good mechanical design it will leak only moderately in the main vessel and a large pressure differential will be maintained. Modeling indicates than a reduction by a factor 100 of the degasing background pressure can be achieved in the reactive zone. Such a machine was designed by Solems in collaboration with a plasma processing machine company NEXTRAL. The main features of the machine lay out are shown in Fig. VIII. Note that the reactive gas lines can be independantly flushed before being injected in the reactor and this even during a deposition process. The first two machines of this kind will be delivered in October 87 and will be immediately tested for the deposition of photovoltaic devices and thin film transistor.

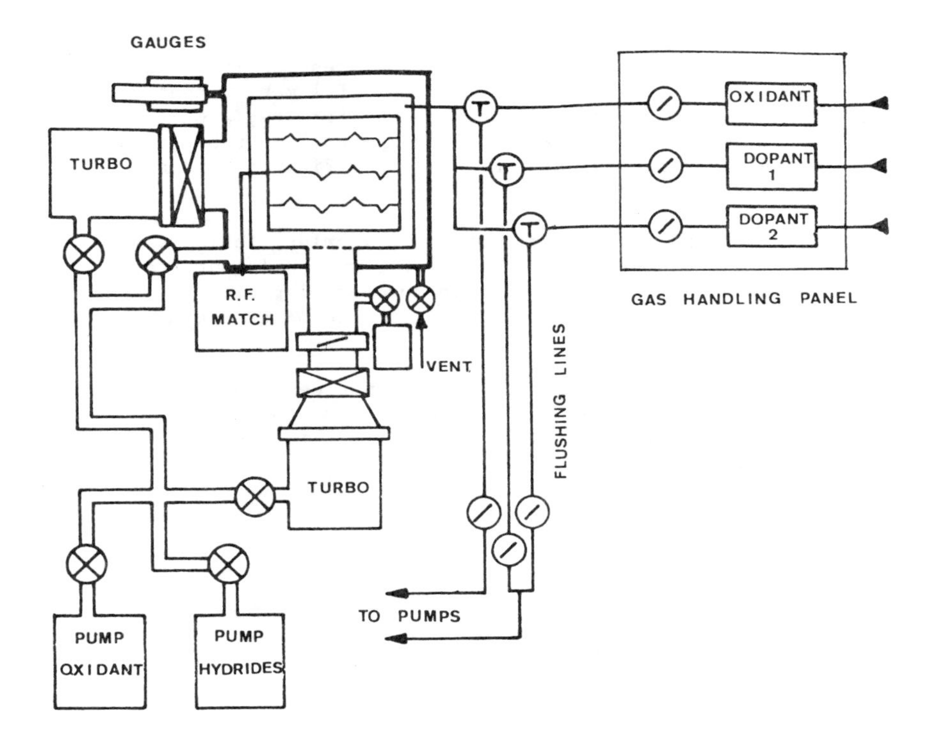

Fig. VIII. Schematic diagram of the SOLEMS/NEXTRAL machine with double pumping and flush lines.

4. CONCLUSIONS

The SOLEMS design for a plasma deposition reactor has three original features, a gas tight double substrate plasma box, a fine mesh central electrode, a laminar flow for reactive gases. This design was extensively tested and evaluated for the production of 30 x 30 cm^2 photovoltaic panels and also for its ability to perform as well on 1 x 0.5 m^2 substrates.

In exploring the large area deposition one limit was found due to finite wave length effect, it becomes sensitive at 13.56 MHz for antenna size of the order of 1 meter. Another limit is related to gas compositional uniformity, a problem in which the laminar gas flow scheme compares favourably to diffusion distribution but remains less efficient than the porous electrode concept.

A plasma process based on nitrogen trifluoride etching was selected as a dry etching process for easy and labour reducing maintenance of the deposition chamber. Typically a chamber is cleaned in one quarter of the deposition time.

Finally an improved plasma box concept was designed where a double pumping system should reduce degasing contamination by 2 orders of magnitude. Two machines of this kind will be operational before the end of 1987.

DEVELOPMENT OF THE SCIENTIFIC AND TECHNICAL BASIS FOR INTEGRATED AMORPHOUS SILICON MODULES

PHYSICS OF THE PIN DEVICE

Contract number: EN 3 S - 0057 - D (B)

Duration : 36 months 1.3.87 - 28.2.90

Total budget : DM 5.428.609,36; CEC contribution DM 2.714.334,68

Head of project: Prof. Winterling for coordination of the activities of the various German partners
Dr. Kniffler for activities of MBB

Contractor : MBB, Energy and Process Technology

Address : 9, Hermann-Oberth-Straße, D 8011 Putzbrunn, West-Germany

SUMMARY

In the first part an overview on the main work done under part B of the joint research and development programme is given. In the second part experiments started at MBB are described.
In particular data on the fill factor are reported for glass/TCO/pin/Al cells prepared with a dopant profile within the central layer. It is found that the fill factor can be enhanced by introducing a phosphorus profile while it is decreased by introducing a boron profile.
The photo capacitance of glass/TCO/pin/Al cells is found to be enhanced by a constant boron doping of the central layer. Both experimental findings can be qualitatively understood by considering the internal electric field near the p^+-i interface.

1. In the first part of this report an overview will be given on the tasks being involved in Part B of the joint research programme as listed below:

B. PHYSICS OF THE PIN DEVICE

B1. Measurement of the internal field

B11. Electro reflectance: an alternating field is superimposed on the internal field and the reflectance modulation is detected by lock-in technique. CNRS, UNIVRSITY OF PARIS, PROF. G. DE ROSNY.

B12. Secondary electron energy measurement in scanning electron microscopy of the fracture edge of a pin junction. UNIVERSITY OF KAISERSLAUTERN, PROF. SCHMORANZER.

B2. Traps and recombination

B21. Time of flight of carriers. When adding delayed reverse field, this technique gives also some access to the recomination rate. CNRS, LGEP, DR. BAIXERAS, DR. D. MENCARAGLIA.

B22. Transient capacitance in the dark: density of gap states - CNRS, DR. MENCARAGLIA, and under illumination: shallow states. MBB, DR. KNIFFLER.

B23. Metastable states in a space charge limited structure. CNRS, ECOLE POLYTECHNIQUE, PROF. I. SOLOMON.

B24. Investigation of recombination losses in a-Si solar cells depending on surface passivation. UNIVERSITY OF KONSTANZ, PROF. BUCHER.

B25. Evaluation of interface losses in particular close to the p/i interface by optical excitation - MBB, DR. GORN / PROF. WINTERLING and by EBIC, ECOLE POLYTECHNIQUE, DR. B. EQUER.

B3. Interfaces

B31. Alloy composition and interface interdiffusion profiling by AUGER and SIMS. S.C.K., PROF. P. NAGELS. LEIS technique for interface analysis complementary with SIMS. UNIVERSITY OF EINDHOVEN, PROF. H. BRONGERSMA.

B32. Analysis of the reactivity of tin oxide and ITO with a silane plasma by in situ ellipsometry, Kelvin probe and chemical analysis. CNRS, ECOLE POLYTECHNIQUE, DR. B. DREVILLON AND DR. J. PERRIN.

B33. Thermal instability of the p+/SnO_2 interface. SOLEMS, DR. TRAN QUOC.

B4. Influence of doping profiling, mainly within the central layer

B41. Doping profiling and contamination control by SIMS. CNRS, MEUDON-BELLEVUE, DR. MARFAING, IMEC, DR. VANDERVORST AND S.C.K., PROF. NAGELS.

B42. Silane purity control. All laboratories will subcontract a common chemical analysis laboratory where silane cylinders could be controlled in relation with the properties of the resulting cells.

B43. The influence of gas phase dopant profiling on fill factor and collection efficiency. MBB, DR. KNIFFLER.

In the following we briefly mention work which is not reported separately at the contractor's meeting and which is carried out at associated laboratories.

B21: Time of flight experiments are presently carried out using Schottky diode structures; once the experimental procedure is fully established the method will be applied to pin structures; in addition a surface photovoltage experiment is being built up so that the diffusion length of holes can be studied by 1988.

B24: In a first step MIS cells are made with the oxide layer produced by low-temperature plasma oxidation.

B32: P. ROCA studied the dark I-V characteristics of pin diodes deposited on a SnO_2 or a Chromium substrate. The dark reverse current is much higher in the case of the SnO_2-coated substrate indicating the already known p^+-i interface contamination by tin.

B41: It was shown by experiments on the ARCAM reactor that B_2H_6 can be thermally decomposed at a small rate already at temperatures of 200° C (CNRS).

In the following we report on activities carried out at MBB during the first six months of this research contract.

2.1 The triode concept

A triode-type electrode configuration allows to spatially separate the plasma volume - in which the gas decomposition occurs - from the deposition volume. A reactor was modified to adapt an electrode set up including a grid as displayed schematically in Fig. 1.

In a first step a series of experiments were carried out to study the relation between the deposition rate and geometrical parameters: firstly the distance d_s grid-substrate was varied at fixed distance d_p and secondly also the thickness of the plasma sheet d_p was varied at a fixed grid-substrate separation. In general it was found that the deposition rate decreases with increasing grid-substrate separation; at higher plasma power densities, however, the plasma did penetrate the grid and reached almost the substrate. This unexpected behaviour made it necessary to check the electrode arrangement. Some parts of which have been redesigned and are reconstructed.

2.2 Doping profiling within the i-layer

A serious problem in pin a-Si:H photocells is the cross contamination of the central i-layer by phosphorus or boron arising from the doped layers. This cross contamination can limit the performance of the pin cell [1].

While our previous experiments have been mainly concerned with a spatially homogeneous doping of the central layer the experiments reported here are aimed to study the influence of a controlled doping profile on the performance of the cells, in particular on the fill factor. These experiments could possibly also verify whether a dopant cross-contamination could be compensated by an intentional counter-doping profile.

The cells used in this study were of the type glass/TCO/pin/Al. The pin layer sequence was prepared in a three-chamber system using a glow discharge of CH_4/SiH_4 or SiH_4, respectively, a gas pressure of $\sim$0,15 Torr and substrate temperatures of $\sim$250° C.

The central layer with a thickness of $\sim$ 0,4 µm was deposited in a time $\leq$ 30 minutes. Its oxygen content was shown by SIMS-profiling to be about $\sim 10^{19}$ cm^{-3}. This low value was mainly achieved by the use of loading and deloading chambers.

The doping profiles were of the type shown in the inset of Fig. 2. Both boron and phosphorus profiles were applied. The vppm-values given in Fig. 2 refer to the maximum gas phase dopant concentration applied during deposition of the central layer. The experimental values of the fill factor as summarized in Fig. 2 show that boron doping - whether with increasing or decreasing profile - causes a decrease of the fill factor in comparison to its value without any (intentional) doping. On the other hand, phosphorus doping can slightly enhance the fill factor when applied with a concentration profile increasing towards the n^+-layer.

This slightly enhanced fill factor, however, is accompanied by a decrease of the open-circuit voltage, so that the resulting efficiencies (between 7,5 to 8 % for test areas ~ 0,1 cm^2) are about the same as in the case of no central layer doping.

It should be remarked that the decrease of V_{oc} induced by the phosphorus profiling follows the trend already observed in earlier experiments studying spatially homogeneous phosphorus doping [2].

2.3 Influence of space charge

Capacitance studies of pin diodes have been started with the aim to get information on the space charge within the central i-layer.

The first preliminary study was made on a series of glass/TCO/pin/metal diodes (test area ~ 0,07 cm^2) prepared in such a way that during deposition of the central layer an very weak amount of a B_2H_6/H_2 mixture was added to the SiH_4 feeding gas.

The dark capacitance was measured in dependence of the bias voltage at a frequency of 100 KHz. At zero bias it was found to decrease with increasing B_2H_6-doping of the central layer.

On the other hand, the photo capacitance of the diodes, measured at 1 KHz and using a signal light of wavelength 400 nm, was found to increase with increasing B_2H_6-doping of the central layer as displayed in Fig. 3. This finding can be understood by considering the mobile charge density close to the p^+-i interface in dependence on the internal electric field. B_2H_6-doping of the central layer is expected to lower the field near the interface so that the mobile charge density can be increased and correspondingly a larger photo capacitance can be observed [3].

Finally first experiments to determine the diffusion capacitance near V_{oc} were carried out.

2.4 Carrier losses near the p^+-i interface

In pin diodes with a good i-material a main loss mechanism can arise from a back diffusion of electrons to the p^+ window layer. This back diffusion is working against the drift motion induced by the electric fields at the p^+-i interface.

The behavior of the light induced carriers close to the p^+-i interface can be probed by measuring the dispersion of the open circuit voltage V_{oc} as already reported and discussed at the Sevilla conference [4]. It was suggested that the observed dispersion $\Delta V_{oc} = V_{oc}(blue)-V_{oc}(red)$ can be used as a qualitative tool for studying the electric field in the neighbourhood of the p^+-i interface. In particular it is expected that the dispersion V_{oc} is a measure of the "interface" field strength relative to the averaged field strength across the central layer. Our data on V_{oc} as displayed in Fig. 4 are consistent with this expectation.

In future this measuring technique can be used for comparative studies of the electric "interface" field in dependence on modifications of the graded layer at the p^+-i interface.

REFERENCES

[1] N. Kniffler, G. Mück, G. Müller, M. Simon, and G. Winterling in BMFT-Photovoltaik-Statusbericht 1984, ed. by Deutsche Gesellschaft für Sonnenenergie, p. 260

[2] M. Gorn, N. Kniffler, and G. Winterling in Proceed. 7th EC PV Solar Energy Conference, Sevilla, 1986 eds. A. Goetzberger, W. Palz, and G. Willeke, D. Reidel publ., p. 412

[3] R. S. Crandall, Appl. Phys. Lett. 42, 45 (1983)

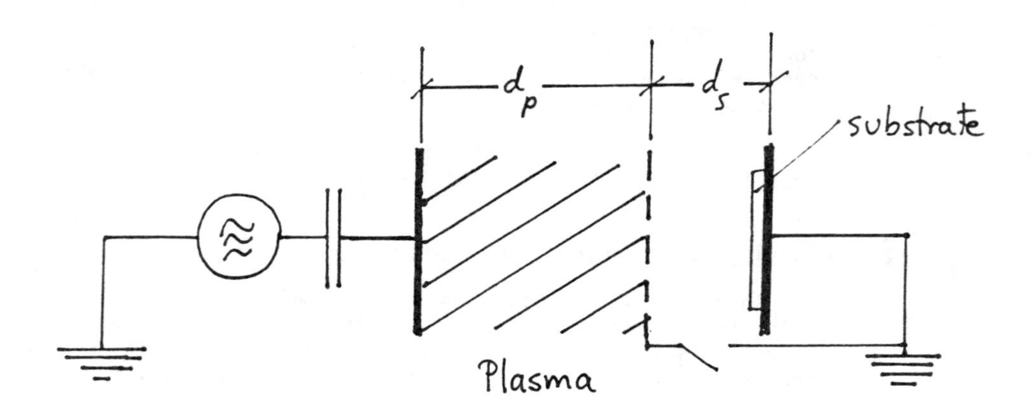

Fig. 1: Schematics of the triode electrode configuration

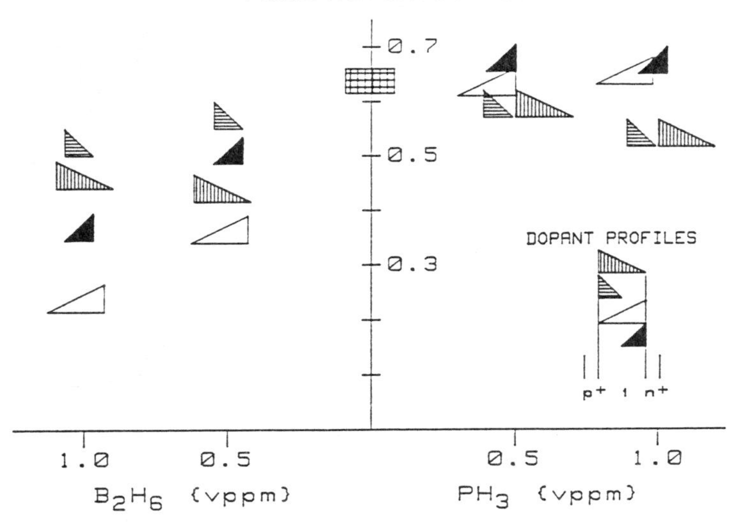

MAXIMUM GASPHASE DOPANT CONCENTRATION DURING i-LAYER DEPOSITION

Fig. 2: The fill factor of glass/TCO/pin/Al diodes for various dopant profiles introduced during deposition of the central layer

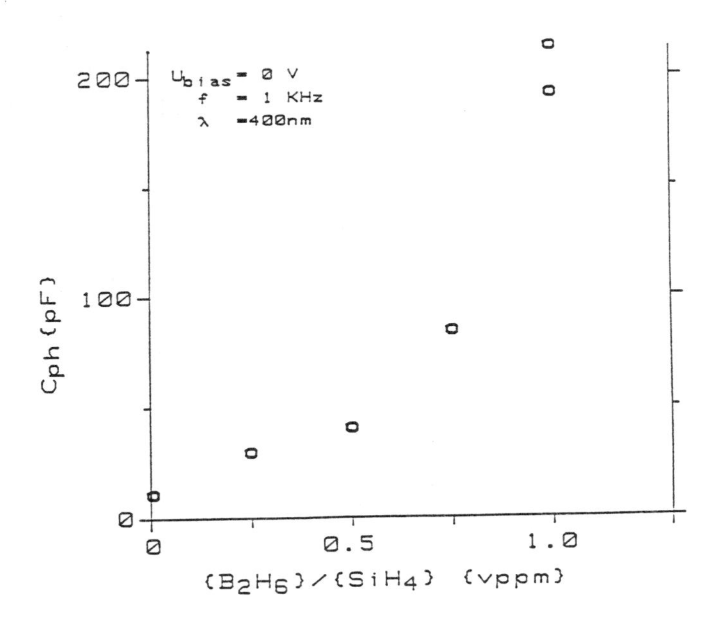

Fig. 3: The photo capacitance of glass/TCO/pin/Al diodes in dependence of a weak boron doping of the central layer

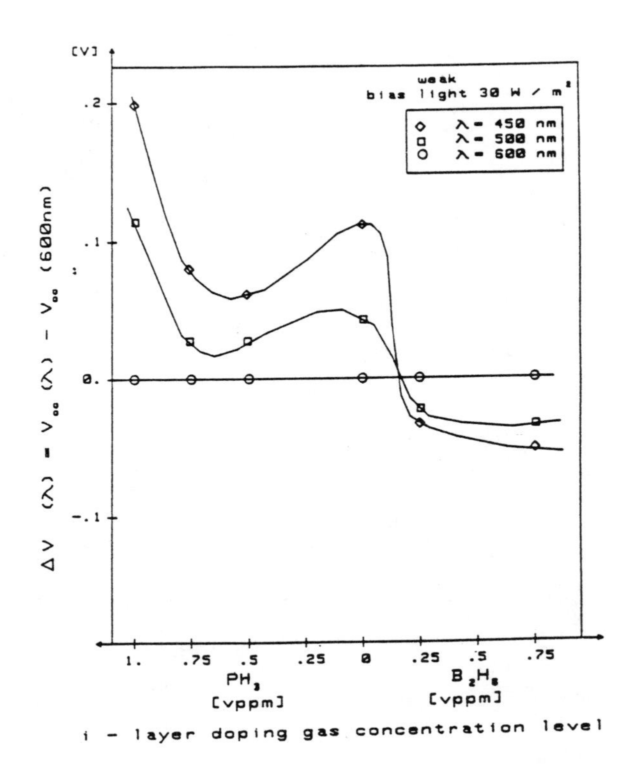

Fig. 4: Measured data of the dispersion V_{oc} as a function of the gas concentration level of B_2H_6 or PH_3 during i-layer deposition

DEVELOPMENT OF THE SCIENTIFIC AND TECHNICAL BASIS FOR INTEGRATED AMORPHOUS SILICON MODULES

IMPROVEMENT OF THE DEVICE STRUCTURE

Contract number : EN3S/0050

Duration : 24 months 1/1/1987 - 31/12/1988

Total budget : ECU 412000 CEC contribution : ECU 206000

Head of project : G. Willeke

Contractor : IMEC

Address : Kapeldreef 75, 3030 Leuven, Belgium

Summary

A brief summary of the main results concerning part C of the common research programme is given. At IMEC the HomoCVD method has been developed as a promising alternative for the deposition of the front window and the p^+-i interface of an a-Si solar cell. In a first series of experiments growth rates comparable to glow discharge deposition have been obtained. Undoped HomoCVD a-Si has been grown with properties similar to good quality gd-material, even in the presence of metallic impurities which have led to a new substrate holder design. Thin films (d>300Å) of n^+ gdμc-Si have been prepared at substrate temperatures as low as 200°C. Growth rates of up to 0.25 μm/hr have been obtained with conductivities between 1 and 20 $(\Omega cm)^{-1}$. In another development gdμc Si films have been grown by diluting silane in He. The use of low temperature screenprinted contacts as back metallisation has been investigated. Contact resistance is the limiting factor, but on n^+ gdμc-Si values as low as 7 Ωcm^2 have been obtained, using polymer based Ag/Ti and Ag/Mo pastes.

1) Introduction

The first part of this report gives an overview of the third part of the common research programme, namely the device structure improvement and briefly summarises the main achievements obtained so far in this field. A more detailed account of the different topics involved will be given elsewhere at this contractors' meeting. In the second part a detailed description of the research work carried out at IMEC is presented.

2) Improvement of the device structure

Part C of the common research programme contains the different topics as follows:

C1. Development of the glass substrate with a transparent electrode
Research on nonconventional substrate top layer (titanium oxide, others...). ST. GOBAIN, DR. O. MARIS in collaboration with several other groups.

C2. Increase of short circuit current by optical improvement of the cell
C21. Enhancement of the red response by structural optimization. ST. GOBAIN, DR. MARIS in collaboration with ECOLE POLYTECHNIQUE, PROF. I. SOLOMON.
C22. Enhancement of the red response by a diffusive silk screen printed paste as a back contact. IMEC, DR. G. WILLEKE
C23. Alloys for optical improvement - S.C.K., PROF. P. NAGELS in collaboration with ECOLE POLYTECHNIQUE, PROF. I. SOLOMON.

C3. Improvement of the fill factor
C31. Analysis of the TCO contact. MBB, DR. N. KNIFFLER and IMEC, DR. WILLEKE.
C32. The rear contact. Evaluation of a microcrystalline n+ layer. MBB, DR. KNIFFLER, IMEC, DR. WILLEKE and CNRS, GRENOBLE, DR. D. JOUSSE.
The different techniques for back contact deposition will be tested and compared: thermal evaporation (SOLEMS), e-beam evaporation (MBB), magnetron sputtering (MBB and SOLEMS).

ST. GOBAIN has focused its attention on the development of a diffusive TCO coated glass for outdoor power applications. The deposition of indium tin oxide by powder pyrolysis on texturized glass has been identified as the most promising technique. Best results are obtained by chemical etching of the glass but this technique has still to be further optimized, in order to obtain sharper pyramids which are required for optimum light trapping. One of the advantages of this method is that the classical deposition techniques of TCO on flat surfaces can be utilised.

Work on topic C21 has mainly been carried out by SOLEMS, where an interference back mirror has been developed, consisting of a properly designed SnO2 layer sandwiched between the back n+-layer and the back metallization. By chosing the correct thickness of the TCO layer, the reflection loss at the SnO_2/metal interface is reduced and the reflectivity is enhanced in the red by interference. This structure has been optimized for a flat substrate, since texturized material of high quality (Solarex, Asahi) has only recently become available.

Work on alloys for optical improvement at MOL had to be postponed, due to the installation of the 6 chamber PECVD reactor, on which work has concentrated. Prof. Solomon's work on the SiC alloy has led to the identification and preparation of 'carbonated' a-Si:H, a high quality SiC material prepared at low power densities compatible with a-Si solar cell technology. Work is now in progress on a-SiCN compounds for use as buffer layers and optical improvement.

2) Work carried out at IMEC

During the first 9 months of the contract period, work at IMEC has focused at the following topics:

A21. HomoCVD - An alternative technique for the front window, P+/i interface
C22. Enhancement of the red response by a diffusive silk screen printed paste as a back contact
C32. The rear contact. Evaluation of a microcrystalline n+ layer

A detailed description of the results obtained is given below. Other topics which have not received as much attention yet are as follows:

B41. Doping profiling and contamination control by SIMS.
C31. Analysis of the TCO contact.

The use of the SIMS technique for impurity control is scheduled for the end of 1987, when the newly developed techniques and materials (HomoCVD, n+ μc-Si, screenprinted back contact etc.) will be included in complete solar cell structures. Work on topic C31 has started by structural, electrical and optical characterisation of the different available TCO coated glasses, the main interest being on texturized material.

HOMOCVD a-Si for the front window

This research activity aims at the development of an alternative technique for the preparation of an improved p+ front window and p+i interface. The so called HOMOCVD technique is based on the pyrolytic (thermal) decomposition of the source gases, whereby a cooled substrate holder ensures the proper hydrogen incorporation in the grown films. In comparison with the standard glow discharge technique the substrate environment is ion free and superior electronic properties of low temperature a-Si:H films have already been demonstrated [1]. In the period under consideration the experimental set up consisting of a conventional Tempress (GS) LPCVD reactor including the necessary modifications to allow for a cooled substrate holder has been installed and started up.

A first series of experiments has focussed on growth rate optimization, since rather low growth rates r<20Åmin-1 are normally obtained. The deposition parameter space has been systematically explored for optimum r. Fig.1 shows the growth rate as a function of gas temperature T_g with the other parameters kept constant. r increases with T_g up to about 700°C, above which it decreases because of silane depletion on the hot reactor walls. In addition, an increasing deposition rate was observed with increasing total gas pressure, increasing gas flow rate and decreasing substrate temperature. A typical growth rate obtained during these studies was 55.4Å/min using the following deposition parameters: 700°C gas temperature, 80°C substrate temperature, gas pressure (SiH4) 0.62Torr and 130sccm gas flow rate. This growth rate is comparable to glow discharge values and therefore demonstrates the potential of the HOMOCVD technique. Much larger growth rates were obtained at even higher gas pressures where the onset of gas phase nucleation was observed. Films grown under these conditions had a very bad morphology similar to some nonoptimised PVD or PECVD layers. By carefully avoiding these phenomena, very uniform films with good morphology and growth rates up to 91Åmin-1 have been obtained.

In addition some optical and electronic properties of undoped HomoCVD a-Si:H films have been studied as a function of deposition parameters. The best film in terms of photo- to darkconductivity ratio has been obtained at 700°C, 250°C, 0.61Torr and 130sccm. A room temperature dark conductivity of 1.8E-10 Scm-1 was observed, as well as a conductivity activation energy of .78eV, a room temperature photoconductivity (approx. 1E16 cm-2sec-1 white light) of 2E-5 Scm-1, an optical band gap (Tauc plot of optical transmission data) of 1.75eV and a growth rate of 40.6Å/min. This data is very similar to state-of-the-art glow discharge material and therefore demonstrates again the potential of the HomoCVD technique. Films deposited at T_s=60°C showed an optical gap of 2.0eV. Fig.2 shows the photoluminescence intensity as a function of temperature for a HomoCVD film grown at 80°C and a PECVD film deposited at 230°C. It is interesting to note that I_L for the two films is comparable at low temperatures. At room temperature the signal is still observable in the HomoCVD film, unlike the glow discharge one, demonstrating its good electronic properties.

SIMS analysis of some of our films showed a large concentration (1E18 cm-3)of metallic impurities such as Cr and Mn, possibly originating from the stainless steel substrate holder exposed in certain parts to the rather high gas temperature of about 700°C. At present a modified substrate holder is being installed to overcome this problem. It is interesting to note that the above good electronic properties have been obtained with the metal contamination.

Evaluation of a microcrystalline n+ layer as a rear contact

As a standard procedure n+ microcrystalline silicon thin films are included in high efficiency a-Si single junction solar cells. Apart from obvious benefits such as reduced series and contact resistances and reduced 'dead layer' optical absorption, certain drawbacks include high power density and strong dilution deposition conditions, leading to defect creation and small growth rates. In addition a minimum thickness of above 500Å is normally required as well as high substrate temperatures (>300°C) [2], which can lead to a possible degradation of the underlying solar cell structure typically grown at somewhat lower temperatures (200-250°C). The present work aims at overcoming these problems by developing a large growth rate, low power density and low substrate temperature process compatible with state-of-the-art solar cell technology.

In a series of experiments carried out in a Plasma Technology PD80 glow discharge system, the silane dilution in hydrogen was systematically varied between 1 and 100%. Fig.3 shows the room temperature conductivity of the films thus obtained as a function of the dilution used. It appears that films grown with a dilution of SiH_4 (containing 1% PH_3) in H_2 of up to 20vol% are microcrystalline with a typical conductivity - approximately independent of dilution - of 3 Scm^{-1}. Above 30% dilution the films appear to be amorphous. As expected, growth rates increase with decreasing dilution and values up to 0.25 um/hr - comparable to state-of-the-art a-Si material - have been obtained.

The conductivity of these films increases with thickness and 'bulk' conductivities of about 10 Scm^{-1} are reached for d>2000Å. n+ microcrystalline layers as thin as 500Å have been prepared, whereas thinner films appear to be amorphous like. This result is encouraging, as 500Å thick n+ back contacts can be tolerated in a single junction a-Si solar cell structure. In a recent experiments carried out in a Nanotech glow discharge system, even thinner microcrystalline n+ layers (d=300Å) have been prepared with conductivities up to 20 Scm^{-1}.

In fig. 4 the room temperature conductivity of the above n^+ microcrystalline films is plotted against the substrate temperature. It appears that the conductivity decreases with decreasing substrate temperature, although this effect may be masked by a thickness dependence, since layers grown at lower T_s were somewhat thinner. In any case films grown at temperatures as low as 240°C are still microcrystalline. In a recent experiments carried out in the Nanotech system, layers grown at 200°C were microcrystalline.

In addition low power density (factor 2 increased as compared to state-of-the-art a-Si deposition) microcrystalline films have been prepared. In another development, gdμc-Si thin films have been grown by diluting silane in He. This result could be particularly important when the conventional reducing hydrogen ambient should be avoided, i.e. for a gdμc-Si film to be grown on SnO_2.

Enhancement of the red response by a diffusive silk screen printed back contact

This work aims at the evaluation of screenprinted back metallization in conjunction with a possible red response enhancement of the a-Si solar cell [3].

The screen printing technology offers advantages as compared to the conventional Al evaporation technique in terms of throughput and simplicity (contact patterning included in the process) and therefore presents an interesting alternative. In crystalline silicon solar cell technology screen printing is quite commonly used; it in this case a high temperature process. Our present work requires the development of pastes which can be cured (polymerized) at about 200°C, compatible with a-Si:H technology. Other requirements for a screen printing back metallization process are as follows :

- Contact resistance with the (typically) n+ layer below about 1 Ohm cm2 corresponding to about 1% relative power loss. In tandem cells a somewhat higher value can be tolerated as current densities decrease.
- Sheet resistance less or comparable to that of the TCO layer (typically 10 Ohm/square).
- Low contact resistance with the TCO layer (typically below 10^{-2} Ohm cm^2), depending on the current density in the interconnection.
- Good adhesion and long term stability.

Screenprinted contacts made from polymer based pastes consist of metal particles of different size and shape embedded in a polymer matrix which ensures structural stability of the metal film and its adhesion to the substrate. Ideally a regular array of small sized spheres would have to be considered in an interconnecting polymer matrix. Even in this case the contact consists of a number of 'point contacts' as compared to a complete coverage in case of evaporated or sputtered contacts. A 'back-of-the-envelop' calculation shows the dramatic effect of this situation. Considering a closed packed layer of identical (hard) spheres of radius R and an electrical contact distance h (over which the sphere is supposed to make electrical contact; this effectively is a tunneling distance), the fraction of the active contact area can be described as

$$\frac{\pi}{2\sqrt{3}}\left[2\frac{h}{R}-\left(\frac{h}{R}\right)^2\right] \quad \text{and for } R \gg h \quad \frac{\pi}{\sqrt{3}}\frac{h}{R}$$

For typical values of R=1μm and h=20Å, only about 0.4% of the total contact area carries electrical current. Evidently the use of small sized spheres and a soft metal increases the active area. To make thesituation worse, in most cases the metal particles are flakes of irregular shape and size. SEM fracture edge studies and infrared reflection have provided experimental evidence for a small effective contact area of our screen printed contacts. At present capacitance studies are being carried out on screenprinted contacts deposited on oxidized cr-Si wafers, in order to determine the effective contact area as a function of the metal and particle size. The above effect is twofold. First of all the contact resistance is calculated on the basis of the total area and secondly because of lateral current flow between the point contacts, the sheet resistance of the n+ a-Si layer of a solar cell - which is typically larger that 20MΩ/square - increases the effective contact resistance.

The lowest contact resistances on n+ a-Si layers are obtained by evaporating or sputtering Mg, Ti and Al. This can be attributed partly to the low metal workfunction of this materials, which ensures a small barrier height and therefore a small contact resistivity. In addition the ability of the metal to reduce a possible interfacial layer such as a native oxide might be important. From the data of the standard free energies of the oxides it can be deduced that these metals tend to reduce SiO_2 even at room temperature. The resulting metal oxides are more conductive than SiO_2 due to a possible nonstoichiometry, including metallic interstitials as well as different electrical properties. For instance tetragonal TiO_2 has a refractive index in the visible of between 2.5 and 2.8, more like for instance the semiconducting SiC (2.65) than the insulating SiO_2 (1.5).

Commercially available polymer screenprint pastes are mainly based on Ag. With such pastes and similar self made ones it was impossible to obtain low contact resistances, probably for the above mentioned reasons. In order to check the influence of the fixed charge concentration in the semiconductor on the contact resistance - heavy doping is known to improve contact properties - the contact resistance of Ag based screenprinted contacts was measured on crystalline Si as a function of doping. It was found that R_c decreased with doping but even for a carrier concentration of 2E20 cm-3 the contact resistance was limited to 100 Ωcm^2 (fig.5). We found that contact resistances could be drastically lowered by forcing a 5A/cm2 current through the contact. However the low value of R_c was not stable and increased with time. SEM and Auger studies indicated that the original decrease of R_c was not due to a breakdown of a polymeric interfacial layer but possibly due to a native oxide breakdown, which is reformed with time.

Fig. 6 shows the contact resistance R_c obtained on different materials including n+ a-, gdμc- and thμc-Si thin films. The latter have been obtained by thermally crystallizing n+ a-Si thin films at about 700°C, leading to material similar to gdμc-Si but with somewhat larger grain sizes and conductivities. It appears that pastes based solely on Ag are not suitable for a-Si solar cells. However, adding Mo and Ti to the paste leads to a significant decrease of R_c. Values as low as 7 Ωcm^2 have been obtained so far on thin layers (d=500Å) of in situ grown gdμc-Si. On thermally crystallized n+ a-Si films, which have an even lower sheet resistance, R_c values below 1 Ωcm^2 have been observed.

The conductive path in polymer based pastes is formed by a series of individual particles embedded in a polymer matrix, i.e. a to some extent porous environment. A non-conductive surface layer, caused by oxidation for instance, can therefore lead to a dramatic increase of the sheet resistance. Pastes based solely on metals which tend to oxidize rather quickly like Al, Ti and Cr led indeed to high sheet resistances. Contacts made from Ag/Mo and Ag/Ti showed values below 1 Ω/square, which is at least an order of magnitude below that of a typical TCO layer. Sheet resistance is therefore not a critical parameter.

The contact resistance of the above Ag/Mo and Ag/Ti contacts on 25 Ω/square SnO_2 Glaverbel substrates has been found to be about 1E-2 Ωcm^2, sufficiently low for practical a-Si solar cell applications.

A possible red response enhancement of an a-Si solar cell can be envisaged as follows. The use of a texturized glass substrate leads to a diffuse scattering of the light entering the solar cell. The screenprinted 'point contact' back side metallization provides effective light trapping due to the small critical angle of reflection at the a-Si/air interface [4]. This effect should be observed by comparing the near infrared total reflection intensity of complete solar cell structures using screenprinted back metallization in comparison with evaporated contacts.

3) Conclusions and future work

In the field of device structure improvement several advances have been made. ST. GOBAIN, in close collaboration with SOLEMS has developed a process for chemically etching glass surfaces in order to obtain a well defined surface texturisation for diffuse TCO coated substrates. This method still has to be further optimized for better light trapping. SOLEMS has prepared an interference back mirror, consisting of a SnO_2 layer sandwiched between the n^+ and metal films. With this structure an average gain of about 30% in the red part of the spectrum has been obtained relative to a similar conventional structure. This technique will be applied to texturised substrates as soon as they become available. As a result of Solomon's work at the ECOLE POLYTECHNIQUE, the deposition conditions of high quality a-SiC alloys for optical improvement of the solar cell structures have been identified. Work will continue on a-SiCN alloys. At IMEC, the preparation of low substrate temperature microcrystalline silicon - compatible with a-Si solar cell technology - has been demonstrated. On thin layers of this material, screenprinted contacts of Ag/Ti and Ag/Mo have shown contact resistances Rc as low as 7 Ωcm^2. A further reduction of Rc is expected from optimization studies. Future work will focus on the incorporation and evaluation of n^+ gdμc-Si layers in combination with different solar cell back metallizations, including screenprinted contacts. The HomoCVD technique has been developed as a promising alternative for the deposition of the front window and p^+/i interface. Further studies will aim at the optimization of the i-layer quality as well as the preparation of doped films. In addition the interaction of the growing film with different TCO coated substrates will be investigated. Complete solar cell structures including the above new materials and techniques will be realized and optimized. The spectral response of these cells as a function of impurity profiles will be measured and correlated with deposition conditions.

Acknowledgements

The author would like to acknowledge Quian Z.M., A. Van Ammel and H. Michiel for developing the HomoCVD technique, K. Baert, M. Honore and A. Janssens for the work on the low temperature screenprinting technology.

References

1 - B. Scott, Alternative methods of a-Si:H growth, IMEC summer course 1984, unpublished

2 - G. Willeke, Ph.D. Thesis, Dundee University 1983, unpublished

3 - K. Baert, G. Willeke, M. Honore, W. Vandervorst, J. Roggen, J. Nijs, Proc. 19th IEEE PVSEC, May 1987, New Orleans

4 - K. Baert, G. Willeke, M. Honore, J. Roggen, J. Nijs, to be presented at 3rd PVSEC Kobe, Japan 1987

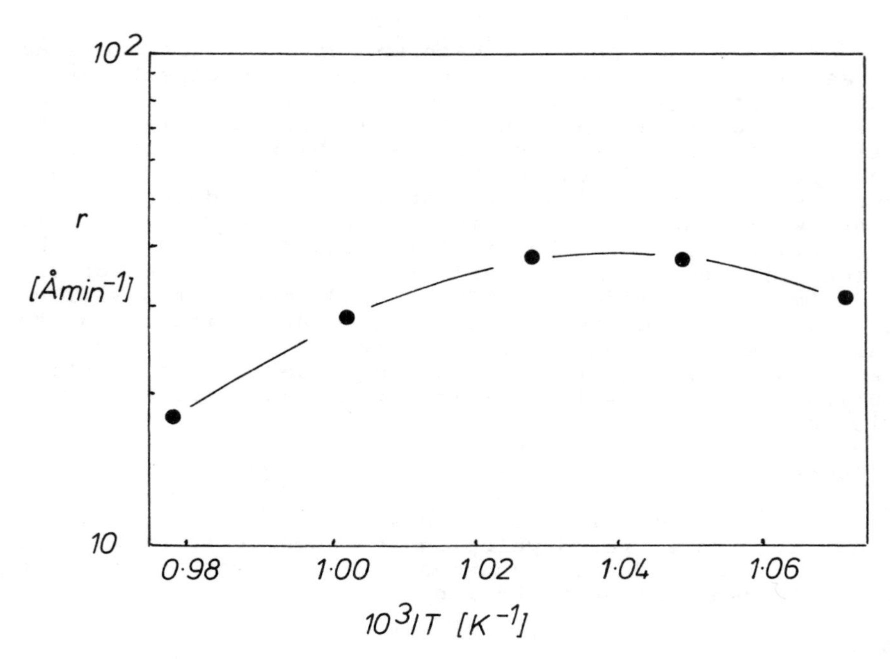

Fig.1: Growth rate r of HomoCVD films versus inverse gas temperature (486mTorr,80sccm,T_S=60°C).

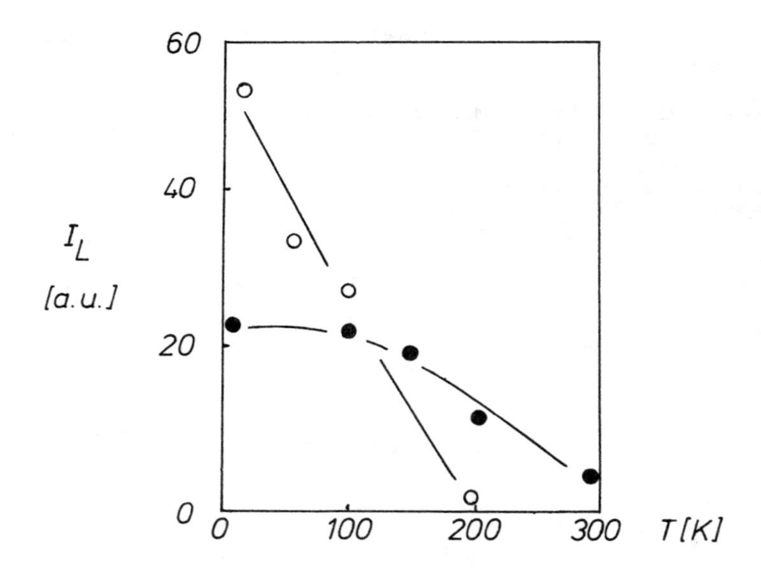

Fig.2: Photoluminescence spectra of HomoCVD- (full circles) and glow discharge films (open circles).

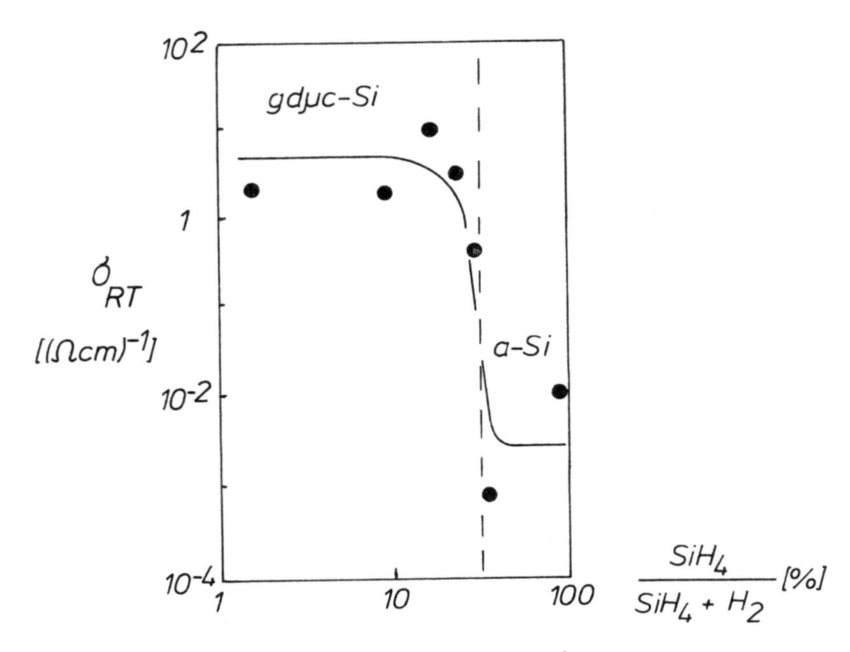

Fig.3: Room temperature dark conductivity σ_{RT} of glow discharge microcrystalline silicon thin films as a function of hydrogen dilution

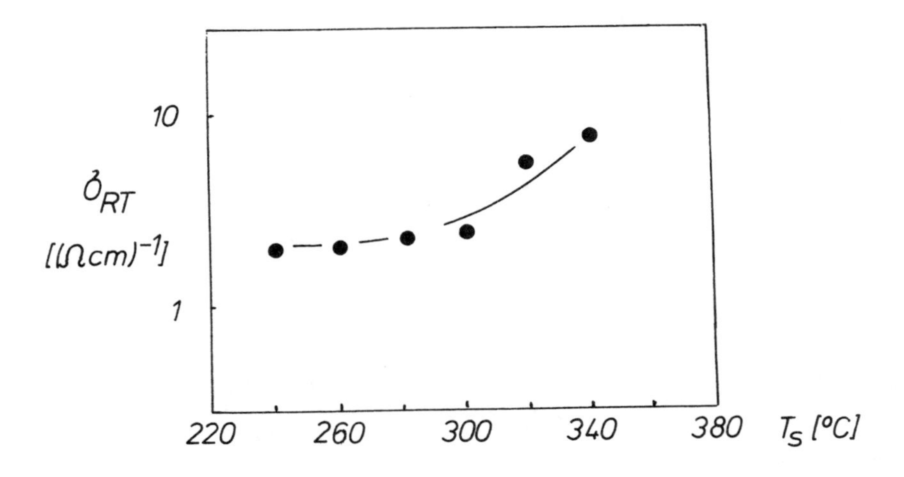

Fig.4: Room temperature dark conductivity of gdµc-Si thin films as a function of substrate temperature T_s.

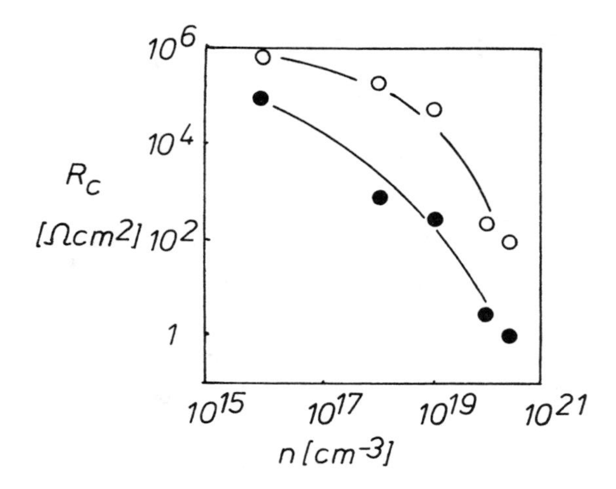

Fig.5: Contact resistance R_c of Ag paste contacts on cr-Si substrates of different carrier concentrations n. Open circles: virgin contacts. Full circles: after $5A/cm^2$ for 30sec.

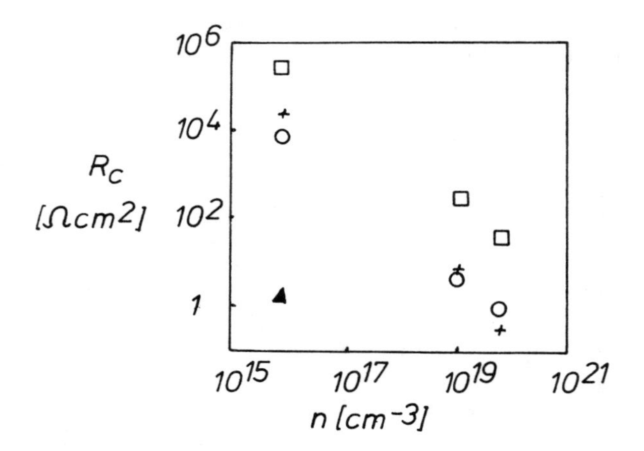

Fig.6: Contact resistance R_c for different substrate materials. From left to right: n+ a-Si, gdμc-Si and thμc-Si. Most likely values for the carrier concentrations n have been taken. Open Squares: Ag, plusses: Ti/Ag, open circles: Mo/Ag, full triangle: evaporated Al.

DEVELOPMENT OF THE SCIENTIFIC AND TECHNICAL BASIS FOR INTEGRATED AMORPHOUS SILICON MODULES.

Description of the research work carried out at S.C.K./C.E.N.

Contract Number : EN3S-0051-B(GDF)

Duration : 30 months 1 July 1986 - 31 December 1988

Total Budget : BEF 19.756.000 CEC Contribution : BEF 9.878.000

Head of Project : Prof.Dr. P. NAGELS, Physics Department

Contractor : S.C.K./C.E.N.

Address : Boeretang 200
B-2400 MOL
Belgium

Summary

During the last six months, the R&D efforts at S.C.K./C.E.N. were concentrated on four topics.
(I) By optimizing the deposition conditions, an efficiency of 7.7 % (cell area = 0.125 cm^2) was achieved for solar cells with a glass-SnO_2-p^+ a-SiC:H-i-n^+ a-Si:H-Al structure.
(II) A new multi-chamber PECVD system was installed in our laboratory and brought into operation.
(III) A phenomenological model for fitting the I-V characteristics of a-Si:H solar cells was developed. Solar cells with different p^+-layer thicknesses were analysed using this new model. It was shown that the mobility-lifetime product $\mu\tau$ was severely influenced by the impurity inclusion in the intrinsic layer during the p-i-n deposition sequence.
(IV) Analysis of the performance decay of cells degraded by current injection led to the conclusion that at least two mechanisms were operative. Temperature studies suggested that cell stability could be greatly increased by raising the operating temperature. A field distortion near the p-i interface was observed for cells degraded by current injection.

1. Preparation of amorphous silicon films and solar cells

At S.C.K./C.E.N., four plasma-enhanced CVD reactors of own construction are in operation since four years. During the last six months, the development of heterojunction solar cells of the type glass-SnO_2-p^+ a-SiC:H-i-n^+ a-Si:H-Al in these systems was continued. By careful optimizing the deposition conditions, we achieved an efficiency of 7.7 % (cell area = 0.125 cm^2). Experiments were started to increase the cell size. An efficiency of 3 to 5 % was obtained for integrated solar cells with an area of 100 cm^2. A low cost mechanical patterning technique was used to pattern the constituent layers of the cell.

A new multi-chamber PECVD system for the preparation of amorphous silicon solar cells was installed in our laboratory recently. Figure 1 shows a general view of this system. It was constructed at Pfeiffer (FRG) in close collaboration with S.C.K./C.E.N. The system consists of six separated chambers, namely a load-, three reaction-, a dc magnetron sputtering- and an unload chamber. The main advantages of such a machine are obvious: better control of the impurity contamination in the intrinsic layer and higher production throughput. The direct coupling of a dc magnetron sputtering chamber to the plasma CVD reactors allows us to deposit the metallic back contact without breaking the vacuum, so that oxygen absorption of the a-Si:H layer can be avoided. Substrates of 10 x 30 cm^2 in area can be transported in both directions. Therefore, the machine is also well suited for the fabrication of multi-junction solar cells. The distance between the electrodes can be varied between 2 and 7 cm. Each chamber is equipped with its own pumping system, consisting of a turbomolecular pump for the high vacuum and a roots vacuum pump for the process vacuum. The total system can be pumped down to a very low base pressure ($\cong 10^{-7}$ mbar). Each process chamber is equipped with 3 MKS mass flow controllers for the automatic control of the process gasses. The operating conditions are microprocessor controlled.

During the last six months, this new system was built up in the laboratory. The chambers and their peripherals were thoroughly tested. The rf discharges were controlled under argon plasma. Experiments with silane and silane mixtures will start in the beginning of September. We will report on the film quality of the individual layers and on the characteristics of amorphous silicon solar cells prepared in the multi-chamber system at the next contractor's meeting.

2. Alloy composition and interface characterization

P-type a-SiC:H and i a-Si:H layers were grown on doped tin oxide to study the chemical reduction of this transparent conductive oxide in the plasma. This work was performed in close collaboration with the Eindhoven University of Technology. Preliminary results can be found in their progress report.

3. Curve-fitting for amorphous silicon solar cells

The photovoltaic performance of an a-Si:H solar cell is essentially dominated by the mobility-lifetime product of the photogenerated carriers and the properties of the interfaces between the different layers composing the cell, as described by the "variable minority carrier transport model" (1). Based on this model, we proposed a new phenomenological model for fitting a-Si:H solar cell I-V characteristics (2). It introduces a voltage-dependent photocurrent I_{ph} given by:

$$I_{ph} = qGl_c \; (1-\Omega \exp(-\frac{qE}{kT} x_c)) \; (1-\exp(-\frac{L}{l_c}))$$

with G = generation rate (number of electron-hole pairs excited per unit volume and per unit time), L = thickness of the intrinsic layer, l_c = collection length = $\mu\tau E$, E = electric field = $(V_b-V_a)/L$, V_b = built-in potential, V_a = applied bias voltage, $\mu\tau = \mu_n\tau_n + \mu_p\tau_p$, mobility-lifetime product of holes $(\mu_p\tau_p)$ and electrons $(\mu_n\tau_n)$, $\Omega = S/(S+E)$, S = front side surface recombination factor and x_c is related to $\mu_n\tau_n/\mu_p\tau_p$. Substituting the voltage dependence of I_{ph} in the single diode model, one gets an excellent fit for the I-V characteristics in the 3rd, 4th and 1st quadrants (as long as $V_a < V_b$).

Using this model, we analysed glass-SnO_2-p^+ a-SiC:H-i-n^+ a-Si:H-Al solar cells with different p^+-layer thicknesses. Results were reported at the 7th EC Photovoltaic Solar Energy Conference at Sevilla (2). We found that the mobility-lifetime product $\mu\tau$ varies as a function of the p^+-layer thickness. We believe that this behavior of $\mu\tau$ is related to the cell fabrication of p-i-n junctions in a single chamber system. The p^+ a-SiC:H layer is deposited prior to the i-layer and acts as the source for boron and carbon incorporation in the i-layer. The amount of boron in the i-layer becomes larger with increasing p^+-layer thickness. Okamoto et al. (3) reported that $\mu\tau$ is strongly influenced by the number of boron and phosphorus atoms in the intrinsic layer. The largest $\mu\tau$ products were obtained when a-Si:H is doped with a small amount of boron. This explains the observed behavior in $\mu\tau$ when increasing the p^+-layer thickness. This effect can be avoided in a separated chamber system. A comparative study between a-Si:H solar cells fabricated in a single and separated chamber PECVD system will be performed in the near future.

4. Degradation by current injection

Glass-SnO_2-p^+ a-SiC:H-i-n^+ a-Si:H-Al solar cells (efficiencies between 6 % and 7 %) were degraded by current injection during 210 min. at 300 K and 326 K. Cells were injected with a forward current of 70 mA (cell area = 0.125 cm^2). At regular time intervals, the forward bias was removed for measurement of the photovoltaic performance. Figure 2 shows the changes of the normalized conversion efficiency as a function of time under forward current injection for a typical cell. The data indicate that cell degradation can be considerably reduced by operating the cell at higher temperatures. Tawada (4) reported that the surface temperature of amorphous silicon solar cells, when installed outdoors, could become higher than 70°C. So, our result suggests that cell degration will be greatly suppressed when operating outdoors. However, one must also take account of the thermal degradation when cells are operated at higher temperatures. One type of thermal degradation is caused by the diffusion of the metal atoms (like Al or Ag) from the backside electrode. Fukada et al. (5) reported that when a-Si:H solar cells were exposed to high temperature conditions, the electrode metals diffused into the a-Si:H layers. After a certain period of thermal treatment, a complete deterioration of the device was observed. This effect is absent in our experiment due to the short period of forward current injection.

The efficiency decrease in figure 2 can be fitted by a double exponential, as shown by the solid line. This suggests that at least two mechanisms are reponsible for the degradation. The I-V characteristics for the cell operating at 300 K were analyzed before and after current injection using curve fitting. The results are shown in Table 1. The increase of the built-in potential is remarkable. If we calculate V_b in the high injection region, however, no change is observed. This difference can be explained by the fact that the calculation in the high injection region probes the average electric field without providing any detailed information about the internal potential distribution. Since we take account of the influence of the interfaces on the photovoltaic performance, our model is very sensitive to field distortions near the interfaces (especially the front side interface). Thus, our results suggest that the internal electric field is peaking near the p/i interface. We believe that this field distortion arises from the build-up of a space charge in the i-layer during degradation. One of the main assumptions when deducing our model is a uniform internal electric field. As already discussed, this condition is not fulfilled in this experiment. So, care should be taken when interpreting the other results from Table 1.

Table 1. The effect of current injection on the fitting parameters of a glass-SnO_2-p^+ a-SiC:H-i-n^+ a-Si:H-Al solar cell (area=0.125 cm^2, L=420 nm). Operating temperature is 300 K.

Injection time(min.)	R_{sh} (ohm)	R_s (ohm)	I_o (pA)	n	V_b (V)	$\mu\tau$ (cm^2/V)	S (V/cm)	x_c/L
0	7.0×10^4	57	12	1.65	0.85	3.7×10^{-8}	3.6×10^4	0.16
210	7.0×10^5	62	53000	2.91	1.26	1.6×10^{-9}	1.8×10^6	0.29

5. Future work

The main part of the research work will be devoted to the starting-up and the optimization of the multi-chamber PECVD system. Films will be studied using the normal optical and electrical measurements. P-i-n solar cells will be characterized by measuring the I-V characteristics under simulated sunlight and the spectral dependence of the photocurrent as a function of the applied bias voltage.

The study of the reduction of the TCO layer and the interaction between the TCO and p^+ a-SiC:H layers will be continued in close collaboration with the Eindhoven University of Technology. The main emphasis will be on the variation of the deposition parameters of the p^+-layer. Reactively sputtered ZnO will be studied as a diffusion barrier.

References

(1) S. Nonomura, H. Okamoto, H. Kida and Y. Hamakawa, Jap. J. Appl. Phys. 21 (Suppl. 21-2) (1982) 279.

(2) J. Smeets, M. Van Roy and P. Nagels, Proc. 7th EC Photovoltaic Solar Energy Conference, Sevilla (1986) 539.

(3) H. Okamoto, H. Kida, S. Nonomura, K. Fukumoto and Y. Hamakawa, J. Appl. Phys. 54 (1983) 3236.

(4) Y. Tawada, Abstracts of SERI/DOE Amorphous Silicon Subcontractors Annual Review Meeting (1985) 149.

(5) N. Fukada, J. Takada, M. Yamaguchi, K. Tsuge and Y. Tawada, Technical Digest of the International PVSEC-1 Kobe (1984) 229.

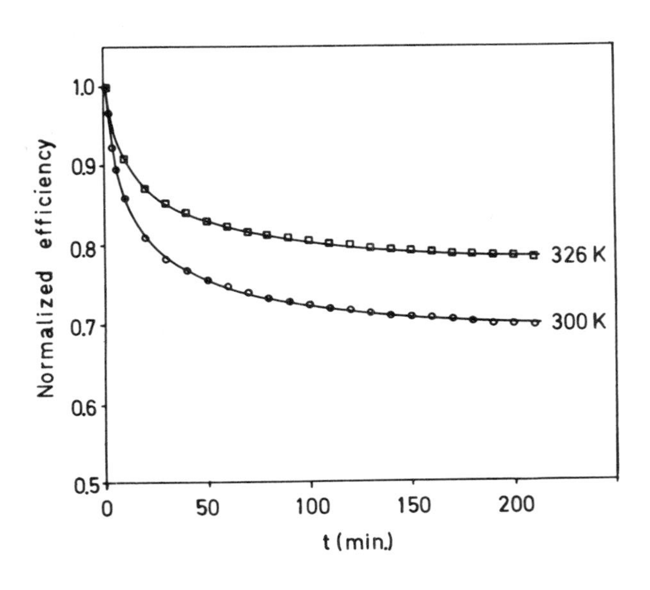

Figure 2. Normalized conversion efficiency versus the time of forward current injection

DEVELOPMENT OF THE SCIENTIFIC AND TECHNICAL BASIS FOR INTEGRATED AMORPHOUS SILICON MODULES

AES ANALYSIS OF THE SnO_2/a-Si:H INTERFACE

Contract Number: EN 3S/0052
Duration : 24 months 1 July 1986 - 1 July 1988
Total Budget : CEC Contribution: Dfl. 227,500
Head of Project: Prof.dr. H.H. Brongersma
Co-authors : A.S. Verlinde, Dr. H.J. van Daal
Contractor : Eindhoven University of Technology
Address : P.O. Box 513
5600 MB Eindhoven
The Netherlands

SUMMARY

The investigations presented here are concerned with reduction and diffusion phenomena at the interface between SnO_2 and amorphous silicon-hydrogen (a-Si:H). These phenomena occur mainly during glow discharge deposition of a-Si:H from silane. The paper deals with Auger electron spectroscopy (AES) profile analyses through the interface made while sputtering with Ar^+ ions. The AES profile measurements indicate that the interface is not sharp. It has a width of about 40 nm which is comparable to that of the a-Si:H layer. When going through the profile from SnO_2 to Si the Sn AES lines show a change being mainly a shift to higher characteristic energies. This shift is attributed to partial reduction of SnO_2. The relative concentration of reduced Sn at the centre of the interface is estimated to be of the order of 10 at%.

1. INTRODUCTION

In the development of solar cells on the basis of amorphous silicon-hydrogen (a-Si:H) one of the problems is still degradation of the interface at the entrance of the cell between the transparant conductive oxide (TCO), usually SnO_2, on glass and the first, mostly p-type silicon (carbide) layer. The degradation is mainly due to reduction of SnO_2 followed by diffusion of Sn into Si occurring during the glow-discharge deposition of a-Si:H from SiH_4 (silane). The present investigations are meant to obtain profile analyses on differently prepared interfaces by aid of several techniques. These comprise low-energy-ion-scattering (LEIS), electron spectroscopy for chemical analysis (ESCA) and Auger electron spectroscopy (AES). Differences in preparation of the interface have mainly regard to changes in the glow-discharge deposition parameters, such as the temperature of the substrate and to application of thin diffusion barriers. This paper reports on a study of the SnO_2/a-Si:H interface by aid of AES.

2. EXPERIMENTAL

The AES experiments were carried out in a Physical Electronic Industries, PHI model 550, spectrometer. The base pressure of the target chamber was typically in the low 10^{-9} Torr range. Profile measurements were done using an Ar^+-ion gun set at 3 keV with a current of 140 nA. The Ar^+-ion beam with a diameter of 0.2 mm, incident at 30° with the sample surface, was scanned over an area of 4 mm^2. AES differential mode spectra were measured with a 9 μA beam of 3 keV electrons with a spot diameter of 0.5 mm.

The sample substrates, prepared by Glaverbel Jumet, consisted of 1 cm^2 glass plates covered with 200 nm fluorinated tin oxide with sheet resistances varying between 12 and 20 Ω per square. Undoped hydrogenated amorphous silicon layers were deposited on the substrates at SCK/CEN Mol (Dr. P. Nagels) by RF glow discharge from SiH_4. The substrate temperature was 250 ^oC, the RF power 2 W, the deposition time 10 minutes, the silane pressure 0.44 Torr, the thickness of the film varied between 40 and 50 nm.

3. RESULTS

An example of an AES depth profile plotted in atomic concentrations of the elements Si, O and Sn versus sputter time is presented in Fig. 1. The top layer consists of 40 nm a-Si:H deposited on a glass supported SnO_2 layer with a sheetresistance of 20 Ω per square. It appears that the interface between Si and SnO_2 is not sharp. It is estimated to have about the same width as the a-Si:H layer, supposing that sputter yields of both materials are not too much different. Furthermore, it appears that the observed ratio of tin and oxygen atomic concentrations in the interface region as well as in the SnO_2 region is about unity. This is probably due to preferential sputtering. This effect, also noticed by other authors, has explicitly been mentioned by Greenwald et al. [1].

The AES Sn and O peaks have been observed to change systematically during the profile measurements. Fig. 2 presents a set of double Sn differential mode peaks obtained in a profile measurement. The lower curves are obtained at the Si-side, the upper ones at the SnO_2-side of the interface. The third curve from below roughly corresponds to the centre of the interface. The observed chemical shift suggests that Sn peaks at the Si-side originate from a mixture of oxidized and reduced tin. In literature the minima of the AES peak of purely metallic (reduced) tin lie at about 429 and 436 eV while the minima of purely oxidized tin are shifted 6 eV to lower energies [2-4]. The observed shift of about 1 eV in the Sn peaks from the centre of the interface to the SnO_2-side suggests a maximum concentration of the order of 10% reduced Sn. A similar result can be obtained from a subtraction and calibration procedure where AES peaks in the mixed state are subtracted from the peak measured in the purely oxidized state. A result of such a procedure in terms of reduced tin concentration (counts/s) vs sputter time, i.e. depth through the interface, is shown in Fig. 3. The top of the curve, when compared to the orginal top-to-top value of about 3000 counts/s., corresponds to about 7 at% reduced Sn.

4. CONCLUSION

AES depth profile measuring results obtained on SnO_2/a-Si:H interfaces, made while sputtering with Ar^+-ions, are probably influenced by preferential sputtering effects. The results suggest that the interface has a width of about 40 nm. The change of the AES Sn line shape during the profile measurements suggests the presence of a maximum of about 10 at% reduced tin. This result needs corroboration from other measurements such as ESCA.

REFERENCES

1. L.A. Greenwald, J. Bragagnolo and M. Leonard, Proc. 19th IEEE Photovoltaic Specialists Conference, New Orleans (1987), paper TR-86-54.
2. Y. Tawada, Ph.D.Thesis, Osaka University (1982).
3. N. Fukada, T. Imura, A. Hiraki, Y. Tawada, K. Tsuge, H. Okamoto and Y. Hamakawa, Jap. J. Appl. Phys. 21, suppl. 21-2, 271-275 (1982).
4. L.E. Davis, N.C. MacDonald, P.W. Palmberg, G.E. Riach and R.E. Weber, Handbook of Auger Electron Spectroscopy, Perkin-Elmer Corp. (1978).

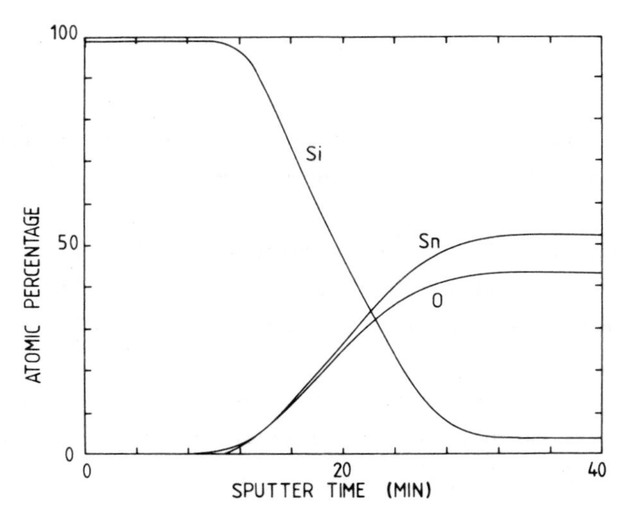

Fig.1 - AES depth profile of the elements Si, Sn and O, in atomic concentration, through the a-Si:H/SnO_2 interface as a function of sputter time

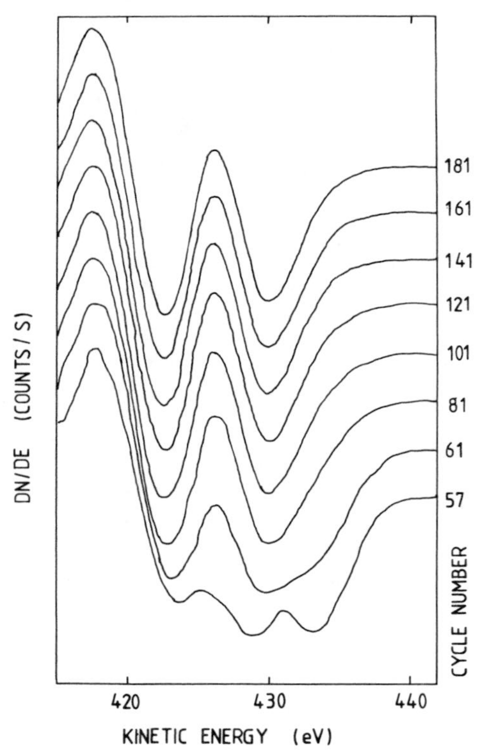

Fig.2 - AES double Sn peaks as a function of kinetic electron energy at different positions in the interface a-Si:H/SnO_2 (all curves have been normalized to the same peak to peak value: the lower curves correspond to the Si side, the upper ones to the SnO_2 side)

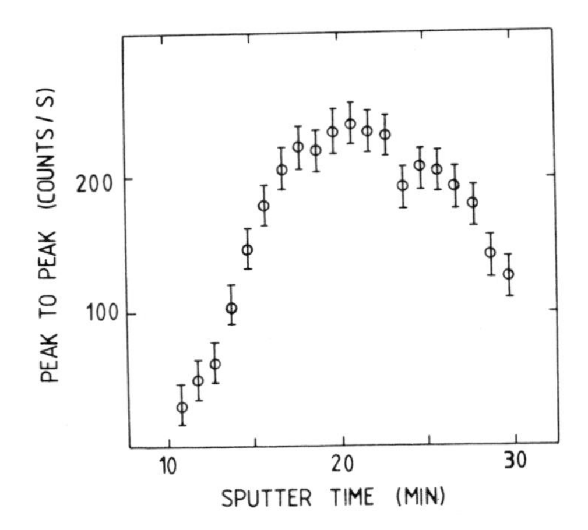

Fig.3 - Variation of the reduced tin concentration in counts/s through the a-Si:H/SnO_2 interface estimated from an AES peak subtraction and calibration procedure

DEVELOPMENT OF SCIENTIFIC AND TECHNICAL BASIS FOR INTEGRATED AMORPHOUS SILICON MODULES

Author : G. de ROSNY

Contract number : EN3S-0054-F

Duration: 36 months - 1 Sept. 1986 - 31 August 1989

Total budget : 4.332.000 FF CEE contribution : 2.046.000 FF

Head of Project : G. de ROSNY, LPICM, Ecole Polytechnique, 91128 PALAISEAU

Contractor : Centre National de la Recherche Scientifique

Address : 15, Quai Anatole France - 75700 PARIS - France

Summary: The present contract is related to the development of industrial production of amorphous solar cells made by SOLEMS in France and M.B.B. in W. Germany. It focuses on basic aspects relevant for CNRS and University research groups that cooperate in the PIRSEM-CNRS " ARC SILICIUM AMOPRPHE".

The following items have been studied:

A - Investigation of the concept of separate discharges in a common vacuum vessel to realize PIN structures. The effect of cross contamination has been studied in details.

B - Dependance of the nature of the substrate on the behavior of PIN diodes. Comparison of SnO_2 and Cr substrates. It is shown that the first one, commonly used industrially, is worse than the other.

C - Study of the contact between a-Si:H and various materials: Cr, SiO_2, SnO_2, χ-Si by optical or electrical methods. The interface is never rigorously abrupt.

D - Investigation of the built-in electric field by various methods: time of flight, EBIC, electroreflectance. These techniques are now under development and a brief report on their status will be given.

A - The ARCAM reactor, made of three separate discharge regions in a common vacuum tank, each one being devoted to make one of the layers constituting the PIN diodes, has proven its ability to make good material and devices. Efficiency of more than 6% under AM1 have been obtained on ≈ 1 cm^2 diodes, without carbon entrance window, back reflector nor optimisation of the antireflection. The comparison with either the single discharge or the saparate vessels concepts has principally to be made in term of cross contamination between the discharge regions.

It has been shown that in the ARCAM reactor the boron may reach the discharge devoted to the deposition of the intrinsic layer. This contamination is solely due to the diborane thermal decomposition on the the discharge walls and on the substrate which are held at 200 - 250 °C. There is no specific effect of the discharge in the region devoted to make the P^+ layer. The boron deposited on the walls is then cycled and partially included in the intrinsic film during its deposition, by an etching effect of the discharge at the same time the deposition takes place.

To evidence the diborane decomposition process, the surface of an intrinsic a-Si:H layer, maintained at its deposition temperature, 200°C, was exposed for 1 hour to a 50 mtorr, 1% B_2H_6-99% SiH_4 mixture. The activation energy of the film decreased from 0.8 eV to 0.66 eV by this procedure. The optical properties of the film were altered, as seen by ellipsometry. There is evidence for a modification of the film surface which becomes granular. The presence of boron has been revealed by SIMS and neutron reactions. This boron is located only at the film surface. If the film is maintained at 100°C, no boron is detected.

The observed effect cannot be completely absent in the case of separate vacuum chambers reactor, as the boron may be transported by the film surface itself. The present test conditions are not in general encountered as such in production reactors, where the residence time of the substrate in diborane ambiant is shorter. However if one seeks for ultra-abrupt junctions, the reported effet may have to be taken in consideration.

B - The ITO-SnO_2/P^+ contact is one important technological goal to master in device realisations. The following experiment stresses the difficulty of the problem. When the same PIN structure is realised on SnO_2 and Chromium substrates, the dark I/V charecteristics are very different the one from the other, with a much better behavior with the Cr substrate, see (fig1, P.ROCA). This observation is to be linked to the evidence, described just below for a reduction of SnO_2 during the film deposition.

C - The contact between a-Si:H and various materials have been studied. The most significant results are the following: When a-Si:H is deposited on SiO_2, there a partial reduction of the oxide and constitution of an interface layer made of a mixture of amorphous silicon and SiO_x, with $x < 2$. On Chromium, there is evidence for a nucleation process at the beginning of the deposition, which may induce columnar structure and caudiflower surface morphology. On SnO_2, there is reduction of the oxide and appareance of Sn. This last interface will be the subject of more elaborated studies, owing its technological importance, the analysis will be also extended to the various commercially availiable ITO's.

The contact between C-Si and a-Si:H has been characterised by internal photoemission (M.CUNIOT), The main result is that, at the interface, the valence bands of the two materials are approximately aligned (Fig.2).

D - Investigations of the built-in potentials. Three methods are at present under development to try to answer the important question of the shape of the built-in potential in devices.

a) Carrier transport analysis by TOF and capacitance methods. These well established characterisations have been improved both experimentally and in their modelisations to analyse the space charge regions. It was shown that most of the previously published data are not accurately interpreted as the effect of the built-in field is not properly taken into account, specially at low temperature. This has led to define procedures to investigate the shape of the space charge regions and also to be in position of interpreting the measurements on thin samples as for instance usual PIN diodes.

b) Electron beam induced current on PIN diodes were shown to be sensitive to the internal field, as the pair creation depth may be varied by changing the microscope acceleration potential. The spatial distribution of the created pairs presents a maximum in the sample volume, which is in contrast to the light induced pair creations, this makes the EBIC method more sensitive than the usual spectral response method. However specific difficulties in the present method have to be solved before being able to extract the internal field profile.

c) Electroreflectance properties of a-Si:H may be used to probe the internal field profile. The experiment, involving Michelson interferometry, high frequency modulation and small signal detection is difficult but is now close to be completely in mastered. An unforseen effect was identified and studied: there is a variation of the film thickness induced by the polarisation of the sample, at present an a-Si:H 1µm thich Schottky diode. The effect, of the order of a fraction of tenth of nanometer, allows in fact the determination of the a-Si:H Young modulus. It is then intended as a by product, to seek for a correlation betweem a variation of the Young modulus and the light soaking procedure used to initiate the Staebler Wronski effect.

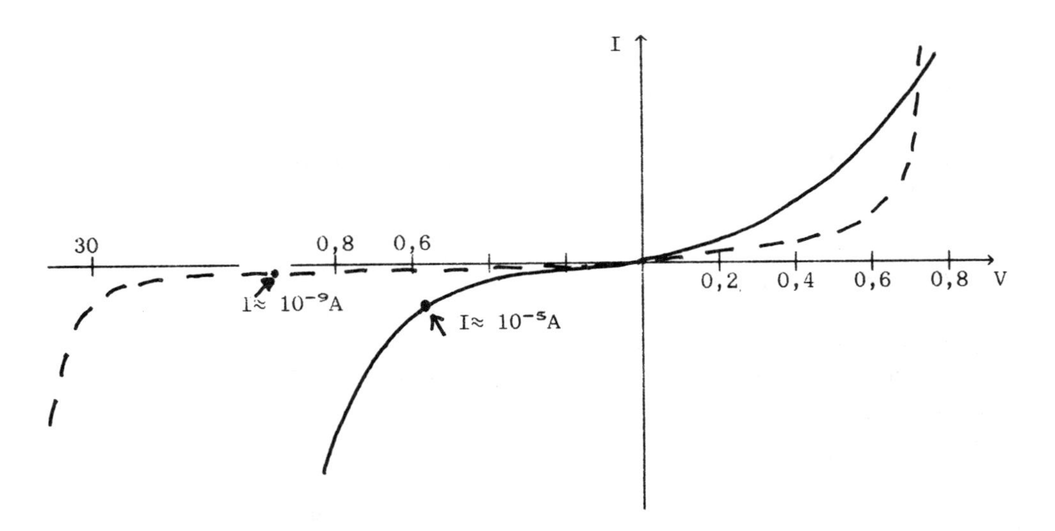

Comparision of the behavior of two identical PIN diodes deposited respectively on SnO_2 (full line) and Chromium (dashed line)

Figure 1

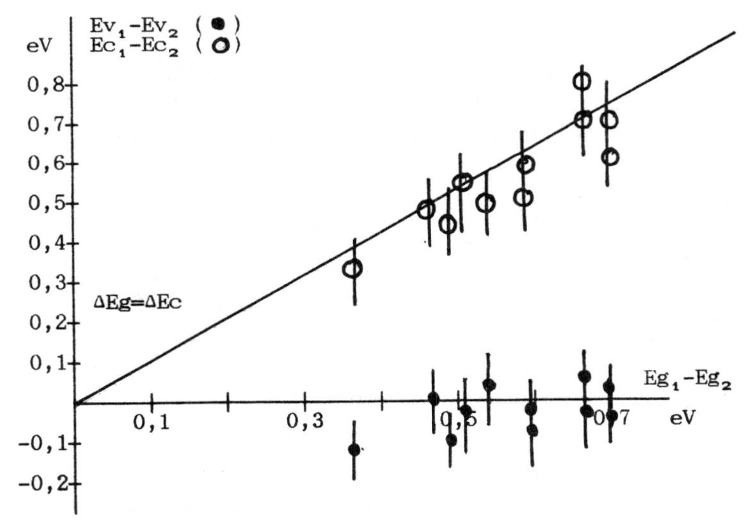

Eg_1 is the a-Si:H optical gap , Eg_2 is the χ-Si gap
Ec_1 and Ev_1 are respectively the a-Si:H conduction and valence band edges
Ec_2 and Ev_2 are the χ-Si conduction and valence band edges

Relative positions of the band edges at the a-Si:H/χ-Si interface as obtained from internal photoemission

Figure 2

DEVELOPMENT OF THE SCIENTIFIC AND TECHNICAL BASIS FOR INTEGRATED

AMORPHOUS SILICON MODULES

LIGHT TRAPPING IN a-Si:H SOLAR CELLS BY TEXTURIZATION OF THE GLASS SUBSTRATE

Author : P. VILATO
Period covered by this report : 31 March-31 August 1987.
Contract Number : EN3S-0055-F (CD).
Duration : 36 months (1 September 1986-31 August 1989)
Total Budget : FF 3 248 000 CEC contribution : FF 1 364 000
Head of project : O. MARIS, Saint-Gobain Recherche
Contractor : Saint-Gobain Recherche
Address : Saint-Gobain Recherche. 39, quai Lucien Lefranc. B.P. 135.
F-93 304 Aubervilliers Cédex (FRANCE). Phone : (1) 48 39 58 00

Summary :

In this work, texturization of glass substrates has been performed in order to trap light in a-Si:H solar cells. Two methods were used : (a) chemical etching and (b) abrasion by high hardness powders. Then these substrates were ITO-coated. Observation was performed by Scanning Electron Microscopy. Samples were also optically and electrically characterized. Chemical etching resulted in pyramidal pits at the glass surface of ≈ 10-20 μm diameter and ≈ 2-5 μm heigh. The mean angle of the slope of the pyramids with the basis of the surface was 14°. Haze ratio at 550 nm was 82%. Geometrical Optics were applied for the description of light scattering. About 1/3 of the light was trapped in this structure. Grains of ≈ 5-25μm were obtained on mechanicaly abraded glass. As no regular topology appeared, Geometrical Optics were difficult to apply. Haze was similar to that found for chemical etching. The proportion of trapped light (≈40%) was larger than for chemical etching but it seems difficult to be improved. Integration of these substrates in a-Si:H cells is being realized by SOLEMS to test suitability of these etching techniques.

1-INTRODUCTION.

In solar cells, trapping light results in increasing optical absorption. (1). This effect can be obtained by texturization of the substrate (glass + Transparent Conductive Oxide electrode) by lithography (2), a TCO layer with large grains (3) or a texturized SiO_2 layer on glass (4). In this work, texturization has been directly performed on the glass substrate, which allows to use our classical TCO deposition technique. In our case, ITO was deposited by spray pyrolysis. The ITO-coated substrates are being evaluated by SOLEMS by integration in a-Si:H cells.

2-RESULTS.

Low ferreous concentration, 1.3 mm thick glass was used in order to minimize light absorption, especially in the red range. Surface etching was performed on one side as it was observed that texturization of both sides led to higher absorption. Two methods for glass etching were used: (a) chemical etching, (b) abrasion by high hardness powders. Then, ITO was deposited on the etched side. The surface of the samples was observed by Scanning Electron Microscopy. Using a He-Ne laser (λ = 633 nm), surface topology was investigated at SOLEMS by light diffraction patterns : sample transmittance and reflectance were recorded as a function of the deviation angle θ (Fig.1). Sheet resistance $R_{\square}$ and spectrophotometric measurements at 550 and 750 nm were also performed. An ITO-coated flat glass substrate was used as a reference (Table I).

2.1. Chemical Etching.

The bath was a solution of LERITE (powder containing ammonium bifluoride) in hydrochloridric acid. Frosting experiments were performed at room temperature. Formation of crystals of ammonium fluosilicate occured at the glass surface. During growth, these crystals protected the underlying glass area, while the other areas were etched. When rinsing after etching, the crystals were removed and their shape remained, giving the surface a characteristic pyramid-like topology (Fig.2a). Pit diameter and height were in the 10-20 µm and 2-5 µm range, respectively. Mean values of optical and electrical measurements are reported in Table I. Reflectance was similar to the value for flat glass substrate but transmittance was slightly lowered by texturization (e.g. 85% to 82% at 750 nm). This was probably due to the light loss by lateral refraction when measuring transmittance rather than an absorption effect. $R_{\square}$ was 15 Ω while it was 10 Ω for the reference. This can be atributed to the fact that, because of the large difference between layer thickness (≈ 230 nm) and the height of the pits, glass was not totally coated by ITO. A simple optical model for light refraction was used in order to geometrically characterize the surface topology from laser beam refraction experiments (Fig.1). A periodic pyramidal structure was assumed (Fig.3). A was the angle between the slope of the pyramid and the flat surface (= the angle of incidence of the light at the glass/ITO interface). B was the angle between the incident and the deviated beams (B = Θ in the transmission mode). In light intensity-Θ plots [I(Θ)], the position of the maximum gave the mean value of B. As $\sin(A+B) = n_g \sin A$ ($n_g = 1.5$ was the glass refractive index), mean value of A was ≈ 14°. In a cell structure, the reflected light by the back electrode is trapped by total internal reflection at the glass/air interface (Fig. 4), i.e. when $n_g \sin 2 A_{tr} = 1$. Then, for a 100 % trapping efficiency, angle A should be at least A_{tr} = 21°. From the I(Θ) plots performed with our structure, the estimated proportion δ of the trapped light at 633 nm was 31%. Similar estimations (Table I) were obtained by spectrophotometric measurements of a texturized sample coated by an Al layer which simulated the back electrode. Sharper pyramids should increase δ Etching solutions of potassium bifluoride in HF were tested with this objective, but no reproductibility and stability of the baths were obtained.

2.2. Frosting by abrasion with high hardness powders.

Abrasion was performed with SiC powders of various granulometries (≈ 5-25 µm). No topology was observed (Fig.2b) and Geometrical Optics were not able to be used Optical and electrical characteristics were similar to those of chemically etched samples (Table I), except haze at 750 nm (73% vs 80.5%) and trapping efficiency δ (41% vs 31% at 633 nm).

3. CONCLUSION.

The I(Θ) plot obtained in the reflection mode after metallization of the texturized side is reported Fig.5. Abraded samples do not exhibit privileged angle of reflection, contrary to the chemically etched samples (reflection centered at 26° (≈ 2 times the A value previously found). Therefore, in spite of its best trapping performance, abrasion is not believed to be as promising as chemical etching, because improvement in δ seems difficult by varying etching condition. Results of tests of the solar cells realized using our both types of substrates are expected to improve or modify etching methods.

REFERENCES.

[1] M.A. Green. "High Efficiency Silicon Solar Cells." Ed. Trans Tech Publ. Aedermannsdorf, Switzerland (1987).

[2] H.W. Deckman and J.H. Dunsmuir. Appl. Phys. Lett. 41 (4), 377 (1982).

[3] H. Ida, N. Shiba, T. Mishuku, H. Karasawa, A. Ito, M. Yamanaka and Y. Hayashi. IEEE Electron. Device Lett. 4, 157 (1983).

[4] M. Misonou, M. Hyodo, H. Nagayama and H. Kawahara. Proc. 18th IEEE Photov. Spec. Conf. Las Vegas. Oct 21-25, 1985. p 925

	550 nm				750 nm				
Samples	T	R	H	δ	T	R	H	δ	$R_{\square}(\Omega)$
Chem. etched	82	9.5	82	33.5	82	7.5	80.5	32	15
Abraded	79.5	10	82	40	80	7.5	73	37	17
Flatglass	86	9	0	0	85	7.5	0	0	10

Table I. Optical (at 550 and 750 nm) and electrical characteristics of ITO-coated glass. T, R, H and δ are transmittance, reflectance, haze and trapping efficiency, respectively, expressed in %. $R_{\square}$ is the sheet resistance.

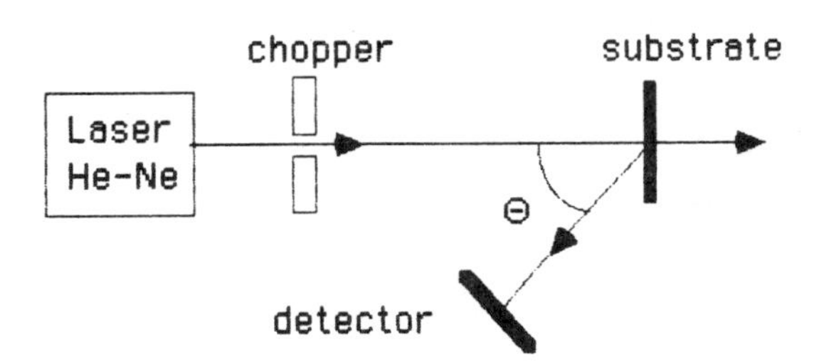

Fig1. Laser beam set-up used to record light intensity I vs θ, either in the reflection mode (as shown here) or in the transmission one.

(a)

(b)

Fig 2a&b - SEM Photographies of texturized glass surface - (a) = chemically etched, (b) mechanically abraded.

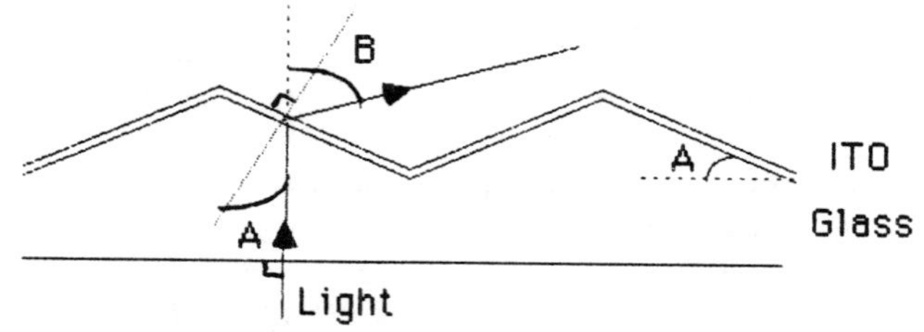

Fig3. Schema of the geometrical model used for the pyramid - like structure. A spatial period is assumed. A = angle of the slope of the pyramids with the flat glass surface, B = deviation of the light through the structure.

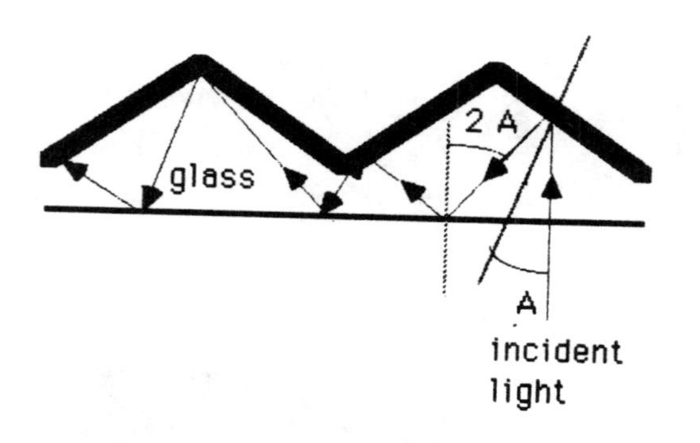

Fig4. Trapping light in a solar cell by total internal reflection of light at the glass / air interface (2A ~ 42° ⇒ A ~ 21°).

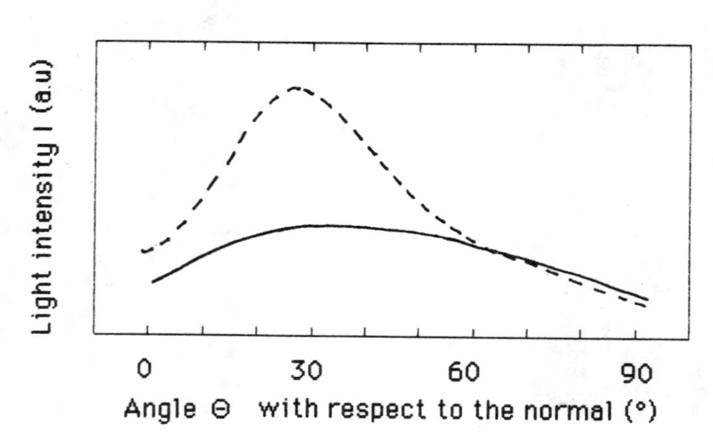

Fig5. Angular distribution I(θ) of the light reflected by
(a) chemically etched glass (dashed line) or
(b) mechanically abraded glass (solid line)
Refraction at the air/glass interface was eliminated by sticking a glass half- sphere on the flat surface of the substrate.

DEVELOPMENT OF THE SCIENTIFIC AND TECHNICAL BASIS FOR INTEGRATED AMORPHOUS SILICON MODULES

PHOTOCHEMICAL PROCESS INDUCED BY A LOW PRESSURE Hg LAMP AND A PULSED ArF EXCIMER LASER FOR DEPOSITING HYDROGENATED SILICON FILMS

Contract number : EN3S/0056

Duration : 36 months

Total Budget : ECU 73 000

Head of Projet : E. FOGARASSY and P. SIFFERT
CENTRE de RECHERCHES NUCLEAIRES - Lab. PHASE
23, rue du LOESS, F-67037 STRASBOURG CEDEX

Contractor : INSTITUT NATIONAL de PHYSIQUE NUCLEAIRE et PHYSIQUE des PARTICULES (IN2P3)

Address : 20, rue Berbier du Mets
75013 PARIS (FRANCE)

Summary :

The growing interest in amorphous materials, in many fields of technology and especially for photovoltaic applications, has stimulated research for new low temperature deposition techniques. Among the new methods, photoassisted processes are particularly attractive. Our laboratory has developed a study on photolytic deposition of hydrogenated amorphous silicon, which consists to dissociate silane gas by using both incoherent and coherent ultraviolet excitation light :

1. a low pressure mercury lamp, providing photons of 253 nm wavelength, which is able to dissociate SiH_4 molecules in presence of mercury vapor by Hg photosensitized reaction ;
2. a high power pulsed excimer laser, operating at 193 nm (ArF) with a typical repetition rate of 60 Hz which is able to dissociate directly SiH_4 molecules by a two-photon excitation process.

This first part of this work dealed with the optimization of the different parameters governing the deposition when using these two different types of light sources. In a second part, it appeared of primary importance to clarify the fundamental chemical reactions involved in these photoassisted deposition techniques by measuring "in situ" the gas phase composition in the reaction chamber during the illumination.

1. COMPARISON OF THE PROCESS INDUCED BY A LOW PRESSURE Hg LAMP AND A PULSED ArF EXCIMER LASER [1]

EXPERIMENTAL AND RESULT

Pure SiH_4 gas sealed in the reaction chamber, at pressures ranging between 0.5 to 500 torr, was irradiated under normal incidence through a suprasil quartz window with two different types of UV excitation sources. The incoherent light was provided by the 253 nm resonance line of a low pressure mercury lamp, giving a light intensity of a few mW/cm^2 at the substrate surface. In this case, the active gas was mixed with mercury vapor from a Hg reservoir usually held at a temperature of 50°C or less. A 20 nsec pulsed ArF (193 nm) excimer laser, operating at a repetition rate of 50 Hz, was used as a coherent light source. The energy density per pulse available close to the surface of the substrate was varied from 5 to 100 mJ/cm^2 by focusing the laser beam with a lens system. In order to evaluate the deposition at ambient temperature on the inner side of the incident window, the film thickness was monitored by optical transmittance at 360 nm provided by a high pressure mercury lamp. At this wavelength, where no photochemical reactions are induced in SiH_4 , the strong absorption coefficient of amorphous silicon ($\alpha \simeq 10^6$ cm^{-1}) is well adapted to follow the initial stages of the deposition (typically $<$ 200 Å).

We report on Fig. 1 the typical transmittances ($I/I_0 = \exp - \alpha x$) recorded as a function of time during the thin film deposition with an ArF excimer laser and a mercury lamp. We can distinguish three main steps which depend strongly on the nature of the exciting light. An extremely short ($\simeq 10^{-1}$ sec) plateau of high transmission ($\simeq$ 100%) with the laser which may be very long (several minutes) when using the mercury lamp, characterizes the initiation time of the dissociation process in the gas phase. The second step, corresponding to the decrease of the transmission following film deposition, may be very sharp (approximately exponential) in the case of the laser. Finally, the transmission decreases to a constant value with a level depending on SiH_4 pressure. Above a few torr, the transmission measured with both the laser and mercury lamp is close to 10% of the initial value. This level corresponds to the deposition of a film of about 200 Å thick.

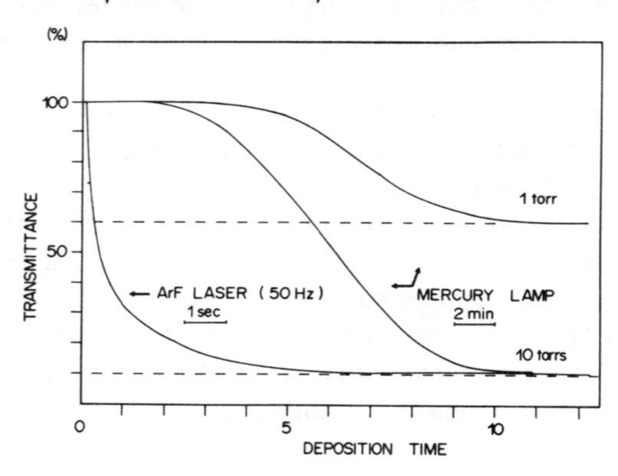

Fig. 1 – Typical transmission at 360 nm through a-Si deposited layers, as a function of irradiation time with laser and lamp (in presence of Hg vapor).

The results of Fig. 2 and 3 show that the laser power and the SiH_4 pressure strongly influence the deposition rates. With a mercury lamp, we measure a deposition rate of about 3 Å/min for an excitation light intensity of $\simeq$ 1.5 mW/cm². This value is about one order of magnitude smaller than the values reported in the literature with light intensities ten times higher [2, 3]. This result confirms the linear dependence on deposition rate with light intensity. By contrast, when using the laser as excitation source, the evolution of the deposition rate with laser energy is strongly non-linear (fig. 2). Below a threshold energy of about 30 mJ/cm², we do not observe any significant deposition. Above this threshold, the deposition rate increases sharply to $\simeq$ 50 Å/sec ($\simeq$ 3000 Å/min). This value which corresponds to the deposition of the first 200 Å is much higher than those already published [4-8] for the deposition of thick layers (> 1000 Å) which did not exceed 600 Å/min.

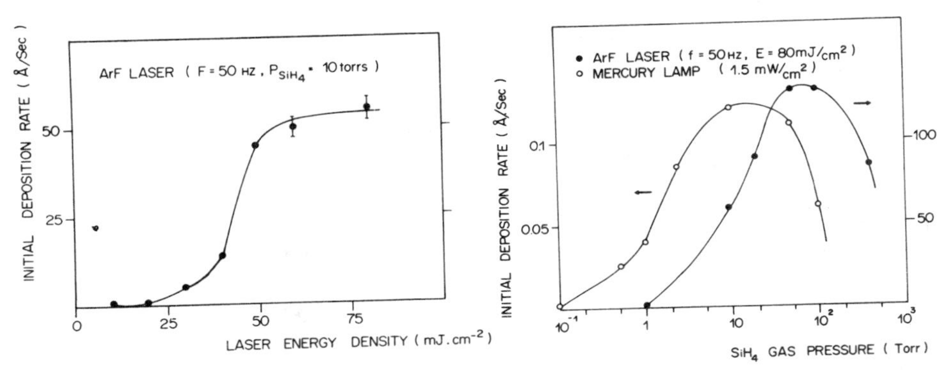

Fig. 2 – Initial deposition rate as a function of ArF laser energy density.

Fig. 3 – Initial deposition rates as a function of SiH_4 pressure for ArF and Hg lamp excitation light.

Finally, when the laser energy exceeds 100 mJ/cm², secondary thermal processes such as recrystallization and evaporation of the deposited layer appear. Film deposition rates also strongly depend on SiH_4 pressure (fig. 3). With mercury lamp, the initial deposition rate begins to increase above a pressure of 0.1 torr, reaching a maximum value around 20 torr, before dropping off rapidly at SiH_4 pressures higher than 50 torr. A similar behaviour is observed with the ArF laser. In this case, the maximum deposition rate is reached at about 100 torr of SiH_4 before dropping off at higher pressures by the formation of large amounts of powdery silicon.

DISCUSSION

The experimental results show that UV photoexcitation of SiH_4 molecules can be achieved through two different pathways which depend on the wavelength and intensity of the light source. With a low pressure mercury lamp, Hg atoms mixed with the SiH_4 are excited by the 253 nm resonance line :

$$Hg\ (^1S_0) + h\nu\ (254\ nm) \rightarrow Hg^*\ (^3P_1)$$

and transfer their energy to SiH_4 molecules by collision :

$$SiH_4 + Hg^* \rightarrow SiH_4^* + Hg$$

With the ArF excimer laser, photons of 6.4 eV (193 nm) are not able to directly dissociate SiH_4 by single photon excitation, because the significant absorption starts only above 7.8 eV (160 nm). However the decomposition of SiH_4 by a two photon excitation of molecular electronic states might occur above a laser threshold energy of about 30 mJ/cm^2 as can be deduced from experimental results of Fig. 2, which shows the non-linear behaviour of the process. In either form of excitation, the nature of the decomposition of photoexcited SiH_4^* molecules into reactive species and intermediate radicals remains a subject of conjecture. Two main primary reactions have been proposed [9, 10] :

(1) $SiH_4^* \rightarrow SiH_3 + H$ $\qquad \Delta H = -\ 91.71\ Kcal.mole^{-1}$ $\qquad (3.98\ eV.mole^{-1})$

(2) $SiH_4^* \rightarrow SiH_2 + 2H$ $\qquad \Delta H = -164.16\ Kcal.mole^{-1}$ $\qquad (7.14\ eV.mole^{-1})$

A third reaction :

(2') $:SiH_4^* \rightarrow SiH_2 + H_2$ $\qquad \Delta H = -\ 59.96\ Kcal.mole^{-1}$ $\qquad (2.61\ eV.mole^{-1})$

sometimes mentioned, appears to be energetically more favourable than reaction (2). In fact, it may not be considered as a primary step since it results from reaction (2) followed by the exothermic recombination of H radicals releasing 104.2 $Kcal.mole^{-1}$.

The quantum yields of reactions (1) and (2) depend strongly on excitation conditions and especially on the wavelength of light source. With the 253 nm (4.88 eV) resonance line, the decomposition of SiH_4 by Hg sensitization is only possible through reaction (1) which needs 3.98 eV. Under these conditions only SiH_3 radicals are created. By contrast the bi-photonic absorption of coherent excitation light from the ArF laser may induce both reactions (1) and (2), leading to the formation in the gas of SiH_2 and SiH_3 radicals. The radical species responsible for film deposition and growth strongly depend on secondary reactions which take place in the gas phase. The secondary reactions more generally considered are [2, 10] :

(3) $H + SiH_4 \rightarrow H_2 + SiH_3$

(4) $SiH_3 + SiH_3 \rightarrow SiH_4 + SiH_2$

(5) $SiH_2 + SiH_4 \rightarrow SiH_2H_6$

At low SiH_4 pressures, reactions between radicals, such as reaction (4) and consequently reaction (5) may be neglected. Then, only SiH_3 radicals resulting from (1) and (3) contribute to the deposition by the Hg sensitization process. This behaviour leads to a linear dependance on the deposition rate with SiH_4 pressure and light intensity. This is in good agreement with results reported on Fig. 3 for $P_{SiH4} \ll 1$ torr. This qualitative description can also be used to explain the UV laser excitation processes. In this case, we must also consider the possibility to form directly SiH_2 radicals by the primary reaction (2) which may recombine with SiH_4 molecules following the reaction (5).

At high SiH_4 pressure, the processes are dominated by the formation of clusters directly in the gas phase that leads to the deposition of powdery silicon. Cluster formation is clearly observed with both incoherent and coherent excitation and could be responsible for the sharp decrease of deposition rate when $P_{SiH4} > 10^2$ torr (fig. 3). In the intermediate range of SiH_4 pressures, interactions between SiH_3 radicals via reaction (4), the formation of Si_2H_6 via reaction (5) and their dissociation through reactions leading to the creation of Si_3H_8 molecules and higher radical species such as Si_2H_4 and Si_2H_5 should be considered.

II. GAS PHASE COMPOSITION DURING THE Hg PHOTOSENSITIZED DECOMPOSITION OF SiH_4

EXPERIMENTAL AND RESULTS [11]

It is of primary importance to clarify the fundamental chemical reactions involved during the mercury photosensitized decomposition of SiH_4 . Different experiments such as mass spectrometry and gas phase chromotography have been already proposed to analyze directly the various reactive species and intermediates photolytically created in the gase phase which are responsible on the film growing. In this work, the gas composition has been determined as a function of irradiation time, by measuring the partial pressure of different species resulting from the dissociation of SiH_4 sealed in the reaction chamber. This simple technique used here consists, in a first step, to lower the temperature of the reaction chamber at 77K to freeze silane and its higher derivates (Si_2H_6 , Si_3H_8 ...), in order to measure the partial pressure of hydrogen. In a second step, after pumping the H_2 molecules, the contributions of SiH_4 and Si_2H_6 are successively recorded by heating the system above their boiling temperature (T_{SiH4} = 161K, T_{Si2H6}= 259K).

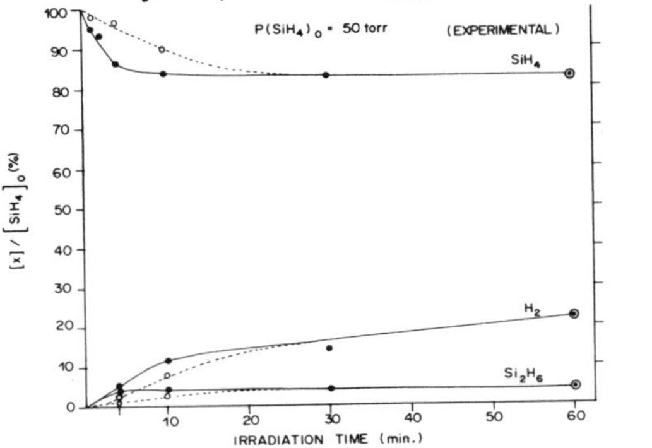

Fig.4 – Gas phase composition as a function of the irradiation time at $P(SiH_4)_0$ = 50 torr. Curve (——) : freshly cleaned cell window ; curve (--) : after 10 minutes, same experiment repeated without any window cleaning.

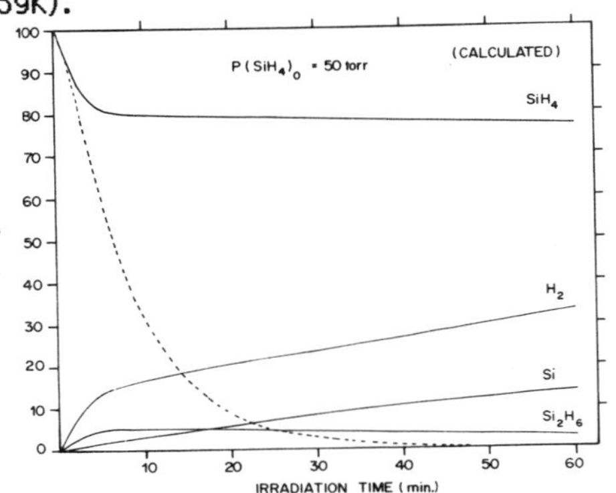

Fig. 5 – Calculated gas phase composition and amount of silicon deposited as a function of the irradiation time ($P(SiH_4)_0$ = 50 torr).

Figure 4 shows experimental results recorded for an initial silane pressure of 50 torr. Initially, the gas phase composition is changing rapidly before reaching a constant value following approximately 10 minutes of irradiation time. This plateau may be related to the existence of a state on chemical equilibrium.

Two arguments support this hypothesis. The optical transmission at 253 nm through the window shows that the equilibrium state is reached before deposited silicon causes complete opacity to the U.V. photons.

DISCUSSION AND MODELING

In order to take into account on this photostationnary state in the modeling, we have to included to the chemical reaction scheme already discussed in the previous paragraph. (1), (3), (4), (5), reverse reactions leading to the regeneration of the SiH_4 molecules. Several reactions have been proposed in the literature [12–14], however it is reasonable to consider only those having the highest rate constants :

(6) $H + Si_2H_6 \rightarrow SiH_4 + SiH_3$

(7) $H_2 + SiH_2 \rightarrow SiH_4$

This scheme must be completed by the reactions describing the deposition on the walls :

(8) $SiH_3 + \text{wall} \rightarrow SiH_x \text{ (solid)} + 1/2\,(3 - x)\,H_2$

(9) $SiH_2 + \text{wall} \rightarrow SiH_x \text{ (solid)} + 1/2\,(2 - x)\,H_2$

The different chemical rate constants (k) corresponding to these reactions and used in our calculations have been detailed in reference 11.

The gas phase composition obtained by using this model and the percentage of Si atoms deposited on the walls are recorded on fig. 5 for initial SiH_4 pressures of 50 torr.

The same calculations have been performed without the reverse reactions (6) and (7). In this case, silane molecules are rapidly dissociated (fig. 5 – dashed line) and completely consummed after a short period of time (approximately 20 mn). This behaviour is not in agreement with our experimental results and justifies a posteriori, the necessity to include reverse reactions in the modeling.

CONCLUSION

In this work, we focused our attention on fundamental photochemical reactions involved when silane gas is sealed in the reaction chamber. These studies will be extended in the future to SiH_4 gas flow conditions in order to analyze the processes induced during the deposition of thick silicon films which are required for device applications.

REFERENCES

[1] C. FUCHS and E. FOGARASSY in "photon beam and plasma stimulated chemical processes at surfaces ", MRS–Boston (Dec. 1986).

[2] J. PERRIN, T. BROEKHUIZEN, in "laser processing and diagnostics II",

edited by D. Bauerle, K.L. Kompa, L. Laude, Les Editions de Physique (1986) p. 129.

[3] N. MUTSUKURA, Y. MACHI, Appl. Phys. B41, 103 (1986).

[4] R.W. ANDREATTA, C.C. ABELE, J.F. OSMUNDSEN, J.G. EDEN, D. LUBBEN and J.E. GREENE, Appl. Phys. Lett. 40, 183 (1982).

[5] M. MURAHARA, K. TOYODA, in "laser processing and diagnostics V39, edited by D. Bauerle (Springer Verlag, Berlin, 1984) p. 252.

[6] A. Yamada, M. Konagai, K. Takahashi, Extended Abstract : laser chemical processing of semiconductor device, presented at the 1984 MRS Meeting, Boston, p. 29.

[7] H. ZARNANI, H. DEMIRYONT, G.J. COLLINS, J. Appl. Phys. 60, 2523 (1986); K. SUZUKI, D. LUBBEN and J.E. GREENE, ibid. 58, 979 (1985).

[8] A. YOSHIKAWA, S. YAMAGA, Jpn. Journ. of Appl. Phys. 23, L91 (1984).

[9] H. NIKI, G.J. MAINS, J. Phys. Chem. 68, 304 (1964).

[10] G.G.A. PERKINS, E.R. AUSTIN, F.W. LAMPE, J. of. Am. Chem. Soc. 28, 1109 (1979).

[11] B. AKA, C. FUCHS, E. FOGARASSY and P. SIFFERT in "photon, beam and plasma enhanced processing" EMRS, Strasbourg (june 1987).

[12] P.A. LONGEWAY, R.D. ESTES and H.A. WEAKLIEM, J. Phys. Chem. 88, 73 (1984).

[13] G.G.A. PERKINS and F.W. LAMPE, J. Amer. Chem. Soc. 102, 11, 3764 (1980).

[14] T.L. POLLOCK, H.S. SANDHU, A. JODHAN and O.P. STRAUSZ, J. Amer. Chem. Soc. 95, 4, 1017 (1973).

DEVELOPMENT OF THE SCIENTIFIC AND TECHNICAL BASIS FOR INTEGRATED AMORPHOUS SILICON MODULES

VACUUM UV PHOTO CVD FOR AMORPHOUS SILICON CARBON ALLOYS

Contract Number	:	EN3S-0058-D (B)
Duration	:	36 months, 1 Oct. 86 - 30 Sep. 89
Total Budget	:	399.000 DM, CEC Contribution: 399.000 DM
Head of project	:	Dr. G.H. Bauer, Institut fuer Physikalische Elektronik
Contractor	:	University of Stuttgart Keplerstr. 7 D-7000 Stuttgart 1

Summary

A vacuum uv photo CVD reactor for the direct photodissociation of SiH_4, Si_2H_6, B_2H_6 and of hydrocarbons has been built up consisting of a deposition chamber, a vuv light source (D_2 discharge lamp) and a gaseous transmission filter for the adjustment of transmission edges. The generation rate of radicals has been calculated using photodissociation cross sections and the spectral distribution of the source in combination with filter transmission. The filter cut off frequency has been continuously shifted across the relevant threshold energies for the formation of different types of radicals by use of O_2 as filter gas at different partial pressures. From simulation of the radical concentration in the reactor growth rates in the range of 0.3 Å per second have been estimated for disilane based films.

1. Introduction

In amorphous silicon pin solar cells commonly the p^+-layer consists of a silicon carbon alloy in order to compensate the reduction in bandgap caused by diborane doping. However, B_2H_6 doping ends up in some more departures of optical and structural film properties of p-a-SiC:H in comparison to undoped films (1). There is great evidence that these effects do not result from the boron incorporated in the films, but from the gas phase/plasma chemistry or from surface reactions in silanes in the presence of diborane or boron hydride radicals (2).

Although the efficiency of conventionally B_2H_6 doped amorphous pin structures have recently reached 12% (3), an improvement can be expected by a reduction of states at the p-i interface decreasing the output current.

Recent activities for the deposition of p^+-layers by photo-CVD methods point towards an improvement of the electronic quality of the p-i interface. Since silanes in the range of $h\nu \leq 6$ eV show only negligible optical cross sections for the direct decomposition and in order to provide a noticeable increase in film growth a deposition system for vacuum-uv-photo-CVD preparation ($h\nu = 6$ eV - 10 eV) has been proposed (4).

2. Experimental Setup

Our deposition system for vacuum-photo-CVD consists of three chambers separated by vuv-transmissive MgF_2-windows ($\lambda \geq 115$ nm).

i) a chamber for the low power D_2-discharge lamp (typically 15 W, Cathodeon Ltd.) flowed with He or Ar for cooling.
ii) a gaseous transmission filter system for the continous adjustment of transmission edges versus wavelength (O_2 at different partial pressures) cutting off the high energetic range of the spectral distribution of the lamp. The filter consists of two MgF_2-windows, spaced 2 cm from each other.
iii) a deposition chamber.

In order to measure the photon flux across the filter at various gas pressures a grating monochromator with photomultiplier is attached instead of the reactor chamber.

3. Experimental results

In a first step the filter has been evacuated to 10^{-5} mbar in order to get the influence of window absorption (Fig. 1). The spectrum is cut off at 133 nm (9.3 eV), so the strong Lyman line at 121.6 nm (10.2 eV) has been absorbed by the two 6 mm thick MgF_2-windows (compare with the upper curve in Fig. 1, which gives an absolute measurement of this lamp type by Key and Preston (5)).

Matching our spectrum to the absolute scale the maximal number of photons contributing to dissociation of SiH_4 and Si_2H_6 can be determined:

$$F = \iint S(\lambda)/h\nu \times \{1 - \exp[-nd\sigma(\lambda)]\}d\lambda d\Omega$$

S -spectral radiance of the lamp
n -gas density
d -light path in the gas
σ -eff. cross section
Ω -sol. angle of the lamp (0.0537 sr)

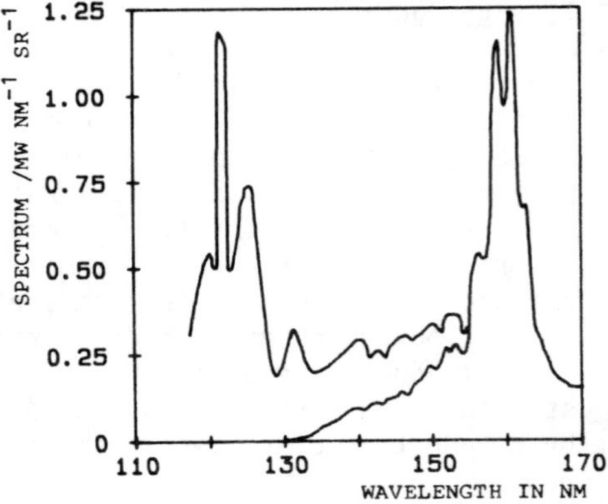

Fig. 1: Spectrum of the D_2-lamp without/ with MgF_2 windows.

With the effective cross sections of SiH_4 and Si_2H_6 (6) we get 1×10^{14} and 6×10^{14} absorbed photons per second resp. with our reduced spectrum. Without MgF_2-windows the number of absorbed photons would be 5×10^{14} for SiH_4 and 9×10^{14} for Si_2H_6.

If the geometric arrangement of the deposition chamber is chosen in a way, that about 10 % of the generated reactive species contribute to film growth on the substrate (1 cm^2) we can estimate in a rough approximation a growth rate of 0.3 Å per second for disilane based films. Compared to conventionally glow discharge deposited films growth rate is reduced by an order of magnitude, but taking into account the thickness of the solar cell window layer ($\approx$10 nm), the additional time spent for film deposition is no severe drawback.

In Figs. 2 and 3 the number of absorbed photons per wavelength is outlined for SiH_4 and Si_2H_6 together with the thresholds for the formation of different types of radicals. We recognize, that different photodissociation processes and ratios of reaction products (radicals) can be controlled and adjusted by a suitable cut off filter for the exciting vuv radiation.

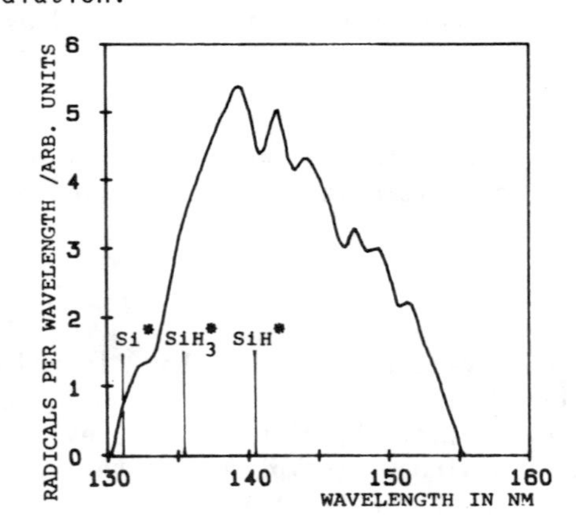

Fig. 2: Absorbed photons per wavelength in SiH_4 (D_2 lamp illum.).

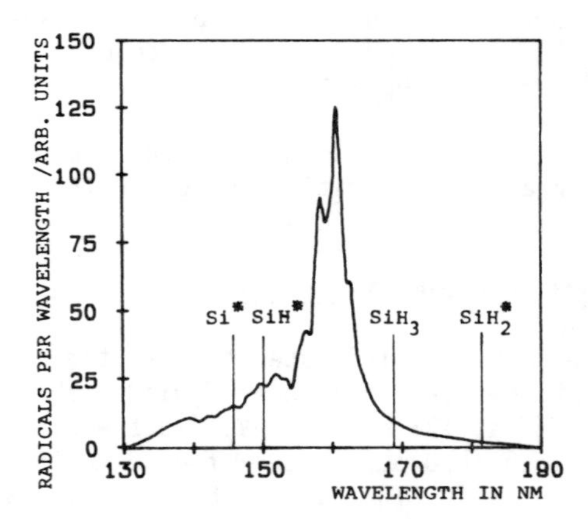

Fig. 3: Absorbed photons per wavelength in Si_2H_6 (D_2 lamp illum.).

Nitrogen has been found to be completely transparent above 130 nm and can't therefore be used as filter gas. Very promising results are obtained with O_2, which has an absorption continuum (Schumann-Runge continuum) in the relevant wavelength region. Below 175 nm O_2 photodissociates to produce one ground state and one excited state oxygen atom (7):

$$O_2 + h\nu \rightarrow O(^3P) + O(^1D)$$

Fig. 4 shows the transmission of the O_2 gaseous filter at different partial pressures. It is clearly demonstrated, that there is a pronounced cut off energy which can be controlled by the O_2 pressure.

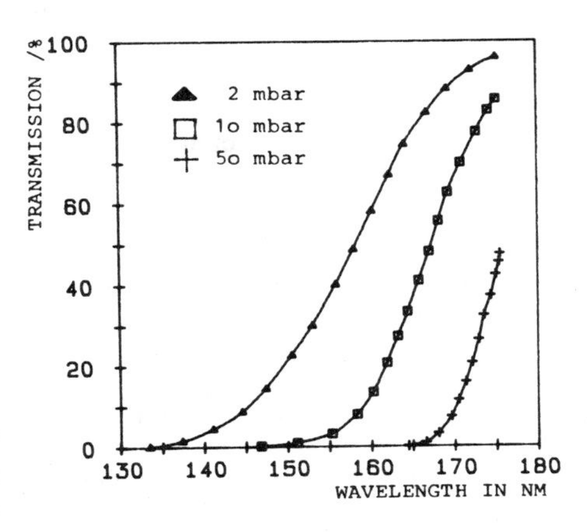

Fig. 4: Filter transmission at different O_2 partial pressures.

4. Calculations

In order to describe the spacial distribution of radicals in the reactor two model calculations have been carried out as borderline cases of the actual deposition chamber geometry:

I. The neutral gas inlet is a small pipe in the centre of the substrate:

The diffusion equations are

$$dn/dt = D\Delta n \qquad \text{and} \qquad J = -D\,\nabla n$$

and the equation of continuity is: $dn/dt + \nabla nv = 0$

n -density of gas (SiH_4 or Si_2H_6)
D -diffusion constant
J -flow rate per unit area
v -velocity of gas

In steady state: $dn/dt = 0$ and $\Delta n = 0$

For a point source the solution of the Laplace equation is well known from electrostatics and can be adopted. The solutions for our semispherical problem are:

$n = Q/2\pi Dr$ Q -total flow rate

$v = D/r$ r -distance from gas outlet

$J = Q/2\pi r^2$

The equation for the density of photogenerated radicals n^* is:

$$dn^*/dt + \nabla n^* v^* = \iint S(\lambda)/h\nu \times n\sigma(\lambda)\exp[-nd\sigma(\lambda)]d\lambda d\Omega - n^*/\tau$$

generation G recombination R

τ -lifetime

With the assumptions of homogeneous absorption ($\exp[-nd\sigma(\lambda)] \approx 1$, $v^* \approx v$) and steady state ($dn^*/dt = 0$) the equation can be solved.
Near the gas outlet the recombination term can be neglected and the density of activated species arises as r, and in the region where the diffusion length is smaller than r, n^* drops as τ/r.

II. Homogeneous density of neutral gas:
The one dimensional diffusion equation (perpendicular to substrate surface) is in steady state

$$-D \times d^2n^*/dz^2 = G - n^*/\tau$$

Assuming homogeneous generation G and the boundary conditions $dn^*/dz = 0|_{z=0}$ (window) and $n^* = 0|_{z=d}$ (substrate) the density distribution is

$$n^*(z) = GD\tau \{1 - [sh(z\sqrt{D\tau})]/[sh(d\sqrt{D\tau})]$$ sh -hyperbolic sine

Taking actual data for diffusion coefficient D, lifetime τ, neutral gas flow rate and sticking propability we calculate growth rates in accordance with the rough estimation from chapter 3.

With help of these simple model calculations the parameter dependency under realistic experimental operation conditions (geometry, neutral gas flow, inhomogeneous generation, different radicals and diffusion constants) can be simulated to give density distribution and flow rates of the activated species in the reactor.

References

(1) H.-D. Mohring et al., Proc. Int. Symp. On Trends And New Appl. In Thin Films, Vol. 1, p. 97 (1987)
(2) M.B. Schubert et al., Proc. 7 EC PVSEC, D. Reidel, Dordrecht (NL), p. 489 (1987)
(3) S. Nakano et al., Proc. 7 EC PVSEC, D. Reidel, Dordrecht (NL), p. 425 (1987)
(4) G.H. Bauer, CEC contractors meeting, Brussels, 1986 in Photovoltaic Power Generation, D. Reidel, Dordrecht (NL), p. 38 (1987)
(5) P.J. Key et al., J. Phys. E 13, 866 (1980)
(6) H. Stafast to be publ. in Appl. Phys. B
(7) K. Watanabe et al., Air Force Cambridge Res. Center, Bedford, Mass., Tech. Rep. No. 53-23, (1953)

DEVELOPMENT OF THE SCIENTIFIC AND TECHNICAL BASIS FOR INTEGRATED AMORPHOUS SILICON MODULES

"PREPARATION OF IMPROVED AMORPHOUS SILICON CELLS WITH p-i-n STRUCTURE BASED ON NEW VOLATILE HYDRIDES AND FLUORIDES OF SILICON AND GERMANIUM"

Contract Number: EN3 S - 0059-D (B)

Duration : 36 months 1 Jan. 1987 - 31 Dec. 1989

Total Budget : Univ. Budget CEC Contr. DM 580.000.-

Head of Project: Professor Dr. H. Schmidbaur,
Anorganisch-chemisches Institut,
Technische Universität München.

Contractor : Technische Universität München.

Address : Anorganisch-chemisches Institut,
TU München, Lichtenbergstraße 4,
D-8046 Garching.

Summary

Synthetic studies have been carried out on a series of compounds with the general formula $(H_3Si)_{4-x}CH_x$ where x = 1, 2, 3, and 4 ("polysilylmethanes"). The first three members of this class of volatile materials are now readily available and are subject to testing in glow discharge and photochemical CVD experiments. None of the compounds is spontaneously inflamable in air, a considerable advantage in handling the substrates. Work on related germanium analogues has started and is in progress. The compounds also have potential for the plasma or photochemical generation of non-conduction surface coatings.

Introduction

Presently the methods of production of amorphous silicon used in photovoltaic devices are mainly based on the decomposition of silane and mixtures of silane with other gases. Materials containing carbon are mostly obtained from CVD of silane/methane mixtures, and diborane or phosphine are used for doping. If desired, fluorine is introduced via volatile fluorides, again of silicon, boron, phophorus or other elements. Higher silanes, like disilane or trisilane, have also been tested.

Attempts to use gases of compounds with more than one or two of these elements present in the same molecules have been limited to only a few species, mainly those commercially available. In the present study, a program has therefore been initiated which is aiming at the preparation of new gaseous compounds containing silicon, hydrogen, carbon, and other elements that can be used for vapour deposition of amorphous silicon and its alloys. The project also considers analogous compounds of germanium.

A few papers on previous work carried out in these laboratories have appeared (1,2,3,4) or are in print (5,6). These publications also give references covering some of the pertinent literature.

Experimental Methods and Techniques.

The compounds prepared for the CVD experiments are handled under inert gas or in a vacuum system under rigourous safety precautions. Their purity is routinely checked by gas and liquid phase infrared and Raman spectroscopy, as well as by gas/liquid chromatography and high resolution mass spectrometry (GC/MS).

Multinuclear magnetic resonance spectroscopy is employed for identification and structure elucidation of new compounds. Elemental analyses are by atomic absorption spectrometry.

Structures of crystalline solids (precursors or derivatives) are determined by single crystal X-ray diffraction, gas phase structures by electron diffraction.

Ionization potentials are measured by UV photoelectron spectroscopy.

CVD experiments, as well as physical characterization and specification of CVD products, are carried out in external laboratories.

Chemical synthesis is by standard procedures of preparative inorganic and organometallic chemistry. Details are given in the publications (1-6).

Results.

Three different classes of compounds containing silicon, carbon and hydrogen have been investigated. Species with a high silicon contents have been given priority, but simple model compounds have received due attention first.

Poly-silyl-alkanes.

The homologous series in which the hydrogen atoms of methane are substituted by up to four silyl groups is one of the most promising classes of materials.

Methylsilane CH_3SiH_3 is readily available through standard routes at low cost. It is stable and not spontaneously inflamable. First results on the plasma CVD products are available.

Disilylmethane $CH_2(SiH_3)_2$ is accessible through two different synthetic routes in good quality and at acceptable cost. Starting materials are non-sophisticated commercial products. The compound is gaseous at room temperature and not spontaneously inflamable, storable at ambient temperature, but more readily decomposed (thermally, photochemically, and in the plasma) than methylsilane. Studies of the plasma CVD products are in progress.

Trisilylmethane $CH(SiH_3)_3$, a new compound, is still not readily available due to the low yields of existing methods of synthesis, but has interesting properties as a CVD precursor molecule. It is sufficiently volatile at ambient temperature and gives products with a high silicon contents.

Tetrasilylmethane $C(SiH_3)_4$ could not be prepared. All experiments carried out in this direction have only given desilylated product. Efforts to avoid this Si-C cleavage are continued.

Poly-silyl-alkanes.

Among compounds with at least one C-C bond, 1,2-disilyl-ethane is an important volatile target molecule. Two methods of preparation have been tested and improved such as to yield the compound at acceptable cost. $H_3SiCH_2CH_2SiH_3$ is very stable at room temperature and not inflamable. It can be purified to high grade and stored indefinitely.

1,3-Disilyl-propane $H_3SiCH_2CH_2CH_2SiH_3$ has similar properties and characteristics, but lower volatility. It has now been made readily available.

Poly-silyl-alkenes.

Vinylsilane and allylsilane are easily obtained from commercially available chlorinated precursors. Their handling needs care and expertise, but presents no fundamental problems though there is more potential danger. Thermal decomposition starts at lower temperatures than with the related alkanes, and photodecomposition is more selective. No experiences with plasma decomposition have been reported. The production of acetylene may be an important special feature.

$$(H_3Si)_2C=C(SiH_3)_2$$

Only very few polysilylated ethenes have been reported in the literature. Tetrasilylethene (Formula) is still an unknown compound, as are other species bearing only SiH_3 substituents. Experiments oriented at the development of synthetic methods for such compounds are continued.

Disilylethyne (disilylacetylene) $H_3SiC{\equiv}CSiH_3$ can be synthesized by convenient routes in acceptable yields. Studies of the CVD properties are in progress.

Hydrosilylation of these alkenes and alkynes is an important route to more sophisticated materials containing other heteroatoms and a higher degree of silylation.

Polygermylmethanes.

Methylgermane H_3CGeH_3 is readily obtained from convenient starting materials. It is stable to water (liquid and vapour), does not inflame in air, and can be stored indefinitely at ambient temperature. Its CVD properties are under investigation.

Digermylmethane $CH_2(GeH_3)_2$ is an analogue of disilylmethane (above). A preparative procedure has been established from precursors generated in the Direct Synthesis using germanium metal. Quantities available are now sufficient for further study. Fluorine derivatisation is an important aspect.

Conclusions

It has been demonstrated that synthetic methods can be found for a whole series of volatile silicon- and germanium-containing compounds of unknown decomposition properties. They are all potential candidates for CVD of amorphous silicon and its alloys. They offer potential for further derivatisation to include heteroatoms for modification of electrical and optical properties. First plasma CVD experiments with these compounds are encouraging and will be continued. It is important that the new materials show advantages regarding handling and safety, and that their cost is not out of proportion as compared to the high purity gases presently in use. Purification to high grade quality is possible.

It should be pointed out that the new compounds also have potential as starting materials for the plasma, thermal, or photochemical generation of surface coatings, e.g. silicon carbide. This aspect will be given some attentionas the work progresses.

Acknowledgement

This work was also supported by Deutsche Forschungsgemeinschaft (Leibniz-Programm) and by Verband der Chemischen Industrie. Prof. Winterling and his group at MBB is thanked for assistance and many helpful discussions. Wacker-Chemie has kindly provided organosilicon chemicals as key starting materials, and Metallgesellschaft has generously supplied alkylating and hydride reagents.

References

(1) J. Ebenhöch, Dissertation, Technical University of Munich, 1987.

(2) C. Dörzbach, Dissertation, Technical University of Munich, 1986.

(3) H. Schmidbaur, J. Ebenhöch, G. Müller, Z. Naturforsch. 42b (1987) 142.

(4) H. Schmidbaur, J. Ebenhöch, Z. Naturforsch. 41b (1986) 1527.

(5) H. Schmidbaur, C. Dörzbach, Z. Naturforsch. 42b (1987) 1088.

(6) H. Schmidbaur, J. Ebenhöch, Z. Naturforsch. in press.
H. Schmidbaur, J. Ebenhöch, G. Müller, ibid. in press.

DEVELOPMENT OF THE SCIENTIFIC AND TECHNICAL BASIS
FOR INTEGRATED AMORPHOUS SILICON MODULES

RESEARCH ON A-Si:F:H (B) ALLOYS AND MODULE TESTING AT IER - CIEMAT

Authors : M.T. Gutierrez, P. Román and L. Delgado

Contract Number : EN35-0091-E (B)

Duration : 32 months 1 August 1986 - 31 March 1989

Total Budget : 89.000.000 Pts CEC contribution: 30.140.000 Pts

Head of Project : Dr. L. Delgado

Contractor : CIEMAT - IER

Address : Avda. Complutense, 22
Madrid 28040. SPAIN

Summary

This report summarizes progress work carried out from 1.04.87 to 30.09.87 by CIEMAT-IER within the frame of the joint european program on a-Si. During this six-month period we have investigated the deposition of fluorinated p-doped films as window layers for solar cells in an attempt to correlate film properties with deposition parameters. It has been found that the use of fluorine prevents the degradation of the bandgap with progressive boron doping, while a dark conductivity increase of up to 7 orders of magnitude can be obtained.
Under the module testing task an outdoor stability test facility as well as some indoor acelerated stresses were brought into operation. Results on two different a-Si module technologies are presented.

1. Introduction.

The present report describes the research effort made by CIEMAT-IER on the following tasks i) material research:p-a-Si:H:F layers, ii) module testing.

In the search for alternative doping processes, alleviating the observed degradation in photoelectronic film properties with conventional (B_2H_6) p-type doping our research focus on fluorinated p-doped layers.

Work on module testing concentrates on outdoor and indoor stability test as an attempt to elucidate photo and thermal degradation mechanisms - in a-Si modules.

Progress work on the subcontracted tasks iii) layer-CVD deposition - and iv) theoretical modelling,will be reported in following progress reports.

2. Description of apparatus and measuring equipment.

Experimental details on the rf plasma reactor, deposition procedure and gases used as well as on the characterization facilities were provided in previous progress reports (1). Thus,only some new details will be given here. Concerning chamber cleaning the use of higher power levels and process pressures in argon diluted gas mixtures lead to some plasma polymerization and powder formation in the chamber. Therefore manual chamber cleaning were sometime needed in addition to the normal cleaning cycles (CF_4 8% O_2).

Besides, more information on the hydrogen bonding structure was obtained by the hydrogen thermal effusion spectra kindly recorded at the University of Barcelona by Prof. J.L. Morenza and Dr. G. Sardin.

Under the module testing task an outdoor module test station was set up. Radiation and electrical module output data are recorded periodically at the same daytime, corrected for module temperature and radiation level and stored in a data logger. A reference a-Si module and a pyranometer are used as reference devices.

3. Results.

3.1. Material research: a-Si:F:H (B) alloys

The variation of dark conductivity σ_d, band gap Eg, deposition rate R, and slope S of the tauc plot as a function of the boron gas fraction into the gas mixture, X is presented in Fig. 1.

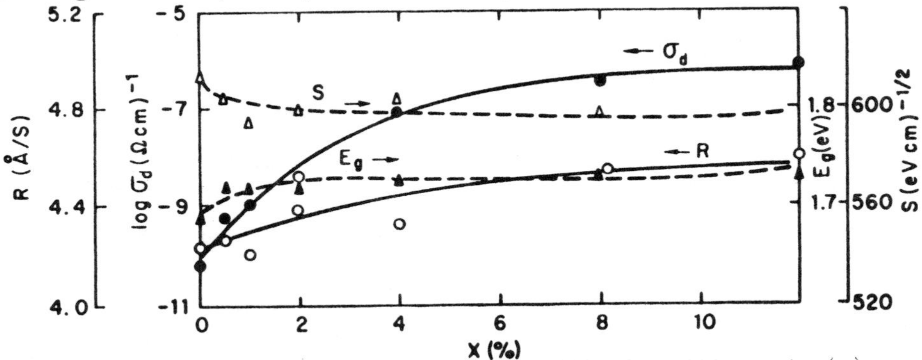

Fig.1. Dark conductivity (•), band gap (▲), deposition rate (o) and slope (Δ) versus the boron gas fraction into the gas phase.

At first typical deposition conditions for pure SiH_4, i.e. total pressure 0,5 T, substrate temperature 250ºC, were used, but similar parameter tendencies were found at 0.75 T and 150ºC. In each case the boron gas flow was varied systematically while SiH_4 gas flow and rf power were held constant at 100 sccm and 18 mw/cm^2.

As expected σd increases up to 4 orders of magnitude as X is increased which means that effective doping is really achieved. However and in contrast with that normally found in diborane doped a-Si C:H films, Eg remains unchanged throughout the doping region (2).

After an initial decrease the slope S remains nearly constant from X=0.05 to X=0.1.The diminution in the slope of the Tauc plot $(\alpha E)^{1/2}$ vs E. is related to film quality degradation (density of defects) as boron incorporation into the film increases(3). Nevertheless, the slope degradation observed here is much lower than that reported for B_2H_6 doped films deposited at the same deposition rate (rf power level).

The effect of substrate temperature (Ts) on σd and Eg is presented in Figure 2. σd shows a maximum ~200ºC, decreasing at higher temperatures. This

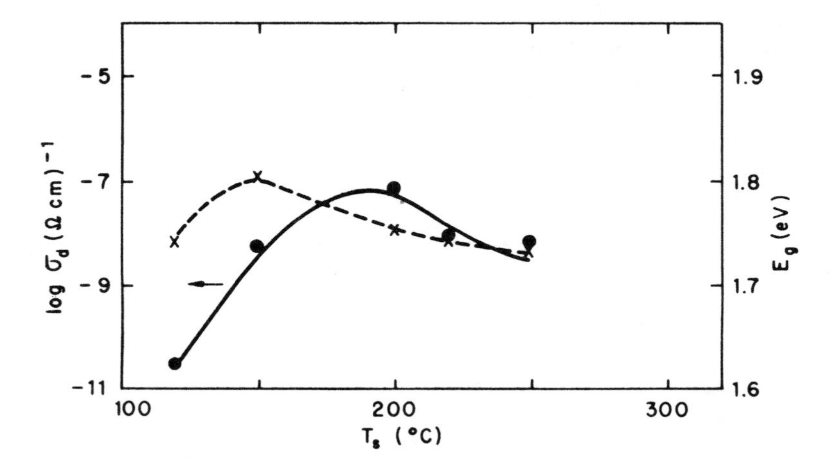

Fig.2. Variation of σd (•) and Eg (X) as a function of substate temperature

seems to indicate that the incorporation of B as electrically active acceptor is enhanced around this temperature. Further deposition at higher total pressure (1T) and lower rf power density (9 mw/cm^2)shows a displacement in the maximum of σd at lower temperatures ~150ºC. In a similiar way Eg starts to decrease at Ts>150ºC. Since it is known that plasma degradation of the transparent electrode (layer on top of which the p-window film should be deposited) starts from ~180ºC, the possibility of depositing high transparent and conductive layers at low temperatures (<180ºC) is a matter of technological interest.

The influence of process pressure P, on σd, Eg, S and R at such low temperature (150ºC) is shown in fig. 3.

From this figure it is clear that at higher total pressure σd, Eg and S increase while deposition rate decreases. This may indicate that despite lower reactant partial pressures at lower total pressures, input power is higher resulting in higher deposition rates and poor film quality (lower S).

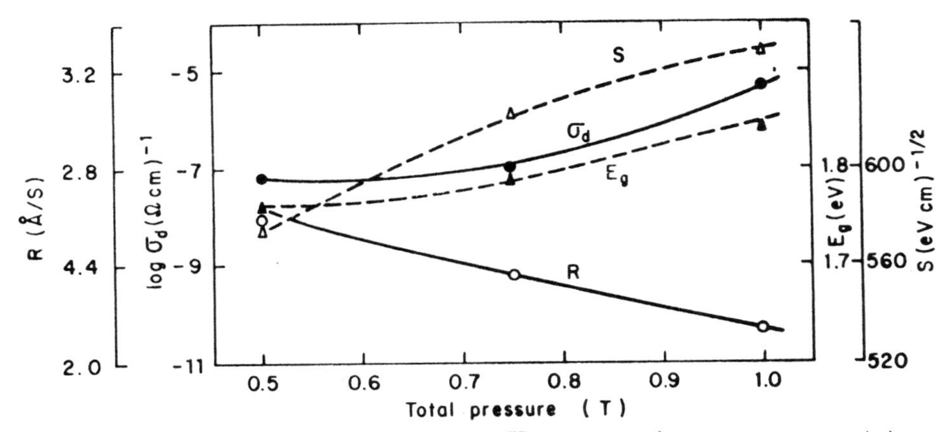

Fig.3. Effect of process pressure on σ_d (●), Eg (▲), S (Δ) and R (o)

The variations in Eg are normally related to the hydrogen bonding structure, thus some attention was paid to the IR and hydrogen thermal effusion spectra. As normally observed in intrinsic a-Si:H films with decreasing Ts from 250 to 150ºC, there is a shift of the 2000 cm^{-1} peak (Si-H stretching) to 2100 cm^{-1} (SiH_2) while the $(SiH_2)_n$ peaks in the 800 - 900 cm^{-1} region emerge. The same tendency is also observed with increasing dopant gas fraction X, process pressure and r-f power density (see fig. 4)

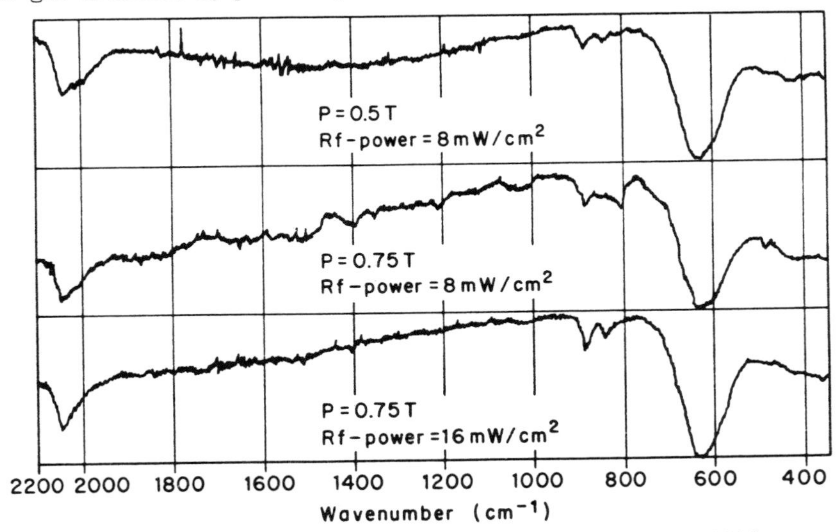

Fig.4. IR spectra of a-Si:H:B films X=0.02, Ts= 150ºC

The total hydrogen content as determined by the IR and thermal hydrogen evolution spectra increase from ~15% at X=0% to ~18.5% at X=2%.

X-ray diffraction patterns of the a-Si:H:F (B) films showed flat spectra characteristics of amorphous phases, without any trace of microcrystallization. Since, it has been reported (4) that in SiF_4/H_2 or SiF_4/SiH_4 - gas mixtures, the presence of fluorine promotes microcrystallization, a typical deposition procedure to obtain μc-films, i.e.: dilution in an inert (argon) gas and higher power levels, was applied. The results are shown in fig.5. where the variation of σ_d, Eg and R with rf-power is presented.

As previously observed in intrinsic a-Si:H films (1) at power levels > 9 mW/cm^2, R decreases with increasing rf-power, probably due to argon ions bombardment onto the films, so that at 80 mw/cm^2, no film deposition occurs.

Contrary to that observed with undiluted gas mixtures, σ_d increases and Eg decreases with increasing rf power from power densities $>$ 8mW/cm^2.

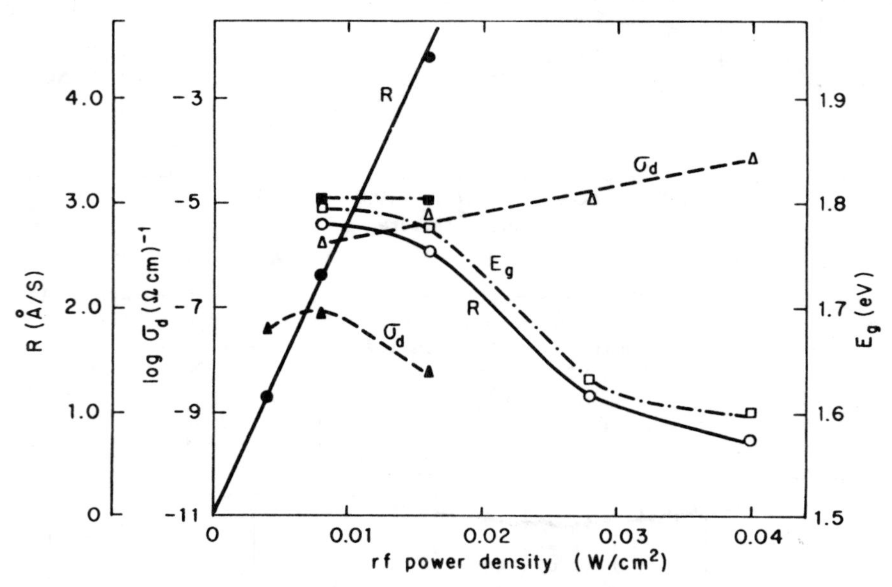

Fig.5. Variation of σ_d (Δ,▲), R (o,●) and Eg (□,■) with rf power density.
Δ,o and □ 150ºC, 1T, SiH_4/A= 10%
▲, ● and ■ : 150ºC, 0.75 T. undiluted gas mixtures

As it was found in films prepared without argon dilution into the gas phase, σ_d increases with increasing process pressure, and no traces of microcrystalline phases could be detected by X-ray difraction analysis. A maximum in σ_d with substrate temperature is found at 250ºC ($\sigma_d \sim 3.5 \times 10^{-3}$ Ω^{-1} cm^{-1}). However the use of argon as a dilution gas is not considered as satisfactory, due to bandgap degradation and deleterious powder formation in the chamber at the higher power levels used. This can be a major drawback while making devices.

3.2. Module testing.

In figure 6 the results of different outdoor and indoor degradation tests on two different amorphous silicon technologies: single cell on glass (module 1) and tandem cell on stainles steel (mod. 2) are presented. The different range of conditions explored: uv+vis light soaking test (B), thermal cycling (C), outdoor sun exposure (D), room light storage (E) and dark storage under short circuit conditions (F) are representative of a real case during qualification.

As can be seen different technologies lead to different deposition patterns influenced by sample history. Tandem cell modules seem to be more stable under outdoor exposure. Most of the power otuput degradation is due to fill factor. A slight recovery in Voc after dark storage is observed in tandem modules.

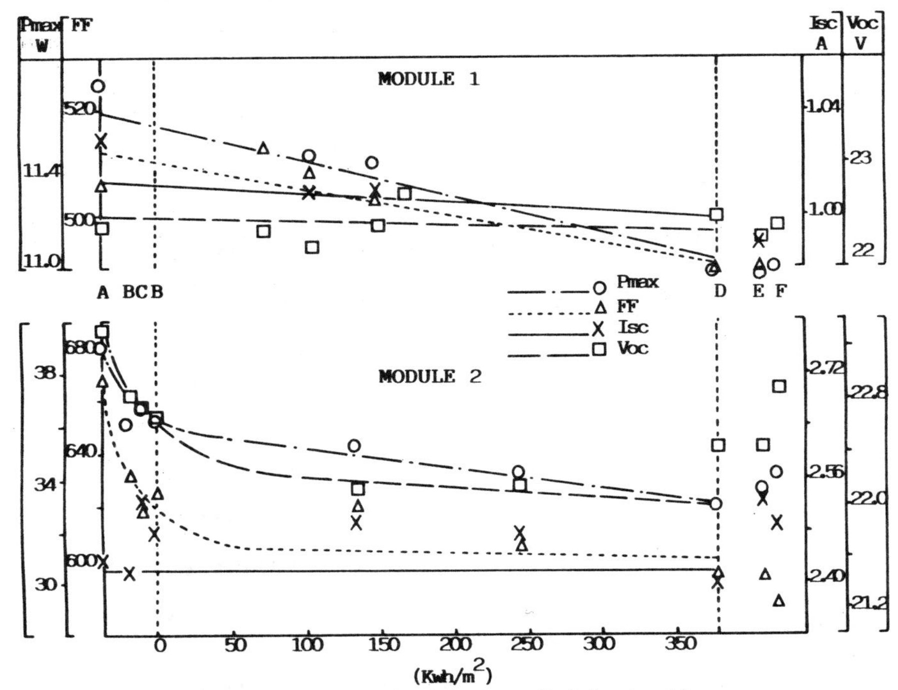

Fig. 6. Outdoor and indoor Module testing

4. Conclusions.

Good quality a-Si:F:H (B) films, as judged by $\sqrt{\sigma d}$ and Eg, can be obtained at low temperatures (150ºC) in a large range of dopant concentration, without deterioration of the band gap. Similar characteristics have been reported in the case of microcrystalline p-doped (B_2H_6) layers deposited at 300ºC (5), in which the nucleation of microcrystallites would enhance both the H bonded concentration (mainly in the dihydride mode, at the surface of the microcrystallites) and the carrier mobility.

The lack of microcrystalline phases in the a-Si:F:H (B) films presented here, points to the role of fluorine, in the gas mixtures used, to explain our results.

With respect to the module testing task the metastable performance of the a-Si modules depends on the sample history and on the measurement method. The prime loss occur in the FF and further work is needed to correlate the loss in FF with material (density of defects,...) or device (series resistance,...) properties.

References.

1. L. Delgado et al "Development of the scientific and technical basic for integrated amorphous silicon modules. Contribution by CIEMAT-IER Progress reports. Octubre 1986. Abril 1987.
2. T. Hamasaki et al; Appl. Phys. Lett., 37, 1084 (1980)
3. F.B. Ellis and A.E. Delahoy. Solar Energy Mater, 13, 109 (1986).
4. H. Matsumura and S. Furukawa. JARECT. Vol. 6, 88 (1983)
5. K. Tanaka and A. Matsuda, JARECT. Vol. 6, 161 (1983)

"LOW GAP ALLOYS FOR AMORPHOUS SILICON BASED SOLAR CELLS"

NARROW BAND-GAP ALLOYS FOR THE IMPROVEMENT OF EFFICIENCY OF AMORPHOUS SILICON-BASED SOLAR CELLS

Contract Number	:	EN 3S-0062-F (CD)	
Duration	:	36 months	1 june 1986-31 may 1989
Total Budget	:	6 361 000 FF	CEC contribution : 6 361 000 FF

Head of Project	:	J.Bullot, Groupe des Matériaux Amorphes, LPCR, Université Paris-Sud, Orsay
Contractor	:	Centre National de la Recherche Scientifique,
Address	:	15 Quai A.France, 75007, Paris

Summary

Progress in the preparation of hydrogenated amorphous silicon-germanium alloys in the "ARCAM" reactor, under various conditions, is described. The main physical properties : optical gap,refractive index, density of defects, Urbach edge parameter, conductivity and photoconductivity efficiency are measured to monitor films preparation. Hydrogen incorporation, which turns out to be the key parameter of preparation, is studied by means of infrared spectroscopy as a function of various deposition parameters. Detailed physical investigations on the alloy a-$Si_{0.82}Ge_{0.18}$:H, prepared in the ARCAM reactor mock-up, known to have good electronic properties, are also presented. They include determination of the density of states at the Fermi level by capacitance measurements, determination of the density of states profile from the mobility edge down to 0.3eV, of the electron mobility in the extended states, electron capture cross section and attempt to escape frequency by the Time Of Flight technique. In addition the Schottky contact prepared from this alloy is shown to have good barrier height and good forward current to reverse current ratio. Finally results on modelling of single and tandem solar cells are given. They show that a tandem cell made of a-Si:H (top layer) and an alloy having an optical gap ≈ 1.4 eV(bottom layer) has an efficiency ≈ 11.5% under AM1. The preparation of such a high-quality alloy by the Glow Discharge technique is our present goal.

1. Introduction

In this paper we report on research performed on narrow band gap amorphous semiconductors in the laboratories of the french ARC groups. In section 2 details on the glow discharge (GD) preparation in the "ARCAM" reactor of of a-$Si_{1-x}Ge_x$:H alloys and on their characterization are given. In section 3 detailed physical studies on the alloy a-$Si_{0.82}Ge_{0.18}$:H, prepared in the ARCAM reactor mock-up -which in previous reports was shown to be a high-quality material- are also presented. Finally in section 4 we describe the results obtained on modelling of single and multi-layer solar cells.

2. Materials research studies on ARCAM-produced SiGe alloys

In previous reports (1) we described results obtained on films prepared in the prototype of the ARCAM reactor. As the design of the GD ARCAM reactor stems from an advanced technology the decision was made to start preparing alloys in this reactor. There are many advantages to learn preparing a-$Si_{1-x}Ge_x$:H films that way : firstly the reactor is fully automated, secondly the possibility is offered to dope n or p the materials and thirdly the multi-chamber design allows complex structures like p-i-n or superlattices to be prepared without breaking vacuum.
Although ARCAM a-Si:H high-quality specimens are now routinely produced (see data on film 71 60 44 in table II), clearly the shift to a-SiGe:H which implies the use of SiH_4+GeH_4+H_2 ternary gas mixtures poses new problems.

The preparation program started december 1986 and so far many different deposition conditions were scanned, a selection of which are presented in this paper (table I). The resulting films were characterized by three fast routine techniques : Optical Transmission Spectroscopy (OTS) to measure the optical gap expressed in this paper as E_{04},i.e. the energy for which the aborption coefficient $\alpha=10^4$ cm^{-1}, Photohermal Deflection Spectroscopy (PDS) which yields informations on disorder (Urbach tail parameter E_0 (meV)) and the density of defects N_d, Photoconductivity (PC) to measure the $\eta\mu\tau$ product (η=quantunm yield, μ=electron microscopic mobility, τ=electron lifetime). Infrared Spectroscopy (IRS) was also used to study hydrogen incorporation in the silicon network when x is increased

2.1. Glow Discharge Deposition Parameters

In addition to the usual deposition parameters : vessel pressure p_e, partial gas flow rate d_i of the ith component of the gas mixture, total gas flow rate , RF power, bias voltage and substrate temperature T_s, in the ARCAM reactor a conspicuous parameter is the pressure inside the chamber p_c where the plasma is held (2). This pressure is quite different from that in the vessel p_e where the three plasma chambers are located. Detailed studies of the influence of the pumping system design upon p_e, p_c at various total flow rates have been carried out with the aim of getting necessary high pressures and low flow rates while keeping a laminar gas flow .

The previous results having shown the beneficial role of hydrogen dilution on film properties, it was decided to systematically dilute germane in hydrogen and we turned to the use of a 10 % germane-90 % hydrogen mixture (dilution r=0.9). The gas parameters are described in table I. Films 70 13 02,70 13 03 and 70 32 41 were prepared in the absence of hydrogen (r=0, d_2=0) whereas all others were synthesized with a 90% hydrogen dilution.

2.2. Characterization of alloy films

Table II shows physical data on SiGe specimens prepared under different conditions. For the sake of comparison we include data on films A (a-Si:H) and C (a-$Si_{0.82}Ge_{0.18}$:H) prepared in the ARCAM mock-up reactor.

The germanium concentration x in the film was measured by Electron Probe Micro Analysis (EPMA). In figure 1 the relationship between the germane concentration in the gas and the germanium content in the solid is shown. It is seen that, as a consequence of the large decomposition cross section of germane with respect to silane (about a factor 2), the Ge content in the film is much higher than that in the gas phase.

OTS data (figure 2) point out the linear decrease of E_{04} when x is increased, a feature already observed by several authors. In this figure we present the x dependence of the Penn gap G_{Penn}-which is a properly weighted average of the valence band to conduction band optical transitions (3).

In figure 3 is shown the dependence of the refractive index n upon the germanium content in the solid. A sharp variation of n is observed when small amounts of Ge are incorporated into the silicon network. A similar behavior is observed for the dark conductivity at room temperature (fig.4) and its activation energy E_a .

The data concerning the Urbach edge E_0 (the open symbols are PDS data and the close ones are optical data) shown in figure 5 reveal a substantial increase of the band tail width above the valence band with x, i.e. an increase in disorder. Typical E_0 values for a-Si:H are of the order of 55 meV. We may anticipate a similar broadening of the conduction band tail; such a correlation has been observed in amorphous silicon-carbon alloys (4). Recent Time Of Flight measurements (5), to be described in section 3, bring support to this point. Therfore we may anticipate that electron trapping should be more efficient in these alloys than in a-Si:H. Experiments along these lines will be undertaken in the future.

2.3. Hydrogen bonding upon incorporatin of Ge atoms into the silicon network

Examination of Table II reveals that apparently the $\eta\mu\tau$ product, which is a measure of charge transport efficiency, is not correlated to the total density of defects N_d measured by PDS : low N_d materials may have poor photoconductivity yields (e.g.films 70 13 02 and 13 03). This might be explained either because PDS cannot measure the density of the defects/recombination centers, located in the upper half of the gap, or because structural inhomogeneities are formed when x is increased as suggested by some authors.

Hydrogen incorporation in the network is a key issue and to gain some understanding we carried out infrared spectroscopy measurements in the 200-4000 cm^{-1} range. The results are as follows.

Firstly, the study of Si-H and Ge-H wagging modes (630 and 570 cm^{-1}) allows the total concentration of incorporated hydrogen, H%, and the ratio R=[Si-H]/[Ge-H] to be measured (fig.6). As seen in table II, H% remains of the order of 6-10 but more importantly R is systematically larger than one, reaching values as high as 3.7 for small x. This means that H attaches preferentially to Si atoms rather than to Ge ones leaving dangling bonds on Ge atoms.

Secondly, stretching modes analysis (1840-2100 cm^{-1}) shows that there are more dihydride than monohydride bonds in these films, SiH_2 being predominant (fig.7).

Analysis of the bending modes allows the isolated SiH_2 bonds to be distinguished from the $-(SiH_2)-$ polysilane network which absorbs at 845 and 890 cm^{-1}. Fig.8 shows the strong absorption at these wavenumbers when the germanium content is increased and when the substrate temperature is decreased. These results are in fair agreement with the H exodiffusion studies (6).
Such observations have far reaching consequences because the presence of a polysilane network indicates that an hydrogen-rich low-density tissue is formed, known to have poor electronic properties. On the contrary films containing only monohydride Si-H bonds are high-quality samples made of a high-density network. This is the situation encountered in the alloy a-$Si_{0.82}Ge_{0.18}$:H (C in tables I and II) in which a single absorption peak at 2000 cm^{-1}, characteristic of Si-H is found .

Clearly the hydrogen incorporation scheme in the silicon network determines the properties of the materials. Incorporation of excess hydrogen is likely due to gas phase polymerization but as demonstrated above (fig.8) substrate temperature also plays a role.

3. Density of states, charge transport and device studies on a-$Si_{0.82}Ge_{0.18}$:H

This GD film was grown from a 50% SiH_4-50% (GeH_4+H_2) mixture in the ARCAM prototype. Characterization techniques have shown that this film has an optical gap of 1.56 eV (E_{04}=1.73 eV), a rather low PDS density of defects and a photoconductivity yield close to that of a-Si:H (table II). Under these conditions it was quite interesting to study in more detail the density of states (DOS) in the gap and charge transport.

3.1. Density of States at the Fermi level

Capacitance (C) and conductance measurements were performed as a function of temperature (7). The curves at a frequency f = 75 Hz are shown in fig.9. From the slope of the linear region of $C^2(dC/dT)^{-1}$ vs temperature, we obtain the density of states at the Fermi level $N(E_F)=6\times10^{16}$ $cm^{-3}eV^{-1}$, a low value for this kind of material in agreement with PDS data. We recall that the DOS at E_F for high-quality a-Si:H specimens is typically $\approx5\times10^{15}$ $cm^{-3}eV^{-1}$.

3.2. Density of states in the conduction band tail and transport parameters

Time of Flight measurements and associated new methods of data interpretation were used to determine transport parameters : microscopic mobility μ, electron capture cross section σ, attempt to escape frequency ν and the shape of the conduction band tail density of states (5). For this purpose a short and strongly absorbed flash of blue light is shined onto the 1-2 µm thick specimen having a Schottky structure. A pulsed field whose duration is smaller than the dielectric relaxation time is applied to the sample to collect charges and measure the drift time.

Several conclusions could be drawn. The main point is that in the range 0-0.3 eV below the mobility edge, the DOS may be described by a single exponential band tail

$$N(E)= (N_0/kT_C)\ \exp(-E/kT_C)$$

with a characteristic temperature $T_C=T_1=470$ K. In contrast in a-Si:H two exponential distributions are found with $T_1=370$ K and $T_2=200$ K (see fig.10). This change in the DOS shape is responsible for the observed drift mobility decrease with respect to a-Si:H. The extended states mobility in the alloy ($\mu=21cm^2/Vs$) is roughly the same as that of a-Si:H ($\mu=33cm^2/Vs$) but σ is about 75 times larger and ν 90 times larger.

3.3. Schottky diodes

The purpose of this work was to test the ability of this alloy to form a good Schottky contact (7). A platinum contact deposited on the film allowed the I-V characteristics to be measured (fig.11) for the forward (solid line) or reverse biased junction (dotted lines) at room temperature. The forward characteristics exhibits a linear shape for V < 0.3 volts. In this region, the current-voltage dependence is of the type :

$$J = J_S\ \exp\ (qV/nkT)$$

J being the current density (q=elementary charge, T=temperature),with n = 1.24 and $J_S = 1.2\times10^{-9}$ A cm^{-2}. Then assuming :

$$J_S = A\ T^2\ \exp(-q\ \Phi_b\ /\ kT)$$

with A = 110 A cm^{-2} K^{-2}, we deduce the barrier height q_b = 0.85 eV, showing the high quality of this alloy. This result has been confirmed from the temperature dependence of the reverse current at a fixed bias, leading to the same activation energy $q\Phi_b$.

The ratio of direct to reverse current at an applied bias of 1V is in the range 2×10^4 to 10^5 (fig. 11) at room temperature.

4. Solar cell modelling

Modelling of solar cells has been performed in a joint program between France and Italy (8). Taking into account different series of meteorological data and using a computer program which allows the chief parameters of single and multi-junctions tandem cells to be changed some interesting conclusions were reached concerning band gap optimization.

The device structure under study is made up by one to three p^+-i-n^+ elementary cells connected in series, with decreasing band gap. The top p^+ layer (front window) has a 2eV gap (a-SiC:H). For each elementary cell, the thickness w_i and the optical properties are taken as those of the intrinsic layer. The following physical model was used : the transport of photogenerated carriers is due to their drift inside the i-layer where the photogeneration rate G is assumed constant, so that the photocurrent is given by :

$$J_{ph} = q\, G\, l_c\, [1 - \exp(-w/l_c)]$$

where

$$l_c = (\mu_n \tau_n + \mu_p \tau_p)\, F = \langle \mu\tau \rangle . F$$

is the carrier collection length and μ and τ are respectively the drift mobilities and the electron (n) and hole (p) lifetimes. For non short-circuit conditions a constant electric field is assumed

$$F = (V_{bi} - V) / w$$

V_{bi} is the built-in potential and V the polarization.

An optimization procedure is used to find the maximum efficiency η, by varying the gap E_{gi} and the thickness w_i of the i-layer.

Results

The maximum efficiencies under AM1 conditions for a single cell and a two-cell stack were examined. For a typical a-Si:H cell (E_g=1.75 eV) and $\langle\mu\tau\rangle$=5x10^{-8} and qV_{bi}/E_g=0.5 an 11 % efficiency may be obtained. The efficiency of a single cell was studied as a function of thickness of intrinsic layer for different values of $\langle\mu\tau\rangle$ on the one hand and as a function of the optical gap for different values of thickness on the other.

More importantly fig.12 shows the iso-efficiency plot for 2-gap tandem cells under AM1. An optimum efficiency of 15.9% is reached for a front layer with E_{gtop}=1.25 eV, W_{top}=0.11 µm and a bottom layer with $E_{gbott.}$=0.70 eV, $w_{bott.}$=0.13 µm.

The conclusions drawn from this work are as follows : the model used does not show any significant variation in the device efficiencies with the spectral distribution of solar radiation. The gain in going from 1 to 2 junctions is significant but there is no advantage in going to three junctions device. It has to be noted that the maximum efficiency (~ 16%) for a 2-layer cell corresponds to $1 < E_{gtop} < 1.5$ eV and $E_{gbott.} < 1$ eV, i.e. very low E_g values but with E_{gtop} ~1.75 eV (a-Si:H) and $E_{gbott.}$ ~1.5eV (a-$Si_{1-x}Ge_x$:H with x~0.4-0.5) an efficiency ~11.5% is still obtained.

5. Conclusions and future

We recall that our project is a three-step program, the first and most important one being the development of reactors and routine characterization facilities . This step should end up in the preparation of high-quality materials allowing detailed physical investigations and device studies to be made.

The present situation may be summarized as follows. The ARCAM reactor is operational and the preparation of a-$Si_{1-x}Ge_x$:H alloys under way.The routine characterization techniques proved to be fast and efficient. At the moment the most important challenge is to discover what the right preparation parameters are and along this line two parallel strategies can and will be followed.

The first one consists in scanning the multi-dimensional parameter space of GD synthesis, systematically varying the correlated parameters to reach the best set and using the routine characterization techniques to asses the materials quality.

The second one consists in the detailed study of materials containing small amounts of germanium, ~1-10%, to understand hydrogen incorporation in the network, the nature of defects : dangling bonds, hydrogen-rich structural inhomogeneities and the correlated changes in the transport properties. Such an approach proved to be extremely useful in the case of silicon-carbon alloys (4). Data presented in this report show that, under the preparation conditions we used, various physical quantities e.g. dark conductivity and refractive index, suffer an abrupt variation when a few germanium atoms are added to the silicon network; but as shown by infrared spectroscopy a high concentration of polysilane bonds is formed when the germanium content is increased. Studies on plasma free radicals (12), in-situ ellipsometric investigations (11), hydrogen exodiffusion (6), Soft X-ray and Photoelectron Spectroscopy (9,13),time of flight (5), dual-beam modulated photoconductivity (10) and other techniques will allow valuable informations to be obtained in this direction.

Solar cell modelling shows that to reach high efficiencies, materials with very low optical gaps, ~1eV, should be prepared. It is not clear at the present time if such materials are within the scope of GD preparation of SiGe alloys. In any case modelling also shows that a tandem cell made of a-Si:H (Eg=1.75eV) and a-SiGe:H (Eg=1.4eV) has an efficiency ~11.5%. Such an alloy can be prepared by GD from silane+germane+hydrogen mixtures and its synthesis remains our target.

Finally we wish to emphasize that collaboration between the european contractors has become a reality and that the workshop on deposition techniques, to be held in november 1987, will certainly be an important step towards a closer collaboration.

References (*)

1. **J.Bullot**, Report n°1,november 1986; report n°2 june 1987
2. **P.Roca i Cabarrocas,** unpublished data
3. **I.Solomon,M.P.Schmidt,C.Sénèmaud and M.Driss-Kodja**, "Band-strucutre and average gap models in a-SiC:H alloys", to be presented at the 12th Int.Conf.Amorphous and Liquid Semiconductors, Prague, august 1987
4. **J.Bullot and M.P.Schmidt,** " Physics of amorphous silicon-carbon alloys. A review", phys.stat.solidi, in press
5. **R.Vanderhagen and C.Longeaud,** "Experimental determination of hydrogenated amorphous silicon and amorphous silicon-germanium alloy electron transport parameters from time of flight experiment", to be presented at the 12th Int.Conf.Amorphous and Liquid Semiconductors, Prague, august 1987
6. **J.L.Morenza**,this conference
7. **J.P.Kleider**, Thesis,Paris, june 1987
8. **D.Chello, I.Chambouleyron and P.Baruch,** "Band Gap Optimization For Amorphous Solar Cells ; A Comparative Study ", Proceedings of the 19th IEEE Photovoltaic Solar Conference, New Orleans, may 4-8 1987
9. **C.Cardinaud, C.Sénèmaud and G.Villela**, "Electronic structure of amorphous SiGe alloys studied by soft X-Ray and photoelectron spectroscopy",J.Non Cryst. Solids,88,55 (1986)
10. **J.Bullot, P.Cordier, M.Gauthier and G.Mawawa,** "Dual Beam Modulated Photoconductivity in hydrogenated amorphous silicon", Philos.Mag.B, 55,599(1987)
11. **A.M.Antoine,B.Drévillon and P. Roca i Cabarrocas,**"In-situ investigation of the growth of RF glow discharge deposited amorphous germanium and silicon films",J.Appl.Phys.,61,2501,(1987)
12. **M.Oria, L.Abouaf-Marguin, A.Lloret, B.Séoudi**, "Plasma free radicals detected by the matrix isolation technique", to be presented at the 8th Int. Symp. on Plasma Chemistry, Tokyo, august 1987
13. **C.Sénèmaud,C.Cardinaud and B.Drévillon,** "Caractèrisation du silicium amorphe hydrogéné et d'alliages silicium-germanium par spectroscopie X",2ème Colloque national sur le silicium,Paris,mai 1987, to be published in "Annales de Chimie"

(*) selected references from ARC publications

TABLE I

Deposition conditions of selected a-$Si_{1-x}Ge_x$:H films prepared in the ARCAM reactor (**)

N°	RH	RG	$d_1{}^a$	$d_2{}^a$	$d_3{}^a$	$d_t{}^a$	$p_t{}^b$	$p_e{}^b$	$p_c{}^b$	W^c
701302	0	8	1.6	0	19	20.6	34	-	-	2
701303	0	11	1.6	0	13.4	15	28	-	-	2
703241	0	9	1.1	0	10.9	12	-	34	65	2
703244	7.3	9	1.1	8	10.9	20	-	27	94	2
703246	7.3	9	1.1	8	10.9	20	-	9	77	2
704085	9	1	1	0	14	15	-	27		2.7
705045	9	1	1	0	14	15	-	44		2.2
705042	9	2	2	0	13	15	-	35		2
705043	9	4	4	0	11	15	-	35		2
705044	9	8	7	0	8	15	-	35		3
716044	a-Si:H		0	0	10	-	-	30	40	2

(a) sccm; (b) mTorr; (c) watts

(**) $d_1 = d(GeH_4+H_2)$; $d_2 = d(H_2)$; $d_3 = d(SiH_4)$

$RG\% = (1-r)d_1 / [(1-r)d_1 + d_3]$; $RH = [rd_1 + d_2] / (1-r)d_1$; $r = 0.9$

TABLE II

Physical data on a-$Si_{1-x}Ge_x$:H films prepared in the ARCAM reactor

N°	RH %	RG	H[a] %	x[d]	R	n	E_{04} eV	E_0 meV	N_d[b] cm^{-3} $\times 10^{17}$	$\eta\mu\tau$[e] cm^2V^{-1}
701302	0	8	6.7	0.44	1.19	-	-	74[b]	0.56	2.9×10^{-10}
701303	0	11	6.2	0.5	1.33	3.65	-	74[b]	0.45	-
703241	0	9	-	0.44	-	3.47	-	91[c] 93[b]	- 4.1	3×10^{-12}
703244	7.3	9	-	0.45	-	3.50	1.50	85[b] 85[c]	- 1.5	7×10^{-11}
703246	7.3	9	-	0.43	-	-	-			2×10^{-11}
704085	9	1	8.8	0.03	3.77	3.48	1.87	75[c] 62[b] 59 58	-	1.8×10^{-8}
705045	9	1	8.3	0.04	1.52	3.1	1.84	69[c] 62[b]	0.9	
705042	9	2	-	0.115	-	3.32	1.80	67[c] 61[b]	2.8	2.7×10^{-7}
705043	9	4	7.1	0.23	2.06	3.37	1.72	84[c] 68[b]	1.5	4.6×10^{-8}
705044	9	8	-	0.38	-	3.71	1.54	87[c] 70[b]	1.1	2.7×10^{-9}
716044	a-Si:H		12	0		3.68	1.92	52	0.2	-
A	a-Si:H		9.6	0		3.68	1.92	66[b]	0.5	1×10^{-5} 4×10^{-6}
C			6.8	0.18	2.77	3.70	1.735	65[b]	1.1	1.2×10^{-5}

a) infrared data. See also H exodiffusion data from Barcelona group ; b) PDS; c) OTS; d) Electron Probe Micro Analysis; (e) : η=quantum yield; μ=electron mobility in the extended states; τ=electron lifetime

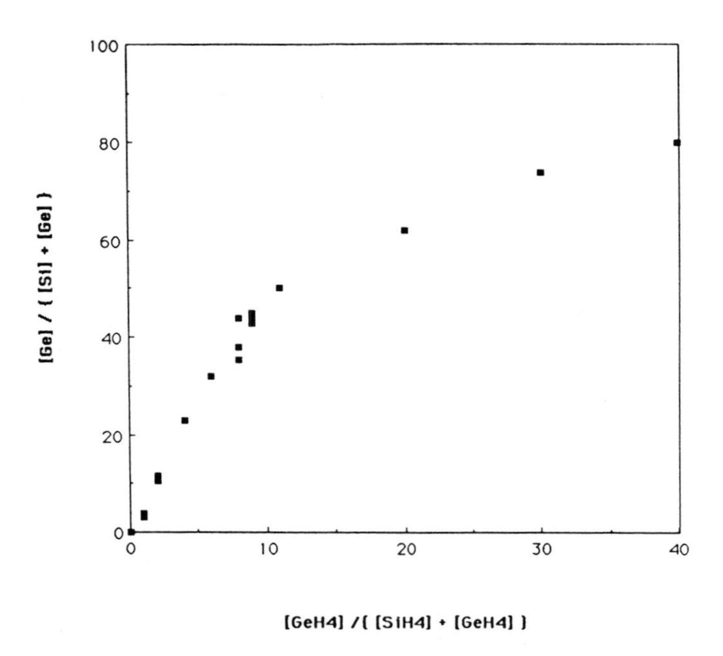

Fig. 1. Film composition vs gas composition in Arcam-produced materials.

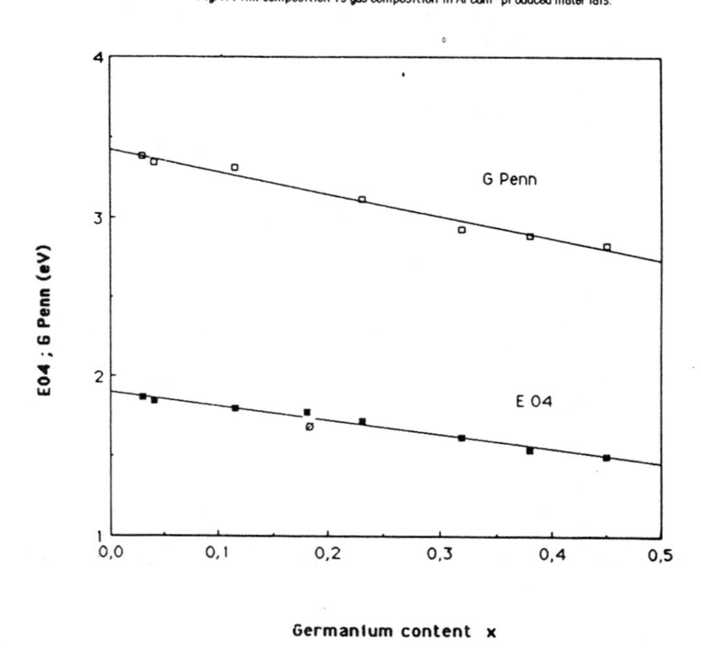

Fig. 2. Dependence of optical gap, E_{04}, and Penn gap on germanium content.

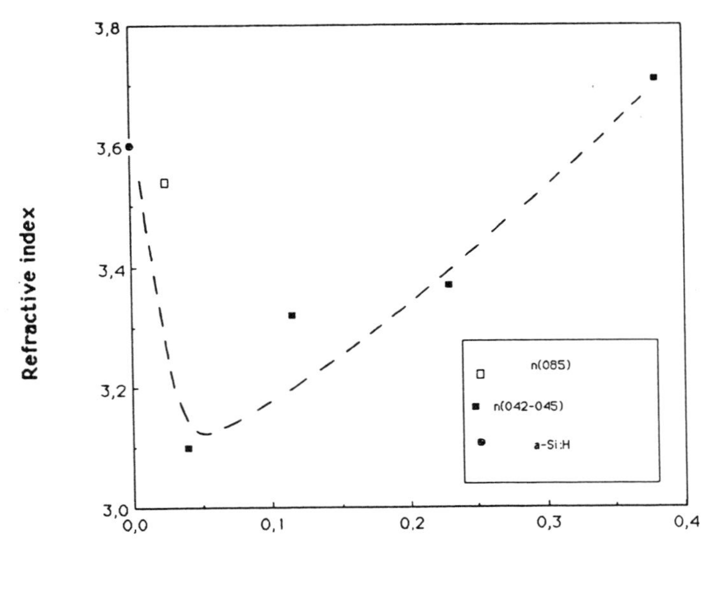

Fig.3. Refractive index of selected SiGe alloys and Si:H vs film composition.

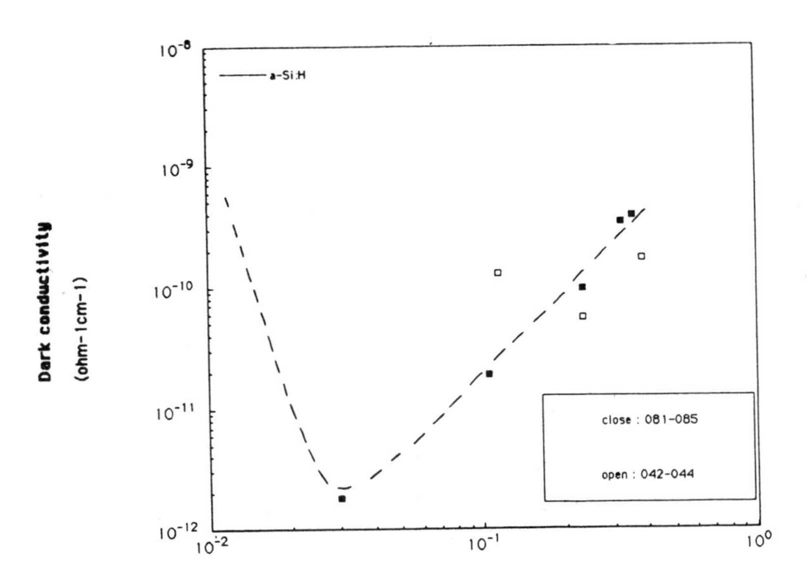

Fig.4. Dark Conductivity of selected SiGe alloys and Si:H vs film composition.

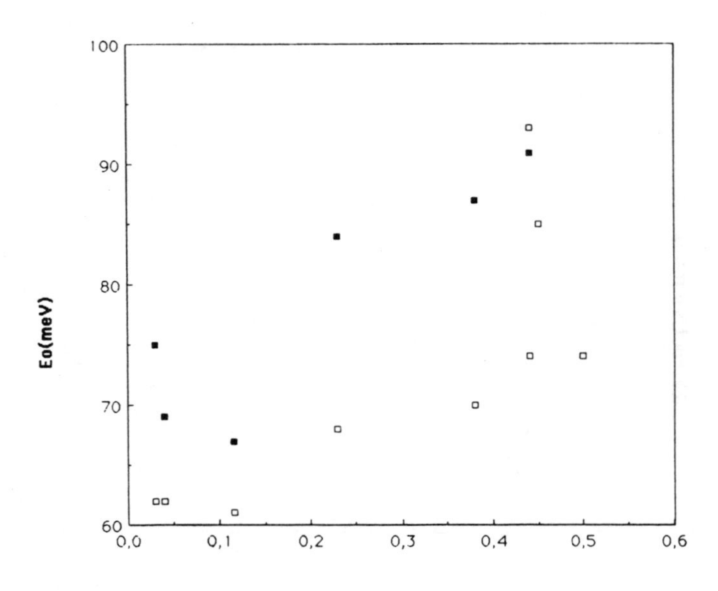

Germanium content x

Fig.5. Dependence on film composition of the Urbach edge parameter of selected SiGe alloys. Open symbols are PDS data,close ones are optical abosrption data.

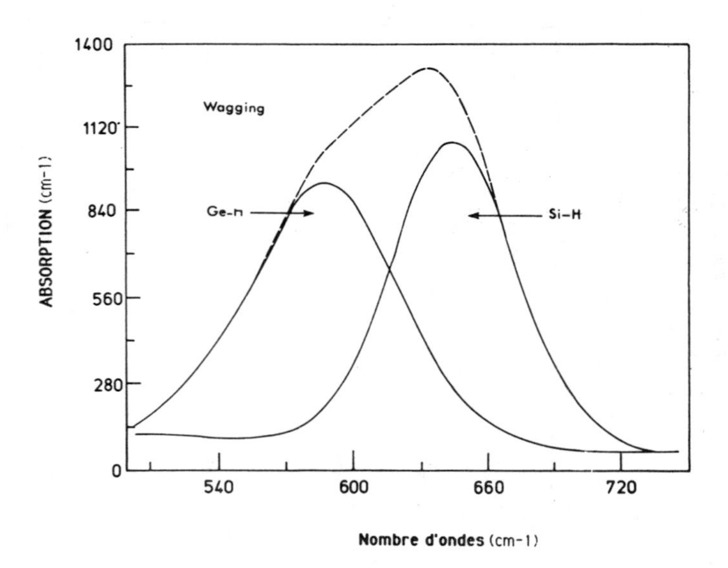

Fig.6. Wagging modes of Si-H and Ge-H vibrations in a typical Arcam-produced SiGe alloy, allowing he total hydrogen content and the ratio R=[Si-H]/[Ge-H] to be measured.

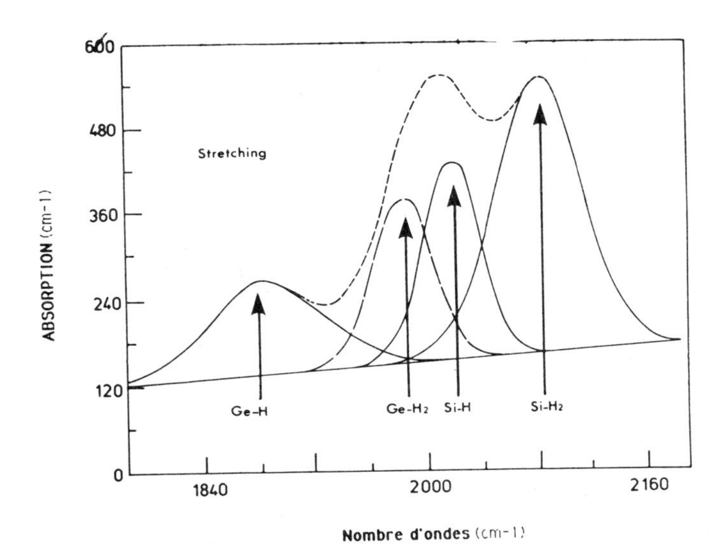

Fig.7. Stretching modes of $Si-H$, $Si-H_2$, $Ge-H$ and $Ge-H_2$. Data are for the same film as in fig.6. The relative amount of monohydride and dihydride bonds can be measured by means of such an anlysis.

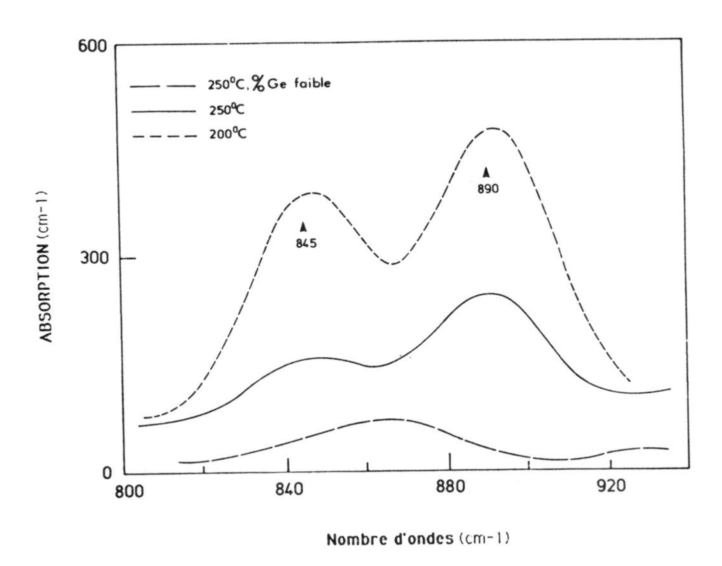

Fig.8. Bending modes of the isolated $Si-H_2$ bond ($845\ cm^{-1}$) and of the polysilane $-(SiH_2)_n-$ network ($890\ cm^{-1}$). Same film as in fig. 6 and 7.

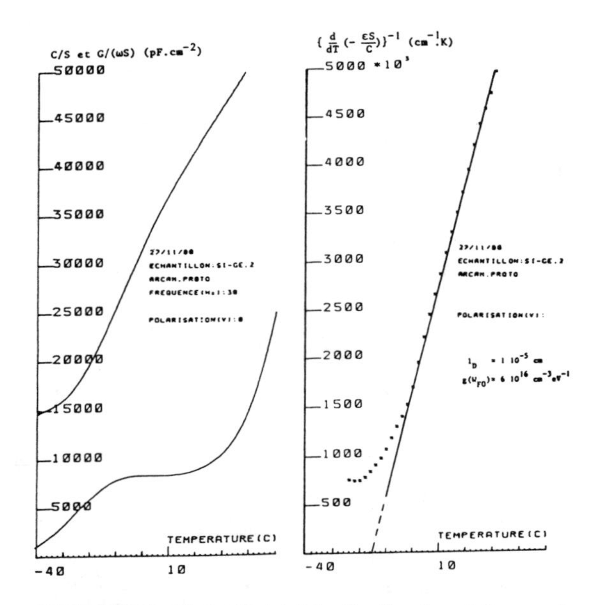

Fig.9. Density of states at the Fermi level of a high quality a-$Si_{0.82}Ge_{0.18}$:H alloy, prepared in the ARCAM prototype GD reactor, measured by the capacitance and conductance techniques.

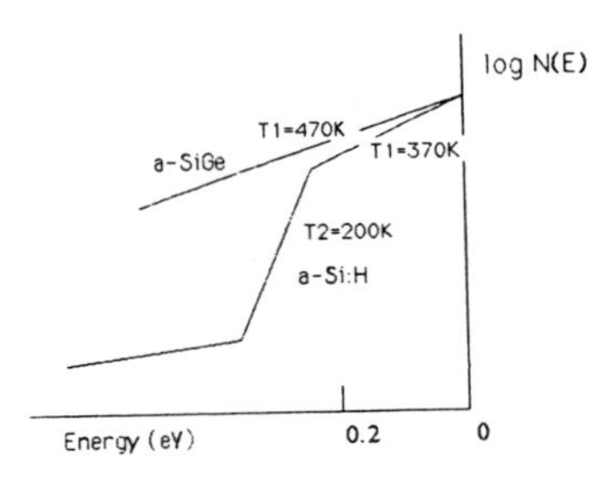

Fig. 10. Density of states profile of the conduction band tail, obtained from Time of Flight data, in the SiGe specimen of fig.9 as compared to a typical a-Si:H film prepared in the same reactor.

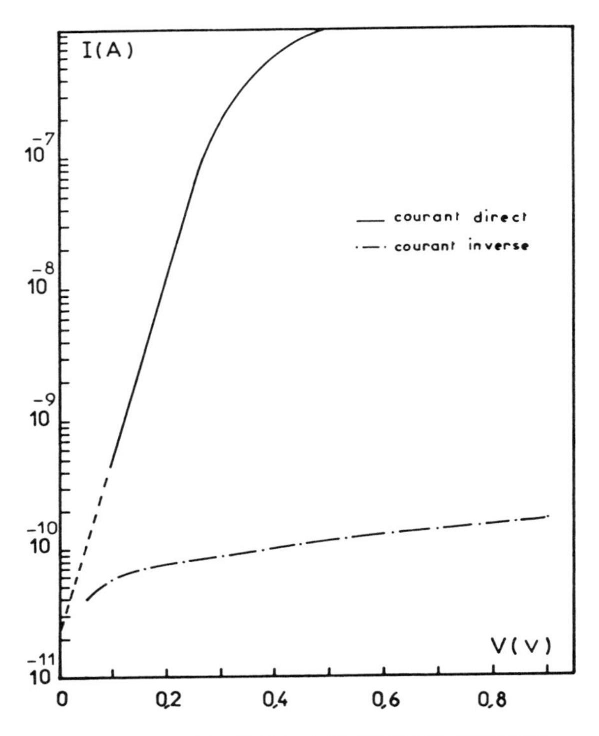

Fig. 11. I-V characteristics of a Schottky diode made of the SiGe alloy of fig. 9 and 10. Heavy line: direct current; dashed line : reverse current.

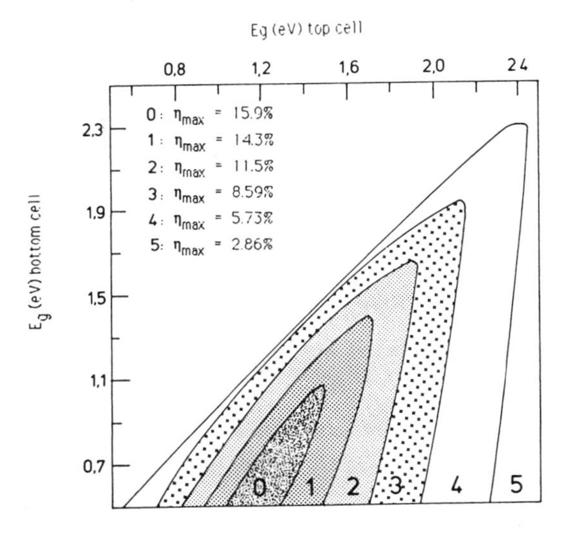

Fig. 12. Results of solar cell modelling : iso-efficiency curves of a tandem cell as function of the optical gaps of the top and bottom layers.

"LOW GAP ALLOYS FOR AMORPHOUS SILICON BASED SOLAR CELLS"

Contract number	:	EN3S-0063-I	
Duration	:	36 months	1 January 1986 - 31 December 1988
Total Budget	:	4.953.000.000 Lit	CEC contribution 684.000.000 Lit
Head of Project	:	Dr. R. Gislon, ENEA	
Contractor	:	Italian Commission for Nuclear and Alternative Energy Sources	
Address	:	ENEA, FARE Via Anguillarese 301 00060 S. M. Galeria (Rome)	

Summary

The most recent results obtained in the framework of the joint program on amorphous silicon based alloys carried on by University of Rome, Eniricerche and ENEA are presented. The principal characteristics of the plasma deposition systems used at the three laboratories are described and the optimum deposition conditions are discussed. In order to reduce sample contamination and to insure a better control of deposition process, many improvements have been introduced on University and Eniricerche deposition equipments. In order to select the best deposition conditions, the influence of several parameters has been studied, including substrate temperature, gas pressure and flow rate, and RF voltage; besides, the role of hydrogen dilution has been carefully investigated. The dependence of amorphous alloys properties on deposition parameters has been systematically tested by means of several techniques. The most important chemical, optical and electronic properties have been measured, including optical gap, activation energy, Urbach tail width. Structural properties of various amorphous silicon alloys have been studied by EXAFS technique. A particular effort has been devoted to the study of the density of gap states and to its dependence on Ge content; despite the intrinsic Ge related disorder, high quality, low gap (1.4 - 1.5 eV) SiGe alloys have been prepared. Finally, a theoretical evaluation of the relationship between alloy properties and multijunction PV cell performances has been carried out.

1.1 Introduction

In order to obtain high efficiency and low cost solar cells, a strong research effort on thin film multijunction devices is in course all over the world. By this kind of devices, both a substantial reduction of absorbing material and a better utilization of sun radiation can be achieved. In this perspective, amorphous silicon appears as a very promising material, due to its optoelectronic properties and to the possibility of preparing amorphous silicon based alloys. The optical gap of these alloys can be tailored and "window" or absorbing layers with the desired optical properties can be produced. Because of their amorphous structure such materials can be easily deposited, usually by glow discharge, in stacked configuration to fabricate multijunction photovoltaic cells. Moreover alloying silicon with low gap materials, such as germanium or tin, absorbing materials optimized for single junction devices (E_g = 1.5 eV) can be prepared.

These considerations justify the efforts devoted to develop semiconducting alloys combining the advantages of a tailor made optical gap with the good electrical characteristics of pure amorphous silicon.

The activities described in the present report are the Italian contribution to a larger joint program carried on by French, Italian, Dutch and Spanish research groups.

The activities here reported have been carried out at the following laboratories:

Physics Department, University of Rome I (La Sapienza);

Eniricerche Laboratories in Monterotondo (near Rome);

ENEA Photovoltaic Research Centre (CRIF) at Portici (near Naples).

1.2 Description of deposition apparatuses and measuring equipments

1.2.1 University of Rome

Two glow discharge deposition systems are available. The first one is based on a 500 W radiofrequency (13.56 MHz) generator HFS 500 of Plasma Therm coupled through an automatic matching network to a couple of coplanar electrodes in the diode configuration. The upper grounded electrode holds the substrate, which can have an area up to 150 cm^2. The reaction chamber is realized by a quartz cylinder substantially larger than the electrodes in order to reduce any contamination from the walls.

The second system, based on a small size ultrahigh vacuum (UHV) reactor, has been realized with standard UHV components. This system is used to grow materials in conditions of extreme cleanliness in order to control the impurities content at very low levels: experimental evidences are reported in literature about the possible connection of light induced instability (Staebler-Wronsky effect) to contamination during the growth phase. It can be mentioned that very good quality a-SiGe hydrogenated alloys have been recently produced by using UHV reactors.

This system is also being used as a preparation chamber connected to vacuum chambers of photoemission spectroscopy apparatus. Samples to be analyzed are transferred from one system to the other under vacuum, allowing the study of samples without the surface contamination due to exposure. By this system both a-SiGe:H alloys and a-Si:H/a-Ge:H etherojunctions have been prepared and studied "in situ".

Several series of a-SiGe:H alloys have been produced in order to investigate the influence of deposition temperature and hydrogen dilution on the optoelectronic properties of the materials. The Ge content x was varied in the range 0 - 0.5 corresponding to optical gaps in the interval 1.8 - 1.4 eV. To investigate the role of deposition temperature T_s was varied in the range 150-400 °C, while keeping constant the other deposition parameters. The effect of hydrogen dilution was studied at T_s=250 °C varying the hydrogen percentage in the feeding mixture from 0 to 95%.

The samples were characterized by infrared (IR) and optical absorption, conductivity, photoconductivity, and photothermal deflection spectroscopy (PDS). Thickness, optical gap and refractive index were determined by optical spectroscopy. The hydrogen content was obtained from IR absorption spectra, by integrating the intensity of Si-H and Ge-H wagging mode. The chemical composition was determined by electron microprobe analysis, performed in the ENEA laboratories of Casaccia Centre. The Ge concentration was found spatially uniform within the sensitivity of the probe (1 at.%).

In order to study the structural properties of silicon based alloys, several samples of hydrogenated silicon-germanium, silicon-carbon, and silicon-nitrogen amorphous alloys were deposited by glow discharge from binary gas mixtures of SiH_4, GeH_4, NH_3, and CH_4 on various substrates, including beryllium, quartz, and silicon wafers (1). The atomic concentration of a-SiN:H was determined from calibration curves obtained b ESCA and Auger measurements; the atomic concentration of a-SiC:H was instead estimated from the optical gap and therefore is less reliable.

The structural investigation was performed by extended X-rays absorption fine structure (EXAFS). The Si K-edge absorption spectra were measured at LURE (Orsay) using the ACO X-ray beam, while the Ge K-edge was determined at the PULS Synchrotron Radiation Facility of Frascati (Italy).

1.2.2 Eniricerche

Silicon-germanium alloys are deposited by glow discharge in a home-made load-lock single chamber reactor, constituted by a deposition chamber and a charging chamber, separated by a vacuum gate. After diffusion pumping and degassing, the typical residual gas pressure of reaction chamber is about 10^{-5} Pa. Samples are loaded in the charging chamber, keeping closed the vacuum gate until the limiting pressure reaches the value of 10^{-5} Pa: by this way, the exposure of the reaction chamber to the contaminant atmosferic gases is avoided as much as possible, with an overall gain in cleanliness and speed of operation.

Reactive plasma is completely confined in the region between the electrodes by means of a mesh electrode which prevents the diffusion of activated radicals to the cold walls of the reactor, where they could condense in form of powder. To further reduce powder contamination, the substrate is held on the top electrode; sample holder is equipped with a guard-ring which improves the spatial uniformity of the plasma.

Further studies have been devoted to the evaluation of the optimum deposition conditions; the equipment has been probed by determining the "Paschen-like" curves of several gases. By the analysis of these measurements a working pressure of about 10 Pa has been selected; this low pressure deposition regime avoids polymer and powder formation and enhances the uniformity of the gas flow inside the chamber. The other deposition parameters were V_{RF}= 1200 V, T_S= 250 °C and flow rate = 20 sccm; the choice of V_{RF} is a compromise between a sufficiently high growth rate (high V_{RF} values) and low defect density (low V_{RF}). Further studies will involve slight variations of flow rate and temperature and a decrease of pressure and RF voltage.

Films are grown by a gas mixture of GeH_4 and SiH_4, both diluted 1:10 in hydrogen, and its chemical composition is controlled by changing gas flow ratio. Full automation of control instruments insures a good sample repeatibility from run to run.

In order to study the Si-Ge amorphous alloys properties over a wide composition range, a set of samples was prepared and systematically investigated by various techniques.
Samples were deposited onto two kinds of substrates: Infrasil quartz for PDS analysis and Dow Corning glass for electrical and optical measurements. Glass substrates are entirely coated with evaporated chromium, except for a central gap 3 mm wide. They are used for electrical measurements in the coplanar buried electrode configuration, for visible and near infrared transmission measurements (through the interelectrode gap) and for the Fourier transform IR reflectance spectroscopy (using the high reflectivity of the chromium layer). The electrical characterization includes dark conductivity and AM1 simulated photoconductivity. By these

techniques all the most important material properties can be obtained : optical gap E_g, static refractive index, activation energy E_a, electronic states disorder parameter E_0, density of defect states N_S, hydrogen content C_H.

Furthermore electron probe microanalysis (EPMA) and/or Auger spectroscopy are systematically used to determine the alloy atomic composition x.

With regard to the evaluation of defect states density, it has to be observed that high quality PDS spectra can be obtained with our equipment by using low absorption Infrasil quartz substrates (Fig. 1).

The integrated absorption plotted in Fig. 1 for some samples is proportional to the density of defect states (2). A computer simulation of various PDS spectra has been performed using a realistic theoretical density of states (3). It shows that the low energy side of the PDS absorption plateau cannot be assigned to bulk volume defects of the films but has to be ascribed to extrinsic surface states due to defects located at the free surface or at the film substrate interface. Their role will not be further discussed in the following.

1.2.3 ENEA Photovoltaic Research Centre (CRIF)

A commercial plasma deposition system, the DP80 model produced by Plasma Technology, is available in the CRIF Laboratory. The system is based on a pyrex chamber and a 400 W radiofrequency (13.56 MHz) power generator coupled, through an automatic matching network, to two parallel electrodes in the diode configuration. Substrates of up to 200 cm^2 are placed onto the lower grounded electrode. Substrate temperature, gas flow rate and gas pressure can be automatically controlled. Three process gas lines (SiH_4, GeH_4, CH_4), including control and purging equipments, have been set up.

A first set of amorphous silicon films has been deposited varying the most important deposition parameters: substrate temperature, gas flow rate and pressure, RF power. By the analysis of sample properties, the best deposition conditions are being selected.

Several techniques for optical and electronic characterization of amorphous films are now available at CRIF Centre in Portici. The density of states can be determined by C-T, C-V and DLTS measurements; these data can be correlated to conductivity measurements results. In order to measure optical absorption and spectral photoconductivity in the range 2-0.8 eV, an optical characterization line has been implemented. By these measurements optical gap, refractive index, absorption coefficient and Urbach tail width E_0 have been obtained. Photovoltaic devices can be characterized by I-V measurements in the light and in the dark.

To determine the hydrogen content of the samples, a FTIR Perkin Elmer spectrophotometer is used; besides, in the ENEA Casaccia Centre, ESCA, Auger and electron microprobe equipments are used to measure atomic composition.

1.3 Results and discussion

1.3.1 University of Rome

Fig. 2 and 3 summarize the main results obtained in the study on the effect of the deposition temperature T_S. In Fig. 2 the optical data relative to the samples described previously ($x=0.37$), are reported as a function of T_S (4). The optical gap decreases while the refractive index increases with increasing T_S, according to what has been found recently for the alloy $x - 0.5$. However, preliminary evaluation of hydrogen content in our samples tends to exclude effects related to H variation. The Urbach tail width, E_0, remains practically constant within the uncertainty of the determination, except for the highest temperature sample. If we assume, as usual, that E_0 is a measure of the disorder in the material, the data point to a certain degree of insensitivity of the structure to the deposition temperature.

Quite to the contrary, T_S markedly affects the photoconductive signal. In Fig. 3 the $\eta\mu\tau$ product is also reported, as derived from the steady-state photoconductivity. There is a broad maximum in the interval 200-300 °C suggesting that the recombination centers have a minimum density for these deposition conditions. A direct determination of the density of these centers is presently underway. As evident from Fig. 3, the rising of the deposition temperature brings about a two order of magnitude increase in the room temperature dark conductivity, which can be presumably accounted for by the reduction of E_g and the concomitant decrease of the distance between Fermi level and mobility edge. Finally, it is worth mentioning that the best samples have a ratio $\Delta\sigma/\sigma = 10^4$ under AM1 illumination (6).

The study as a function of hydrogen dilution is summarized as follows. Fig. 4 shows the absorption coefficient for samples with different x. Just a representative spectrum for each series is reported, since only minor variations in the lineshape and in the low energy absorption were brought about by H_2 dilution. Alloying with Ge produces a sizable increase of α in the low energy part, compared to that of a-Si:H; the variation of Ge content, however, causes no further change in the gap absorption, only determining a rigid shift of the whole spectrum to lower energies. The optical gaps, obtained using a Tauc's plot, were 1.6 eV, 1.48 eV, 1.38 eV for the three series, with $x=0.22$, 0.37 and 0.48 respectively. Using the usual a-Si:H conversion factor, we found gap state densities of 1×10^{15} cm^{-3} for the amorphous silicon films, and of $1 - 2\times10^{16}$ cm^{-3} for the alloys.

The Urbach tail width, as shown in Fig. 5, rises more than 10 meV in passing from a-Si:H to the alloys, with a considerable spread in the values. However, no clear trend as a function of the Ge content or of H_2 dilution can be inferred. Hence, it seems that H_2 dilution has very little influence on the E_0 values.

The most striking effects of H_2 dilution are seen in the electrical properties. In Fig. 6a we report the values of the quantum efficiency-mobility-lifetime product $\eta\mu\tau$, as calculated from steady-state photoconductivity data. As clearly seen, the dilution increases the $\eta\mu\tau$ product in all series. This effect is more evident at higher Ge concentrations. As a consequence, although at a given H_2 dilution a rise in Ge content always determines a drop in $\eta\mu\tau$, samples with higher x but grown at higher H_2 dilution can have better photoconductivities than samples with lower Ge content.

Another parameter often used as an indicator of photovoltaic quality is the ratio $\Delta\sigma/\sigma$. This parameter is reported in Fig. 6b. The photoconductivity measurements were carried out with a flux of 1×10^{15} photons/(s cm^2) at the wavelength for which the absorption coefficient was about 3×10^3 cm^{-1}. We have estimated that the scale of Fig. 6b should be multiplied by a factor ≈ 100 in order to obtain AM1 conditions. $\Delta\sigma/\sigma$ behaves differently from $\eta\mu\tau$. With H_2 dilution this ratio decreases in a-Si:H samples, keeps almost constant in the lower Ge content series, increases considerably for higher Ge content. This fact is related to the features exhibited by the dark conductivity, which increases with H_2 dilution and attains a similar value for all series, including the a-Si:H samples, of about 10^{-8} $(\Omega\ cm)^{-1}$.

Further information on the effects of H_2 dilution was obtained by IR spectra analysis. We found that the total H content doesn't seem to vary appreciably with H_2 dilution, nor does the preferential attachment of H to Si with respect to Ge. The only influenced IR feature (Fig.7a) is the percentage of hydrogen bonded as $Si-H_2$, which decreases with respect to Si-H.

It is known from earlier studies on a-Si:H that the presence of high $Si-H_2$ content is associated with the existence of columnar microstructures; our data, therefore, suggest that H_2 dilution improves the morfology by reducing $Si-H_2$ related inhomogeneities. Furthermore, we observed a decrease in the growth rate as we increased the H_2 content of the gas mixture (Fig.7b). Such a decrease allows the formation of a less strained network and is in agreement with the above interpretation of the IR data. This morphology modification could indeed bring about a narrowing of the conduction band tail, with a consequent rise of the drift mobility and/or an increase of gap states above the Fermi level, with a consequent alteration of the recombination kinetics. In this way the relevant variation in phototransport without a corresponding change in optical properties would be readily explained.

As for the structural investigations by EXAFS technique, the results have been discussed in the previous report. In the following we briefly summarize the main experimental findings.

- In a-SiGe:H and a-SiN:H alloys, all the nearest neighbour distances are independent of concentration and equal to: Si - Si = 2.36 Å, Si - Ge = 2.39 Å and Si - N = 1.74 Å. Quite to the contrary, in a-SiC:H the Si - C bond length varies from 1.80 Å for x=0.5 to 1.89 Å for x=0.9, while the Si -Si bond length slightly increases from 2.36 Å to 2.39 Å.
- A clear evidence of a tendency to form stoichiometric microclusters was found in a-SiN:H at high values of x.
- In a-SiC:H and in a-SiN:H a contribution to the EXAFS signal due to the second nearest neighbours was detected at high values of x. In the a-SiN:H case this second shell is composed of Si atoms, while predominantly of C atoms in the a-SiC:H alloys.

1.3.2 Eniricerche

We observe in our a-SiGe:H alloys the expected preferential attachment of Ge over Si. For all samples we have systematically $x > y$, where x is the film composition measured by EPMA and/or Auger, and y is the ratio of GeH_4 mass flow to GeH_4 + SiH_4 (Fig. 8). The comparison with literature results (shaded area in Fig.8) suggests that the intensity of this phenomenon is a fairly general property of GeH_4+SiH_4 plasmas, unaffected by hydrogen dilution. Mixing H_2 to the discharge seems to play a passive role: it reduces the sample growth rate, but does not affect appreciably neither the main reaction path of GeH_4 and SiH_4, nor the hydrogen content. This is probably due to the fact that the bond strengths of GeH and SiH are quite similar to each other.

The optical gap E_g has been found essentially driven by the atomic composition x. We stress the fact that literature data and two distinct sets of our samples are in quite good agreement with the same relationship(Fig.9). The first set was grown keeping constant pressure, flow rate, RF voltage and temperature and varying the ratio y (dots); in the second set the ratio y and the RF voltage were kept constant while the other parameters were slightly varied: p = 5-10 Pa, Ø=20-40 sccm, T = 250-300 °C (crosses). Fig. 9 proves therefore that E_g is fairly insensitive to the story of the deposition process and is regulated only by atomic composition x. The small scattering of the experimental E_g vs. x data can be well ascribed to underlying C_H variations.

Whereas optical properties depend on x but not on preparation details, the DOS and the photoconductivity are dramatically affected by hydrogen dilution. The elaboration of PDS spectra (Fig. 1) allows to determine the volume density of gap states related to dandling bonds N_S and the electronic disorder parameter E_0 (Fig. 10). We stress the fact that, in spite of the increase of N_S with x, which is an unavoidable by-product of the incorporation of Ge atoms, the density of defects is fairly low for the alloys of photovoltaic interest (x = 0.4 - 0.5, E_g = 1.4 - 1.5 eV) and is much lower than that one of materials grown without hydrogen dilution. Furthermore, the structural disorder of our a - GeSi:H alloys is practically unaffected by the Ge content. A deeper investigation of the detailed role of Ge atoms is beyond our present experimental possibilities. As shown by computer simulation, the low energy plateau of PDS spectra contains the contribution of both Ge and Si dangling bonds, which are nearly degenerate; to count separately Ge and Si related defects, ESR spectroscopy should be used.

The thorough examination of the above mentioned electrical properties provides the following information:

- the room temperature conductivity of the samples doesn't show the variable range hopping contribution usually found in contaminated a-Si:H or in highly defective a-Ge. This result agrees with the low density of dangling bonds deduced by PDS spectra;
- the activation energy E_a of the dark conductivity stands around midgap for all atomic compositions. This is the normal behaviour for undoped samples, whose Fermi level is pinned near midgap at the energy level of defect states. A certain deviation from linearity in the experimental E_a vs x relationship is often observed in a-GeSi:H alloys (Fig. 11) and is probably due to the details of the relative weight of Ge and Si dangling bonds, which possess slightly different energy levels. An attempt to investigate this problem by computer simulation is currently under way;
- the dark conductivity spans over seven orders of magnitude (Fig. 12): this is mainly due to the 0.85 eV decrease of electronic gap;
- despite the density of dangling bonds is raised by the incorporation of Ge, the AM1-photosensitivity of the alloy of PV interest (x = 0.4 -0.5 , E_g = 1.4 - 1.5 eV) is quite high : $\Delta\sigma/\sigma$ = 10^4 (Fig. 12). This involves that the product of the photogenerated carriers is

high, and that such a material is susceptible of doping treatment as well as photovoltaic applications. Doping of this material is planned for the next future.

1.3.3 ENEA Photovoltaic Research Centre (CRIF)

At present, experimental data on silicon based amorphous alloys are not yet available. Both deposition systems and characterization equipments are being tested by preparing and measuring samples of pure amorphous silicon.

Theoretical modeling activity on multijunction cells has been carried out in collaboration with the University of Paris VII. Photovoltaic devices having one, two or three active layers have been computer simulated and their energy conversion efficiency has been numerically evaluated as a function of the band gap, the absorbing layer thickness, and the mobility-lifetime product of photogenerated carriers. Calculations have been carried out under two sets of hypothesis. In both cases no optical and electrical losses were assumed, and an absorption coefficient obeying Tauc's law was supposed. Besides, in the first case, the hypothesis of unit generation probability and collection efficiency has been made and the superposition principle has been applied; in the second case, the model assumes that the photocurrent is given by a drift process and the collection efficiency is a function of both the electric field in the intrinsic layer and the mobility-lifetime product of the excess carriers (7). Assuming a $\mu\tau = 10^{-8}$ cm^2/V, calculations give, as a maximum conversion efficiency of a tandem cell, 21% for the first case and 16% for the second case. Using the less simplified model, the values of the optimized energy gaps for the top and bottom layers come out substantially smaller than in the first case. An energy gap between 1.0 - 1.4 eV for the top layer and an energy gap lower than 1.1 eV for the bottom layer are required in order to obtain a 16% conversion efficiency. Energy gap values obtained by these models are substantially smaller than values currently reported in the literature. Further work to clarify this point is in progress.

1.4 Conclusions

The main results obtained up to now can be summarized as follows:

- Several glow discharge deposition systems, both single and multi-chamber, have been set up at the three laboratories participating to the program. Special attention has been paid to reduce sample contamination and to improve process reliability.
- Several techniques for measuring chemical composition, optical and electrical properties are used to determine alloy composition, hydrogen content, optical parameters, dark and photoconductivity, and the density of gap states. Structural characterization of several silicon based

alloys has been performed by EXAFS technique.

- Low gap (E_g =1.50 eV) device quality a-SiGe:H alloys with ratio $\Delta\sigma / \sigma > 10^4$ under AM1 illumination have been deposited. These materials appear very promising for cell fabrication.

-PDS measurements demonstrate that the best a-SiGe:H samples have a Urbach tail width independent of the germanium content.

- The role of hydrogen dilution in improving photoconductive properties has been demonstrated.

-Finally, on the base of a theoretical model, the optimized optical gaps of a three -junctions solar cell have been calculated.

References

1) A. Filipponi, P. Fiorini, A. Balerna, S. Mobilio and F. Evangelisti, Proceedings of the "4th Int. Conf. on EXAFS and Near Edge Structure", Fontevraud 1986

2) W. B. Jackson and N. M. Amer, Phys. Rev. B52, 5559 (1982)

3) G. D. Cody, "Semiconductors and Semimetals", Vol.21, Part.B, J.I. Pankove Ed.

4) D. Della Sala, M. Nsabimana, M. Capizzi, F. Evangelisti, U. Coscia and M. Vittori, Proceedings of the "7th European PV Solar Energy Conference", Sevilla 1986

5) K. D. Mackenzie, J. R. Eggert, D.J.Leopold, Y. M. Li, S. Lin and W. Paul, Phys. Rev. B31, 2198 (1985)

6) L. Mariucci, F. Ferrazza, D. Della Sala? M. Capizzi and F. Evangelisti, Proceedings of the "12th ICALS", Prague (1987)

7) D. Chello, I. Chambouleyron and P. Baruch, Proceedings of the "19th IEEE PV Specialist Conference", New Orleans 1987

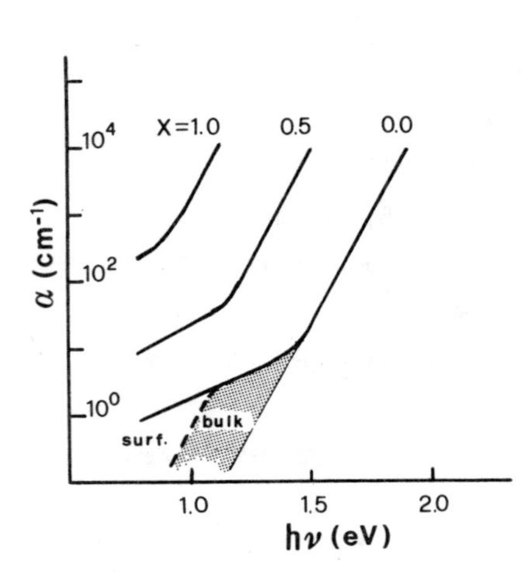

FIGURE 1
Some relevant PDS spectra of a-SiGe alloys obtained in Eniricerche.

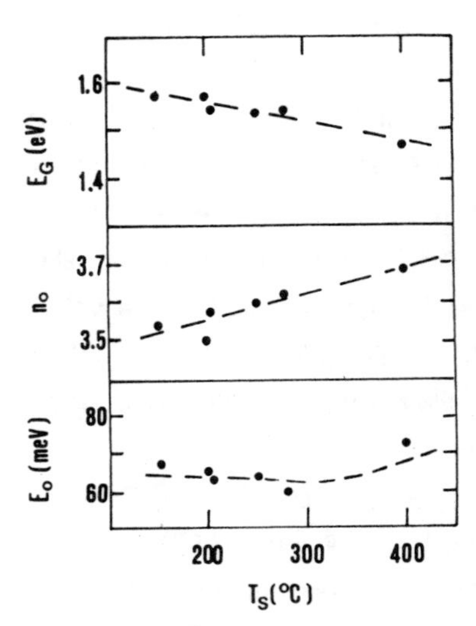

FIGURE 2
Optical gap (E_G), static refractive index (n_o) and Urbach tail width (E_o) vs. deposition temperature.

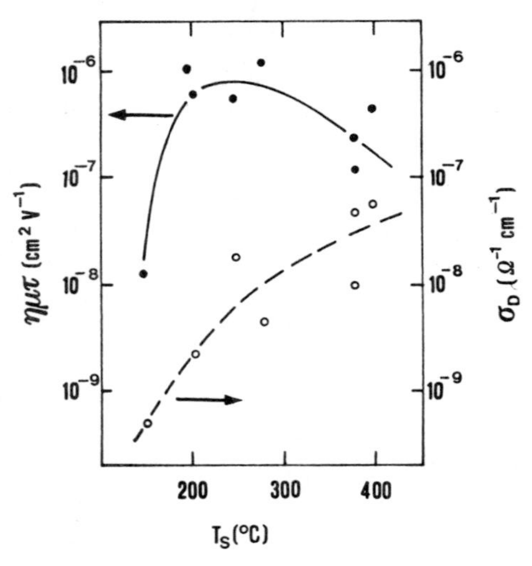

FIGURE 3
Dark conductivity and quantum efficiency-mobility-lifetime product vs. deposition temperature.

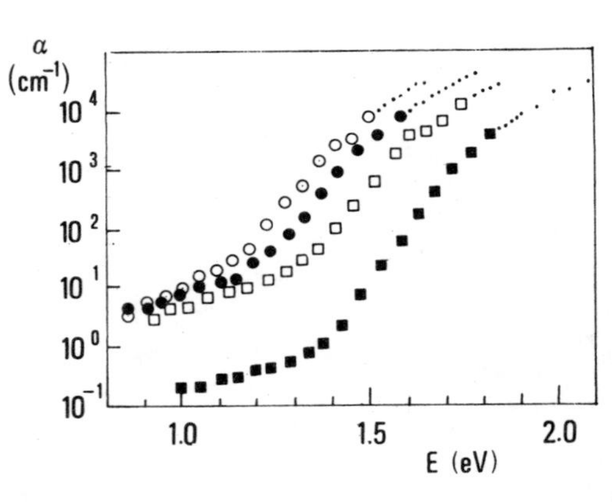

FIGURE 4
Optical absorption vs. energy
[■ x=0; □ x=.22; ● x=.37; o x=.48]

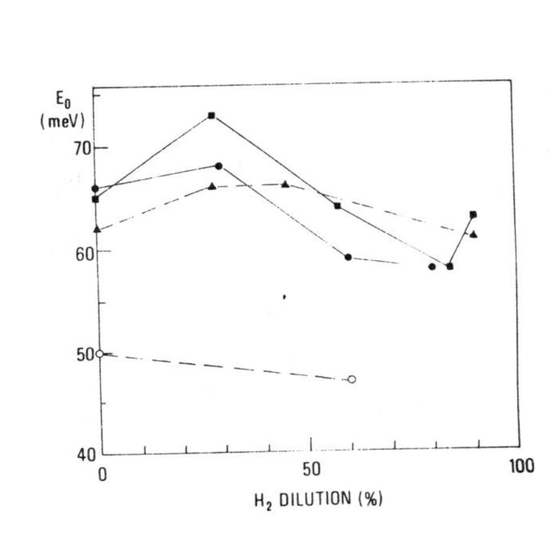

FIGURE 5
E_o vs. H_2 dilution[o x=0; ● x=.22; ▲ x=.37; ■ x=.48]

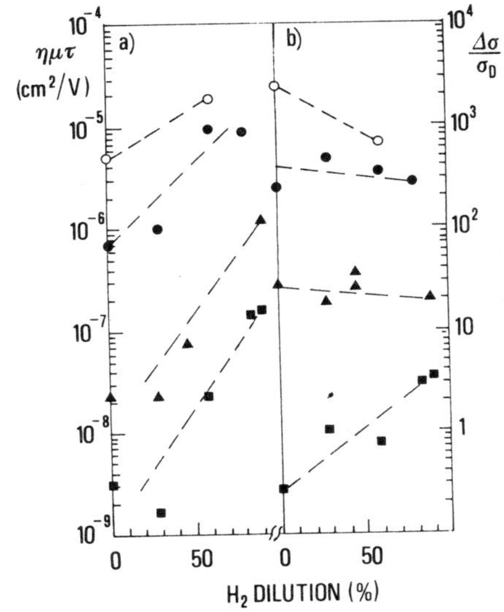

FIGURE 6
Symbol explanation in Fig. 5

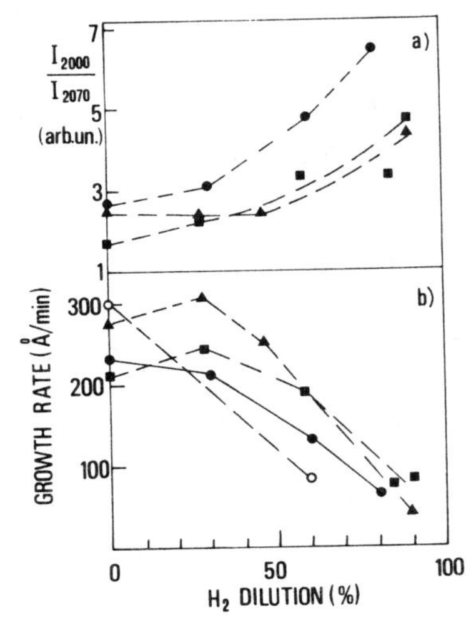

FIGURE 7
Symbol explanation in Fig. 5

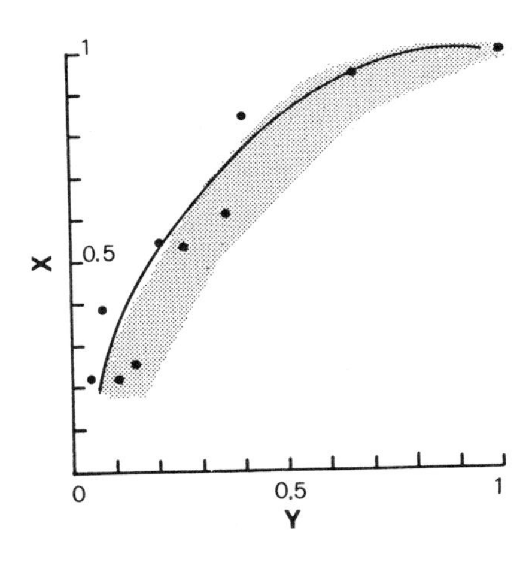

FIGURE 8
Atomic vs. gas mixture composition. Experiment (dots) and literature data (shaded area).

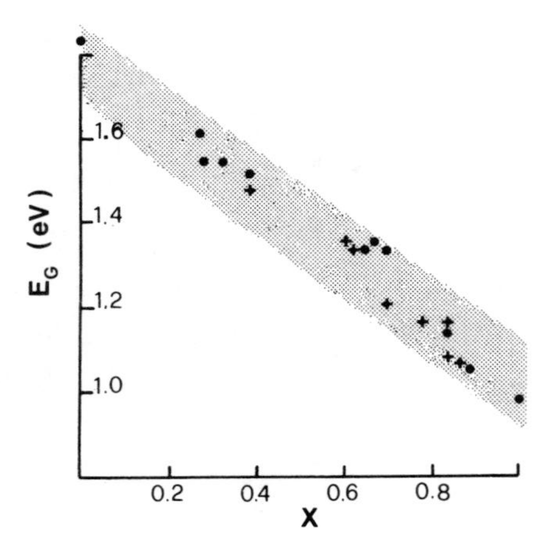

FIGURE 9
Optical gap vs. atomic composition. Experiment (dots and crosses), literature data (shaded area).

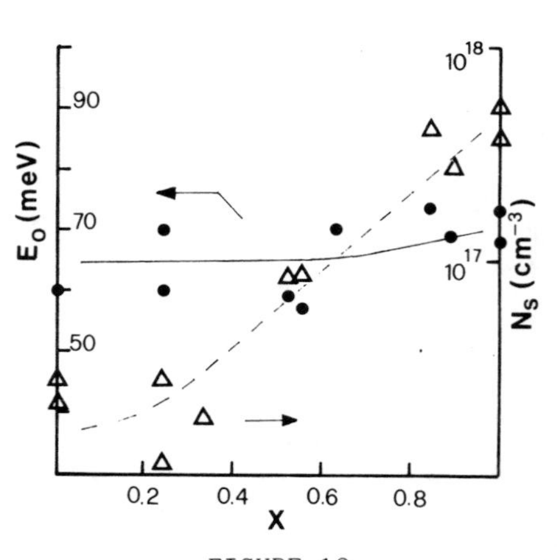

FIGURE 10
Urbach tail width and spin density vs. atomic composition.

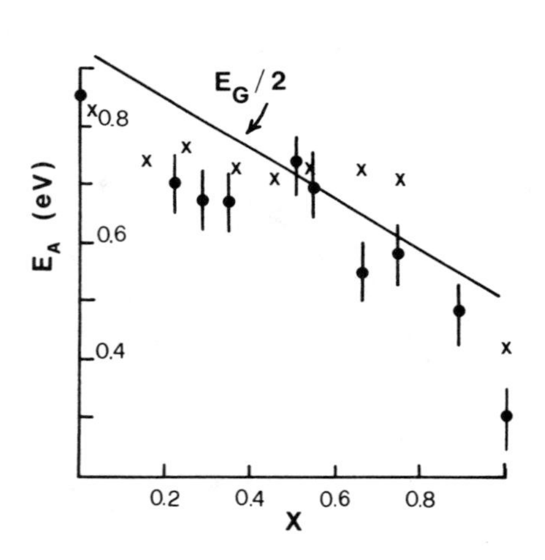

FIGURE 11
Activation energy vs. atomic composition. Crosses: data from Ref. (5).

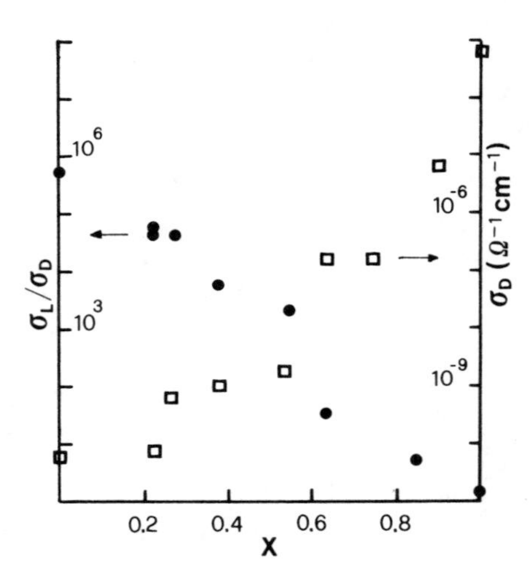

FIGURE 12
Photosensitivity and dark conductivity of a-SiGe alloys.

LOW GAP ALLOYS FOR AMORPHOUS SILICON-BASED SOLAR CELLS

CHARACTERIZATION OF AMORPHOUS SILICON ALLOYS THROUGH THERMAL H EFFUSION

Contract Number	:	*EN3S-0061-E(B)*	
Duration	:	*36 months*	*1 January 1986 - 31 December 1988*
Total Budget	:	*19.500.000 Pta*	*CEC contribution: 13.500.000 Pta*
Head of Project	:	*J.L. Morenza*	
Researchers	:	*G. Sardin, C. Roch*	
Contractor	:	*Universitat de Barcelona*	
Address	:	*Dept. de Física Aplicada i Electrònica Av. Diagonal 645 E-08028 Barcelona*	

SUMMARY

Samples of a-SiGe:H have been analysed through thermal H effusion. By means of this technique we have measured the total and partial H contents. Among the two partial H contents we have evidenced that the presence of a low temperature H effusion (LTE), corresponding to weakly bonded H, is not an intrinsic characteristic, but only a tendency due to the Ge alloying. The later can be overcome, fortunately since the LTE has been identified as rising from weakly bonded H (WBH), correlated to the fall of electronic properties and considered an inductor of metastability.

I. INTRODUCTION

Ge alloying of a-Si:H has been thought to be a convenient way to adjust the energy gap according to the desired application, in particular for the photovoltaic cells, in way to optimize their efficiency to solar illumination. Also it has been reported (1) that increasing the Ge content above 40% would decrease the Staebler-Wronsky effect. In counterpart a shortcoming would be the difficulty in getting a low level of dangling bonds, probably due to the lower H bonding affinity to Ge than to Si (2), which leads to a lost of the H efficacy. Therefore hydrogen related characterizations are of fundamental interest in order to understand the H behaviour and the properties of such an alloy.

The thermal H effusion provides the total H content and presents the adventage of allowing to differenciate between weakly bonded H (WBH) and tightly bonded H (TBH) and to get the percentage of each. In the case of a-Si:H, the WBH is considered to rises from clustered Si-H bonds and/or Si-H2 while the TBH from isolated Si-H. The determination of the WBH has adquired recently much interest since it has been established a direct correlation with the light-induced metastability (3).

II. EXPERIMENTAL

The H thermal effusion has been effectuated by heating up to 750°C, at a rate of 18°C/min, samples of a-SiGe:H located in a quartz tube with a static vacuum of 10^{-6} mbar. The evolution of the hydrogen partial pressure has been measured by a quadrupolar mass spectrometer, Quadruvac Q200 Leybold-Heraeus.

In addition to numerous samples of a-Si:H, seventeen samples of a-SiGe:H grown in the ARCAM reactor of the Ecole Polytechnique (Palaiseau, France) have been analysed. They were deposited on monocrystalline Si substrates within a large range of technological parameters. The silane flux stands in between 3 and 20 sccm, the germane flux in between 0.1 and 0.7 sccm and the H2 flux in between 0.9 and 19.7 sccm. The pressure varies from 22 to 95 mTorr and the RF discharge power from 2 to 5.5 W. The temperature has been fixed at 250°C but for five samples it was only of 200°C. The Ge content of the thin films has been found to stand in between 3 and 50%.

III. RESULTS AND DISCUSSION

The thermal H effusion from the samples of a-SiGe:H has shown most often two peaks corresponding to a low temperature evolution (LTE) and a high temperature evolution (HTE), but in a few cases only the HTE. Figure 1 represents typical thermal H effusion spectra of a-Si:H and a-SiGe:H deposited in similar conditions. It can be observed that the introduction of GeH4 in the RF glow discharge has introduced a peak at low temperature not present in the corresponding spectrum of a-Si:H. Therefore it merges that Ge alloying has favoured the appearence of a LTE. Nevertheless from the set of samples studied no clear correlation between the integrated LTE and the Ge percentage has been found. Any possible correlation may have been hidden by the quite diverse deposition parameters of these samples, introducing structural differences which may have cover the influence of the composition.

The evolution taking place at low temperature (LTE) is considered to proceed from weakly bonded H (WBH) as a consequence of its clustered state, while the evolution at high temperature (HTE) is considered to proceed from tightly bonded H (TBH), due to its isolated state. The LTE is attributed to the effusion of hydrogen almost without diffusion, while the high temperature evolution (HTE) rises from the effusion of hydrogen through diffusion (3). The LTE can be regarded as a quasi-desorption eventhough it takes place in the bulk. This is so because when present a LTE the bulk is then partially porous, i.e. it possesses a low density phase (tissue) characterized by its microporosity. The desorption proceeds from the surface of the microvoids which populate the bulk and often effuses through the low density porous phase almost without diffusion. Instead the HTE is considered to be a diffusive process taking place into the high density phase, i.e. in the grains. Once having reached the grains boundaries the H atoms recombine into H2 molecules which then desorbe out through the low-density porous phase. The differentiation between WBH and TBH stands also on considerations upon the activation energy for bond rupture. In particular the WBH evolves at low temperature because the activation energy for desorption of an H2 molecule needs only 1.5 eV due to an energy gain of 4.5 eV from the recombination of two H atoms (the bonding energy of each Si-H is sensibly 3 eV).

It has been reported by Beyer et al (3) that the LTE, proceeding from WBH, appears upon incorporation of minute amounts of germanium. We have checked if this was an intrinsic behaviour. Hopefully we have found that this was not the case, that such behaviour is primordially dependent on the deposition parameters, and that there is no direct correlation between the WBH and the Ge contents, which implies that the amount of WBH depends rather on the structure than on the composition (% of Ge). Although the incorporation of Ge shows effectively a tendency to favour the appearence of WBH, samples of a-SiGe:H without WBH have been successfully obtained. Therefore the presence of WBH does not correspond to an intrinsic characteristic of a-SiGe:H and can be avoided.

Another interesting result rises from the fact that the two evolution peaks vary in intensity and in temperature, according to the deposition parameters. For most of the samples two well separated peaks have been observed which reflect two differentiated evolutions. Depending on the deposition parameters the maximum of the first peak lays in a temperature range 175°C wide, which starts at a temperature as low as 375°C and varies up to about 550°C, in which later case it becomes confounded with the second peak. The maximum of the second peak itself is situated in between a narrower temperature range of only 50°C, laying between 535 and 585°C. The variation of the evolution peaks has been analysed in order to deduce their structural implications. For this purpose four representative samples have been selected according to the degree of separation between the evolution peaks (fig.2).

The first sample presents only one evolution peak. The second sample presents two well separated peaks, for the third one the peaks are less distant and for the fourth the two peaks are so close that they are almost confounded. The presence of one or two peaks, and the separation between them present a great interest due to the associated implications and can be interpreted in the following way. In the first case the presence of only one evolution reflects an homogeneous network composed of a single phase, which is populated by only one type of H, the tightly bonded one. In contrast the second group present two nettly differentiated peaks, with a temperature intervale of the order of 175°C. The first peak has been associated to weakly bonded H and considered to proceed from a low density, eventually porous phase. In the third case the two peaks are less separated, essentially the low temperature peak has moved to higher temperatures. This indicates that the low density phase has become slightly denser, which induces a diffusive delay. Finally, in the forth case there are still two evolutions which are so close that their peaks are almost confounded. This implies that there are still two phases, but the previously low density one has become denser to the point to behave, from the thermal effusion stand point, almost identically to the high density phase.

V. CONCLUSION

Ge alloying strongly favours the formation of WBH, nevertheless some samples have not shown it. A clear correlation with the Ge content has not merged, probably due to structural differences that may have overcome the influence of the composition. The appearence of two H evolutions points out the presence of two phases, a low density and a high density phases respectively associated to the LTE and HTE. The difference of density between the phases is indicated by the separation between the LTE and HTE peaks while their intensity indicates the respective amount of WBH and TBH, and to some extent the proportion of each associated phase.

In view of reaching good material stability and properties it is primordial to avoid the formation of WBH, which grants poor and unstable electronic properties. Hence it is essential to get a better knowledge of the active parameters leading to the formation of WBH as well as a better understanding of the physical laws that govern this process.

REFERENCES

(1) Y.Yukimoto, in "Amorphous Semiconductor Technologies and Devices", JARECT 6, Y.Hamakawa (ed.), North-Holland, Amsterdam, 1983, p.136.

(2) G.Lucovsky and W.B.Pollard in "The Physics of Hydrogenated Amorphous Silicon II", J.D.Joannopoulous and G.Lucovsky (eds.), Springer-Verlag, Berlin, 1984, p.351.

(3) M.Oshawa, T.Hama, T.Akasaka, T.Ichimura, H.Sakai, S.Ishida and Y.Uchida, Jpn. J. Appl. Phys., 24, L838 (1985).

(4) W.Beyer, H.Wagner and F.Finger, J. of Non-Cryst. Sol., 77, 857 (1985).

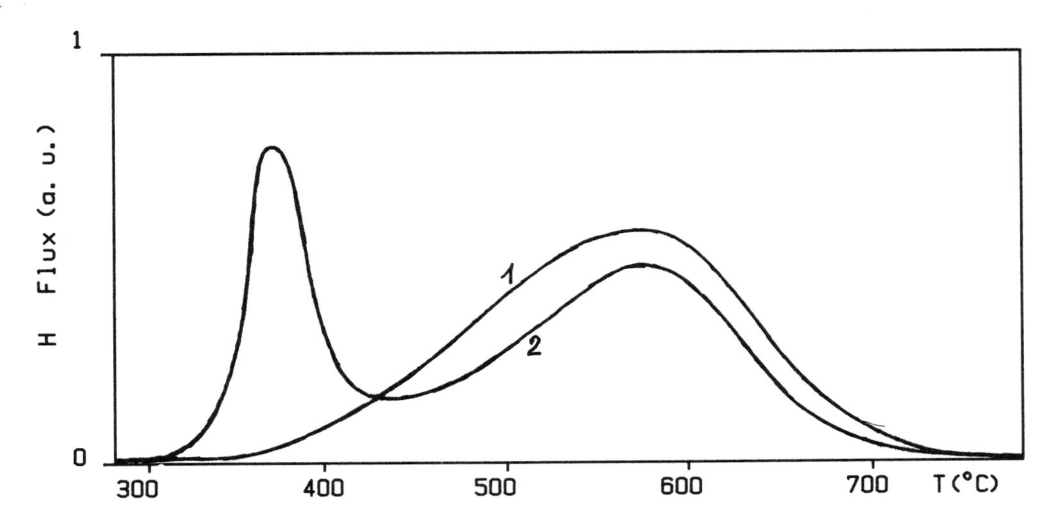

Fig 1 : Thermal H effusion spectra of a-Si:H (1) and a-SiGe:H (2) deposited in similar conditions.

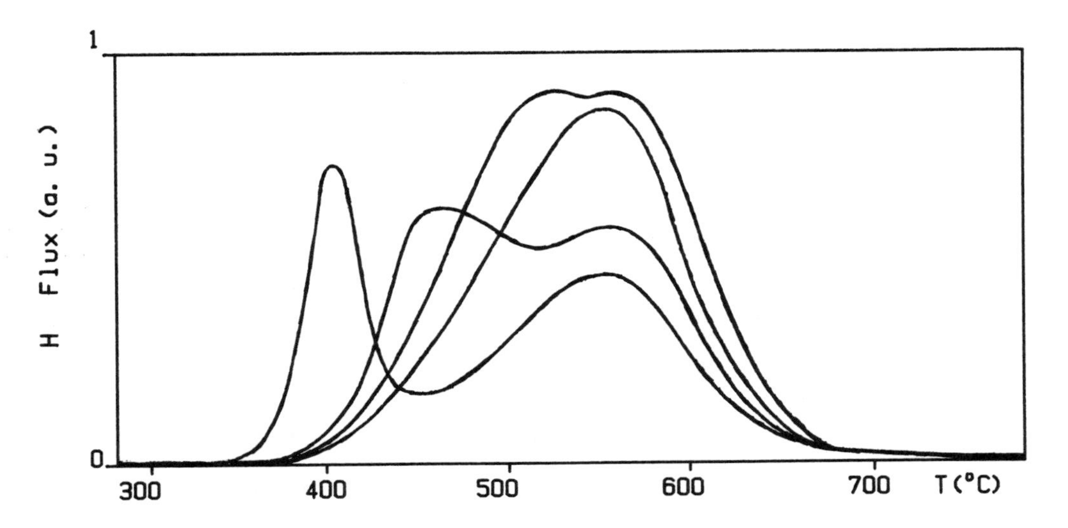

Fig 2 : Thermal H effusion spectra of a-Si:Ge:H deposited in different conditions but at the same temperature (T =250 °C)

LOW-GAP ALLOYS FOR AMORPHOUS SILICON BASED SOLAR CELLS

CHARACTERISATION OF HYDROGENATED AMORPHOUS SILICON BY PHOTO DEFLECTION SPECTROSCOPY

Contract Number : EN3S-0064-NL

Duration : 36 months 1 April 1986 - 31 March 1989

Total Budget : Hfl 2.269.000 CEC Contribution : Hfl 489.000

Author : Dr. J.W. Metselaar

Head of Project : Prof.dr. M. Kleefstra
Laboratory of Electrical Materials

Contractor : Delft University of Technology
Faculty of Electrical Engineering

Address : Mekelweg 4
NL-2628 CD Delft
The Netherlands

Summary

The aim of this part of the research was to establish the deposition conditions to obtain optimal quality amorphous silicon for the development of high efficiency amorphous silicon based solar cells. The sub-bandgap absorption of instrinsic hydrogenated amorphous silicon films deposited by rf glow-discharge from different silane-hydrogen mixtures has been measured by means of PDS. Using a mixture of 55 vol.% silane, 45 vol.% hydrogen and a relatively high rf power level of 36 mW/cm^2 to obtain a high growth rate, yielded films with an improved structural order, as deduced from the slope of the Urbach edge, compared to films grown under the conditions of using pure silane and an rf power just sufficient to maintain the plasma. These conditions are known to yield good quality films, but with the disadvangage of having a low growth rate.

1. Introduction

The rf glow-discharge deposition method of growing hydrogenated amorphous silicon (a-Si:H) at a minimum rf power and with pure silane yields good quality material at a low deposition rate. Diluting the silane gas with inert noble gases deteriorates film properties (1). Using hydrogen as a diluent gas alters the deposition rate and strongly affects the optoelectronic properties (2). The aim of the present work is to establish the influence of the rf power level and the degree of silane dilution with hydrogen on the sub-bandgap absorption of intrinsic a-Si:H. Photothermal deflection spectroscopy (PDS), in cooperation with J.S. Payson from ECD Inc., Troy, Michigan, USA, has been used as a method to analyse the films.

2. Method

The deposition apparatus used is an rf GD reactor. The 13.56 HMz. generator is connected to the upper electrode and the substrate is placed on the grounded lower electrode. The combined leak/desorption rate of the system is 2 10^{-5} torr.l/s. The flow rate, pressure, substrate temperature and electrode distance have been kept constant at 100 sccm, 0.3 torr, 523 K and 5 cm. respectively. Undoped a-Si:H films, typically 1 μm, thick, were grown on Corning 7059 glass substrates by decomposition of a range of silane-hydrogen mixtures (33 - 100 vol. % silane) and at two different power levens (15 and 36 mW/cm^2).

It was necessary to establish the deposition rate (Rg) of a-Si:H to be able to grow films of the same thickness in order to avoid possible thickness dependent absorption. The growth rate of the a-Si:H films formed under the above conditions was determined by optically measuring the transmission and reflection (3). These measurements also calculated the optical gap (E_g) which is generally related to the slope of the Urbach edge E_o (4).

3. Results

The results of the determination of the growth rate, the optical gap and the slope of the Urbach edge as a function of the silane concentration and rf power density are summarized in table 1. The PDS spectra are given in figure 1 and 2. The value of E_o and the tail of the PDS spectra are known to be sensitive to defects in crystalline materials (4). Comparing figure 1 and 2, the sub-bandgap absorption and E_o increase with the rf power level, as expected for films grown with pure silane. Diluting the silane gas with inert noble gases deteriorates film properties (1), more defects are created and the films more or less exhibit a columnar film growth structure. Adding hydrogen of up to 45% lowers both E_o and the sub-bandgap absorption.

This effect is more pronounced for the films grown with the higher rf power level. The layers grown with 55% silane, 45% hydrogen and an rf-power of 36 mW/cm^2 are the films with the lowest defect density, with respect to the value of E_o and the low tail absorption. Conversly, the film grown with pure silane and an rf power level of 15 mW/cm^2 (just sufficient to maintain the plasma) exhibit a higher structural disorder. These conditions are widely known to be able to provide good quality material, however with the disadvantage of having a low growth rate, 1.0 A/s. We have proven that one is able to grow a higher quality a-Si:H by using silane diluted up to 55% with hydrogen. At the relatively high rf power level the growth rate was enhanced by a factor 5 with an additional increase in quality.

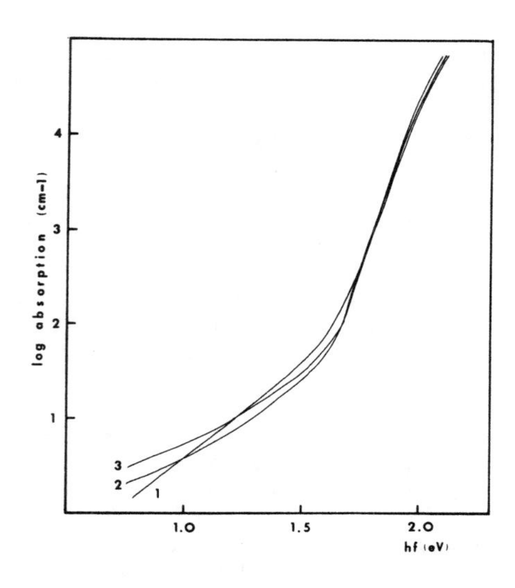

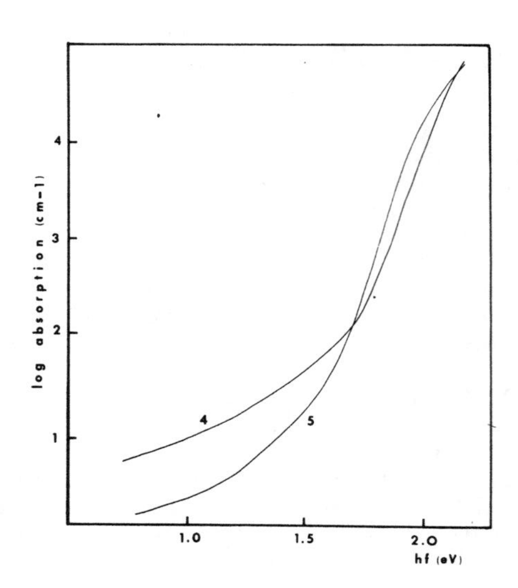

FIGURE 1
Absorption spectra of samples no. 1, 2 and 3.

FIGURE 2
Absorption spectra of samples no.4 and 5.

Table I: Deposition conditions and values of the optical gap E_g and the slope of the Urbach edge E_o for the different samples.

sample no.	$[SiH_4]$ (%)	rf (mW/cm^2)	Rg (A/s)	E_g (eV)	E_o (meV)
1	100	15	1.0	1.68	59
2	55	15	1.4	1.69	55
3	33	15	0.9	1.71	62
4	100	36	5.7	1.74	62
5	55	36	4.8	1.71	52

As mentioned, a relation exists between E_o and E_g (4). A decrease in E_g is accompanied by an increase in E_o. This relation is found only for the films 1 and 2. Obviously the relation between the shifts in E_g and the structural order is not explicit.

4. Analysis of results and comments.

The deposition process of a-Si:H films is not well understood, making it difficult to interpret these results. The benificial role of hydrogen might be explained as follows. The film precursors are SiH_3 and SiH_2 radicals. For SiH_3 the hydrogen desorption from the growing film surface is the growth rate limiting step, while for SiH_2 radicals this is not the case. If the precursors are SiH_3 radicals the deposition process is more like CVD than of like PVD (5). High quality films are known to grow under conditions with SiH_3 as the precursors (6). The SiH_3 and SiH_2 radicals are formed by the following reactions:

$$SiH_4 \underset{2}{\overset{1}{\rightleftharpoons}} SiH_3 + H \qquad (a)$$

$$SiH_3 \underset{4}{\overset{3}{\rightleftharpoons}} SiH_2 + H \qquad (b)$$

The forward reactions are enhanced by the rf power and the backward reactions by the hydrogen concentration. Increasing the hydrogen concentration may lead to a lower SiH_2 concentration as more hydrogen is present to form SiH_3 out of SiH_2. Reaction 4 is favoured over reaction 2. Consequently the concentration of SiH_3 radicals is raised at the expense of SiH_2 and therefore a higher rf power level can be maintained.

Furthermore, hydrogen enhances the surface mobility of absorbed species allowing the network to become more ordered. The surface mobility is higher because more of the dangling bonds, present at the surface of the film, are passivated. These free bonds react very quickly with absorbed species such as SiH_3 radicals which then become immobile. Saturating the dangling bonds with hydrogen makes it possible for SiH_3 to travel greater distances over the surface before being incorporated into the film Thereby energeticly more favourable sites can be occupied resulting in a higher structural order.

5. Conclusions

Using extra hydrogen as a tool to improve the quality of intrinsic hydrogenated amorphous silicon layers, we have shown that it is possible to deposit high quality material by using silane gas diluted up to 55% with hydrogen at a relatively high rf power level of 36 mW/cm^2. The advantages are both that the quality of the films and that the deposition rate will increase. These conclusions are in accord with the earlier conclusion from our density of gap states estimation by the space-charge-limited current method. No definit relationship was found between the structural order as measured by the slope of the Urbach edge and the optical gap of the films.

References

1) J.C. Knights, R.A. Lujan, M.P. Rosenblum, R.A. Street, D.K. Biegelsen, Appl. Phys. Lett., 38(5), 1981, 331.

2) P.E. Vanier, F.J. Kampas, R.R. Cordermann, G. Rajeswaran, J. Appl. Phys., 56(6), 1984, 1812.

3) R.C. van Oort, M.J. Geerts, J.C. van den Heuvel, H.M. Wentinck, Proc. 7th Eur. Ph. Sol. En. Conf., Sevilla, 1986, 480.

4) P. Cody in Semiconductors and Semimetals, vol. 21 B, J.I. Pankove (vol. ed.), A.P., 1984, chapter 2.

5) C.C. Tsai, J.C. Knights, G. Chang, B. Wacker, J. Appl. Phys., 59(8), 1986, 2998.

6) H.A. Weakliem in Semiconductors and Semimetals, vol. 21 A, J.I. Pankove (vol. ed.), A.P., 1984, chapter 10.

KINETIC STUDY OF THE DEPOSITION PROCESS OF GLOW DISCHARGE a-Si:H FOR HIGH EFFICIENCY SOLAR CELLS

Authors : D. Mataras, D. Rapakoulias, S. Kavadias

Contract number : EN3S-0065-GR

Duration : 36 months, 1 April 1986-1 April 1989

Total budget : 39000000Dr CEC Contribution: 29200000 Dr

Head of project : Prof. D. Rapakoulias

Contractor : Institute of Chemical Engineering and High Temperature Chemical Processes

Address : P.O. Box 1239, GR-26110
University Campus-Patras Greece

Summary

We report our Research activity in the first year of a three year plan. Two complementary approaches are used to the problem of a-Si:H deposition mechanism in RF glow discharge. Computer simulation and optical diagnostics results are summarized. Kinetic modelling of the gas phase composition can be used for the prediction of the deposition rate which can be optimized by varying macroscopic parameters.

A theoretic calculation of the SiH_2/SiH_3 radicals relative abundance is presented.

Also experimental work in monitoring ground state radicals is being caried out.

1.1 Introduction

One of the major problems in glow discharge a-Si:H deposition consists in establishing a direct relationship between process parameters and the resulting film quantity and properties.

Such a relation would permit us to operate in optimized conditions, regarding the deposition rate and optoelectronic quality of the film.

The key steps, in this sequence of interconnected processes, are:

(a) The influence of macroscopic parameters to microscopic plasma properties.
(electron density, electron distribution function)

(b) Gas phase reactions mechanism and kinetics, for the identification of main film precursors.
(decomposition cross-section, homogeneous reactions rates)

(c) Plasma-surface interaction and film growth chemistry.
For the determination of the limiting step and the selectivity of the process.

Despite the fact that silane's chemistry has received considerable attention results are often controversial, leaving still many open questions. (1)

In this report we present the work carried out in Plasma Chemistry and new materials laboratory of our Institute, during the first year of a 3-year project.

The aim of this project is to study, along with the collaborating laboratory of the University of Bari, the kinetics of the deposition process, in order to contribute to the elucidation of some of the above mentioned problems.

Our strategic objective consists in enhancing the deposition rate, in conditions when a-Si:H film of satisfactory quality is obtained.

This involves complete comprehension of the deposition mechanism, discovering and control of the bottle-neck processes.

1.2 Description of apparatus and measuring equipment

An important objective of our Research Institute, and of this project also, is to develop the ability of constructing complex installations, for both producing new materials, and using well-established diagnostic techniques, in order to provide fundamental and applied research services.

In this direction we have constructed and assempled a rather complicated set-up in a relatively short period of time.

We have managed to make our apparatus functional despite the fact that a large number of units must operate at the same time, by an extensive use of computer data aquisition and control(Fig.1)

1.2.1 Description of Deposition apparatus and Peripheral devices

The deposition chamber consists of an 16 cm diameter cylindrical stainless steel tank, having four centered quartz windows (4 cm diam.) perpendicular to each other.

Our effort was to construct a quite similar reactor with the one in Bari, in order to have comparable results.

Two parallel round plates of 8cm diameter are used for R.F. and grounded electrodes.

A stainless steel, or an alternative grid, shield are used to confine the discharge at the fixed Rf electrode. Grounded electrode is based on a vacuum tight linear motton feedthrough, in order to control the interelectrode spacing.

Special design feedthrough's are used for thermocouple, power and gas mixture insertion.

Substrate temperature is adjusted with a P.I.D. controller using an in-sight thermocouple while RF sielding techniques are used to prevent temperature deviation (R.F. on-R.F. off = less than 1°C) R.F. power at stable frequency (13,56 MHz) is supplied through a home-built matching network (SWR<1) minimazing the R.F. interference problem.

The system can be pumped down to 10^{-7} torr with a diffusion pumping stack.

Entire deposition apparatus is controlled by a separate unit containing mass flow-meters, electromagnetic valves, downstream pressure control via throttle valve, and digital indicators for pressure and flow rate. (Fig. 3,4)

1.2.2 <u>Description of experimental diagnostics</u>

The main diagnostic techniques used in this project are optical emission spectroscopy (O.E.S) and Laser Induced fluorescence (L.I.F.). (Fig. 2)

Our experimental set-up has the ability of recording both O.E.S and L.I.F simultaneously. Detection system is composed of an 1 m high resolution monochromator (J.Y. THR-1000, 3600 gr/mm) and a wide range 250 cm spactrameter equiped with fast response P.M.T's.

The signal is recovered with an electrometer amplifier (for O.E.S) or with a multiple Boxcar integrator, and processed under digital form in a real-time data aquisition system. A N_2 pumped, tunable dye-laser is used as exciting source.

Complementary diagnostics include Langmuir Probes IR and Vis spectroscopy, dark and Photo-conductivity measurements. (For materials testing).

1.3 First Results

We use two complementary approaches for the investigation of the deposition mechanism.

The first method deals with process simulation of the gas-phase kinetics by computer modelling while the second applies LIF as a non intrusive in-situ technique for monitoring ground-state species which are commonly believed to be responsible for the film growth.

We present here some of our first results.

1.3.1 Modelling

Our first attempt to simulate the deposition process has resulted in a relatively simple kinetic model, dealing mostly with gas-phase chemistry. In order to reveale possible corelations between gas-phase composition and film formation. The obtained results where communicated at the 8th Symposium of plasma chemistry, this September in Tokyo (2)

Three different versions were developed for calculation

of:

I) e^- impact decomposition rate from known effluent has compositions

II) Gas-phase composition and deposition rate

III) Relative abundance of SiH_2/SiH_3 Radicals

The model considers a plug-flow reactor type, obtaining concentration profiles from a general equation of the form:

$$-\frac{dy_i}{dt} = \frac{1}{\rho}\left[W_i - y_i \sum_{i=1} W_i\right]$$

where ρ - molar density

y_i - formation or depletion of component i, taking into account all reactions in which i takes part

W_i - partial mass balance

$\sum_{i=1} W_i$ - The sum of W_i

Thus a system of i equations is obtained which can be resolved by avariable step Runge-Kutta method.

A very simple equation is used, for the moment, for the deposition rate considering that all types of radicals have the same contribution to the film growth.

A small part of the results obtained with this model is presented below.

In Table I are listed predicted values of SiH_4 decomposition constant at given conditions.

P=0.5 torr, flow=100 sccm, Power=10 Watts

Table (I)

Mixture	k_1 (Model)	k_1 (Ref.)
10% SiH_4-90% H_2	2.8 sec^{-1}	1.09 sec^{-1} (Turban)
20% SiH_4-80% H_2	2.2 sec^{-1}	
30% SiH_4-70% H_2	1.6 sec^{-1}	
40% SiH_4-60% H_2	1.4 sec^{-1}	
50% SiH_4-50% H_2	1.2 sec^{-1}	
60% SiH_4-40% H_2	1 sec^{-1}	
70% SiH_4-30% H_2	0.8 sec^{-1}	
80% SiH_4-20% H_2	0.6 sec^{-1}	
90% SiH_4-10% H_2	0.2 sec^{-1}	
100% SiH_4	0.16sec^{-1}	0.23 sec^{-1} (Nolet)

Our results are in better agreement with values reported in the literature, compared to other similar models that have been published.

In Figure 5 is presented the gas composition VS percentage of silane in the mixture.

For mixtures above 65% in silane a significant drop in the decomposition rate is observed, due to the defficiency in energetic electrons.

This causes a similar decrease in the concentration of radicals while disilane fraction continues to increase untill about 80% because of the higher importance of neutral-neutral reactions at these conditions.

We were interested to see the effect of this behaviour to the deposition rate and to correlate it with other plasma parameters. Thus in figure 6 we present the influence of both mixture composition and flow rate on the deposition rate.

The model predicts a maximum for mixtures from 30% to 60% and flow rates as low as 10-20 sccm.

Therefore further development of this model can give the possibility of optimizing deposition rate in conventional a-Si:H deposition gas mixtures.

Another attempt to extract more informations from the model is presented in fig. 7.

For 10% SiH_4-90% H_2 mixture and flow rate 20 sccm.

It results that at this very low flow conditions (depletion regime) SiH_3 is three times higher than SiH_2, but there is no evidence of kinetic elimination of SiH_2. (2)

At higher concentrations this situation changes favouring SiH_2.

1.3.2 O.E.S and L.I.F

A large number of OES spectra were recorded at various conditions, for silane-H_2, silane-He and pure silane, in order to observe their influence on SiH* emission intensity in figure 8 we present a sample OES high resolution sprectrum of the Q branch of SiH ($A^2\Delta$-$X^2\Pi$).

LIF spectra of R.F. Helium plasma were recorded while ground state SiH radical monitoring experiments at various conditions are under development.

By these experiments we hope to get more informations about vacicals concentration profiles between the two electrodes and also at the boundary layer when changing microscopic plasma parameters.(3)

A number of deposited samples have been examined, mainly by dark,photo conductivity measurements,and have demonstrated good electronic quality.

2. Conclusions

Considerable work has been done in this first year of our project mainly in the direction of experimental setup construction and elimination of related problems.

Also some results have been already reported adding new information about the possibility of process simulation and control.

Our initial plan has been proven realistic, so far. However small changes were considered necessary in order to resolve technical problems and to add more reliability to our experiments.

As a general conclusion to our theoretic and experimental work caried out, we believe that an enhancement of the deposition rate is achievable, however complete comprehention of the whole process requires more informations from various points of view.

References

1. P. Capezzuto and G. Bruno, ISPC-8, p.1434 (1987) to appear in Journal of Pure and Applied Chemistry.

2. D. Mataras, D. Rapakoulias, S. Cavadias ISPC-8, p.1484, (1987) (and references there)
3. Y. Matsumi, T. Hayashi, H. Yoshikawa and S. Komiya J. Vac. Sci. Technol. A4(3), (1986), p. 1786.

Figure 1: Information flow chart

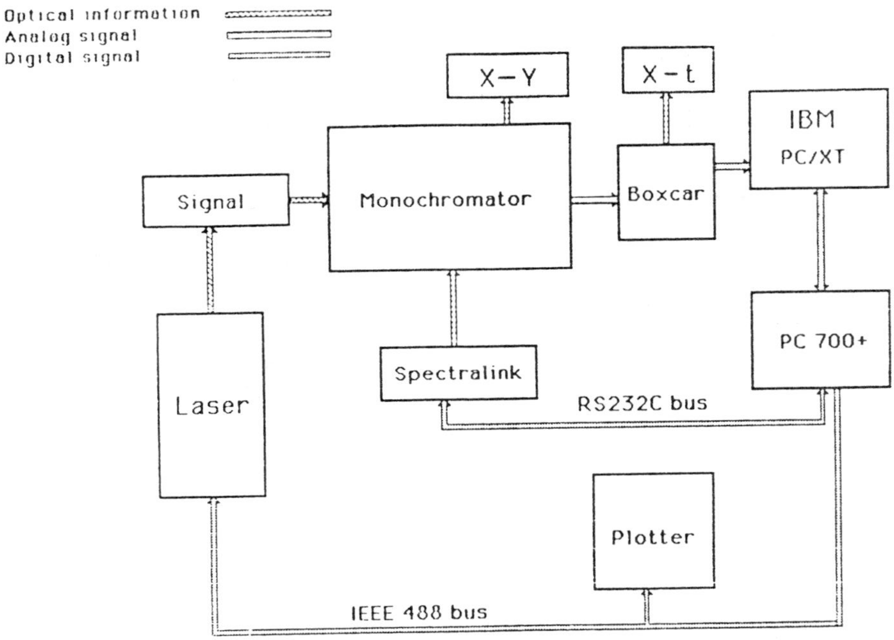

LIF – OES

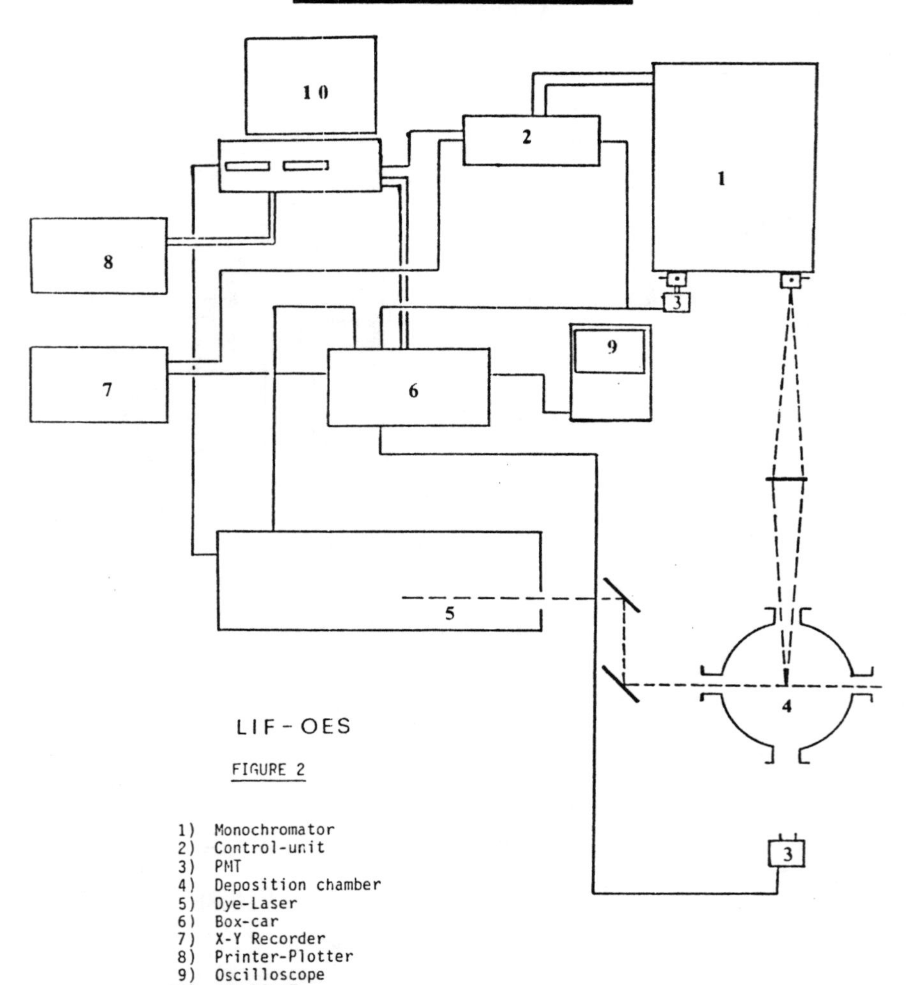

LIF-OES

FIGURE 2

1) Monochromator
2) Control-unit
3) PMT
4) Deposition chamber
5) Dye-Laser
6) Box-car
7) X-Y Recorder
8) Printer-Plotter
9) Oscilloscope
10) IBM-PC/XT

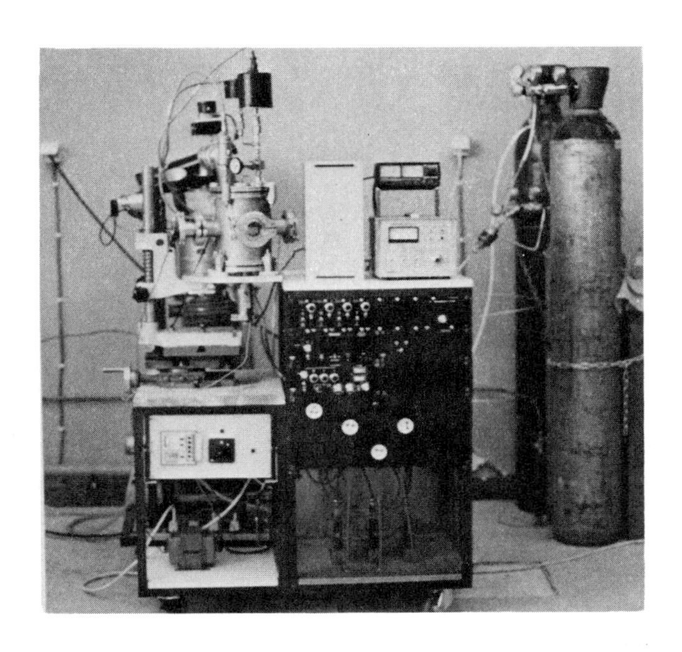

Fig. 3: Glow discharge deposition set-up

Fig. 4: Reaction chamber

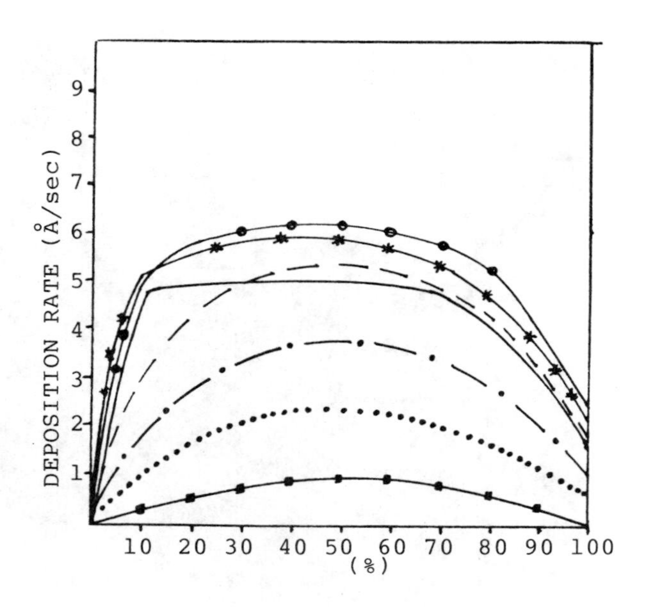

Fig. 5: Gas composition versus silane fraction in SiH_4-H_2 mixtures Conditions: Q=100sccm, p=0.5 Torr W=10Watts

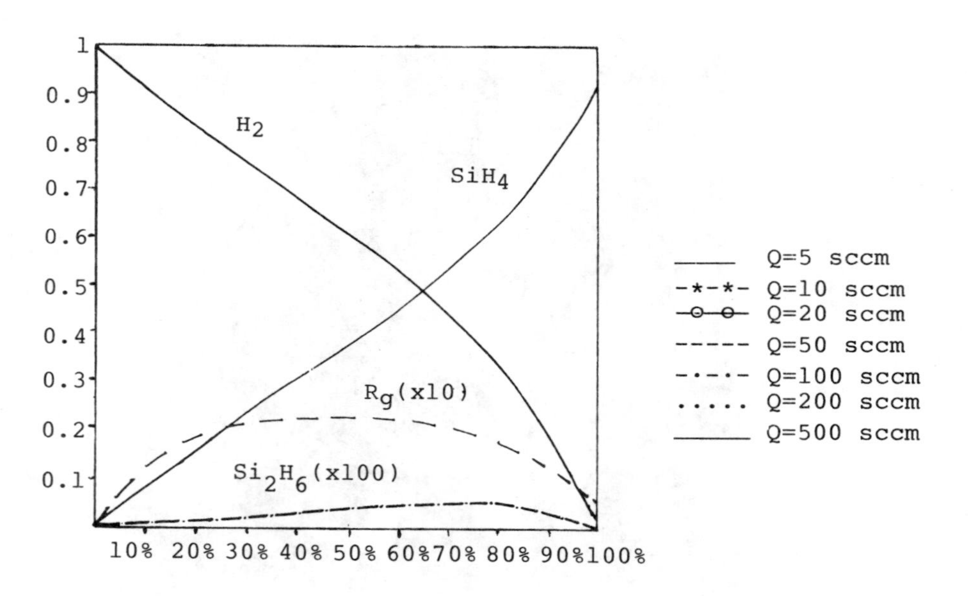

Fig. 6: Deposition rate vs silane fraction at different flow rates. (P=0.5 Torr, W=10W)

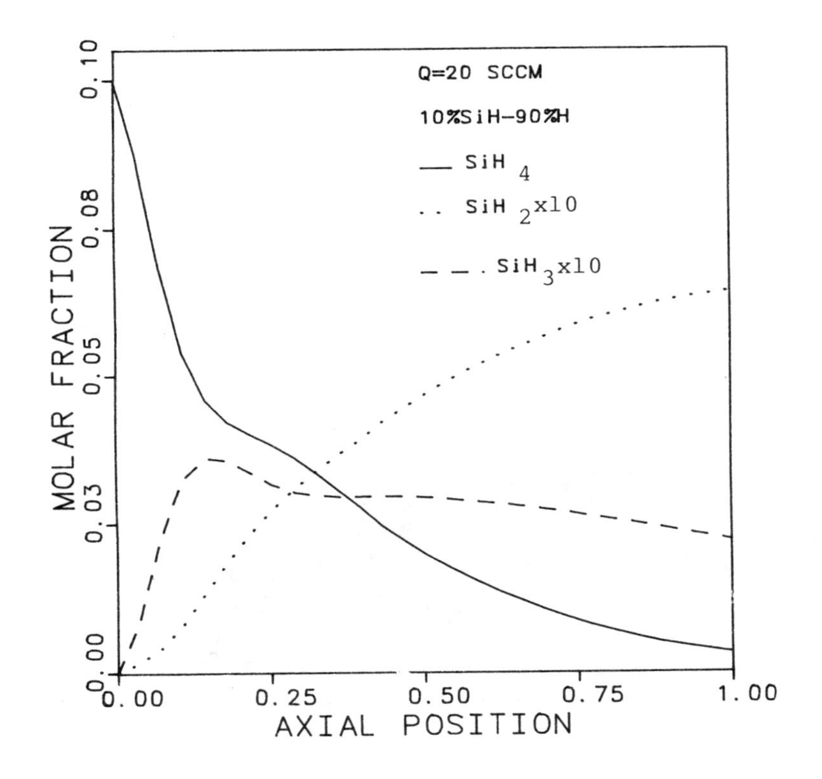

Fig. 7: Gas composition vs axial position in the reactor

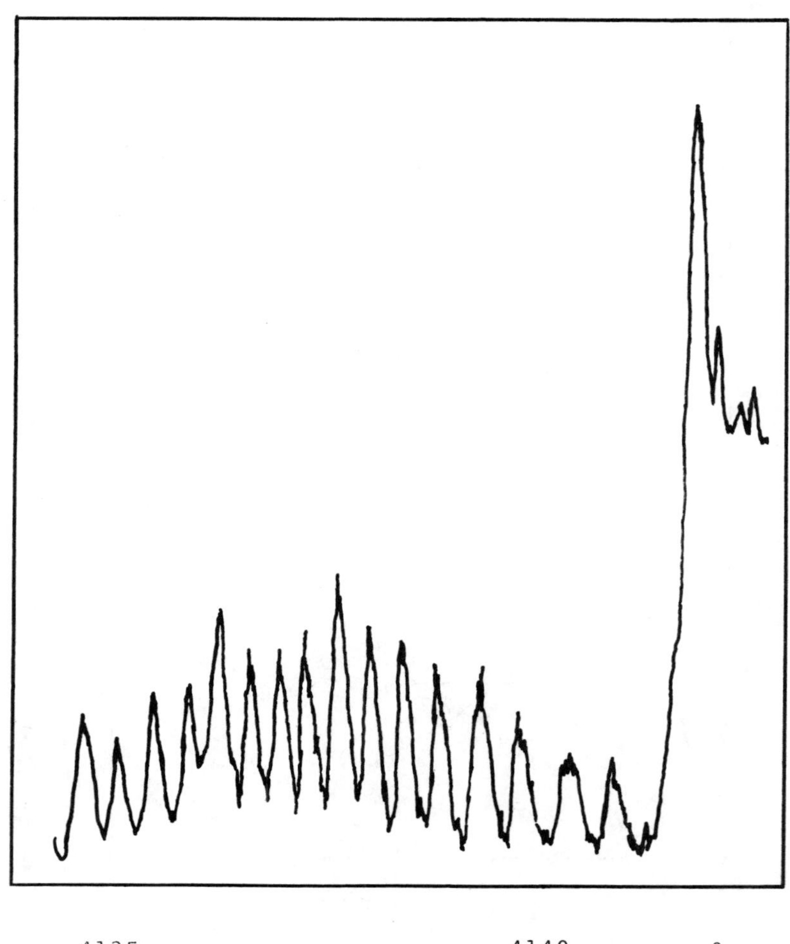

Fig.8 OES spectrum of the Q branch of SiH(0,0) band.

CHARACTERIZATION OF $SiH_4:H_2$ PLASMAS FOR a-Si:H FILM DEPOSITION: OPTICAL EMISSION SPECTROSCOPY AND ELECTRICAL PROBE MEASUREMENTS.

Authors : G. Bruno, P. Capezzuto, and F. Cramarossa.

Contract number : EN3S - 0066 - I

Duration : 36 months, 1 January 1986-31 December 1988.

Total Budget : L. 250.000.000, CEE Contribution: L. 150.000.000.

Head of Project : Prof. F. Cramarossa, Dipartimento di Chimica

Contractor : Università di Bari. Dipartimento di Chimica, Laboratorio Plasma.

Address : Dipartimento di Chimica, Università di Bari
via G. Amendola, 173
70126 Bari, Italy.

SUMMARY

Plasma deposition of amorphous silicon films from silane (SiH_4) is a well established technique to produce photovoltaic good quality material. However, the scientific understanding of the chemical processes involved, as well as their influence on the material properties, is far to be reached. In our opinion some of the unsolved questions regarding the identity of the plasma and plasma/surface processes are related to the absence of meaningful diagnostic techniques on the same system. Therefore, we have studied the deposition of a-Si:H films in r.f. plasma reactor fed with $SiH_4:H_2$ mixtures, with emphasis to plasma diagnostics.
In this paper we report data obtained by optical emission spectroscopy (OES) and electrical probe measurements, under different experimental conditions. The results show that OES and electrical probe techniques are very powerful diagnostic tools, but not enough for a thoroughly understanding of plasma deposition systems. In particular, we have evidenced that the actinometric analysis of the OES data can not be used to obtain relative density values of Si,SiH,H species in the plasma phase.
The ongoing investigation will attempt to evaluate the effect of other macroscopic parameters such as residence time and gas composition, and to improve the plasma characterization with mass spectrometry and laser interferometry. The combined results might give the answer to the still open questions.

1. INTRODUCTION.

Plasma-enhanced chemical vapor deposition (PECVD) is one of the most utilized techniques to prepare Si-based alloys. In particular, r.f. glow discharges fed with SiX_4 (X= H,Cl,F) are used to produce halogenated and/or hydrogenated amorphous silicon films, a-Si:H,X. Continuing research is directed towards the acceleration of the growth rate and improvement of the qualities of these Si-based materials. A part of this effort is the development and application of plasma diagnostic techniques which, in addition to providing a better controll of the plasma process, are likely to lead to improved understanding of the relationships between microscopic parameters and material properties.

The peculiarity of the plasma assisted deposition process is the presence in the gas phase of energetic electrons, ions, atoms and radicals in their ground as well as excited states. In order to clarify the still open questions on the plasma deposition of a-Si:H from SiH_4 (other deposition gases can be treated in similar manner), the relative roles of all these species has to be assessed (1). To this task, optical emission spectroscopy (OES) and electrical probes measurements are successfully applied to provide information concerning the energy and density of the plasma electrons, as well as information on atomic and molecular species.

In this paper we will report a comparative analysis of the results obtained by OES and electrical probe measurements during the deposition of a-Si:H film from SiH_4:H_2 mixture. The effect of the experimental macroscopic parameters such as r.f. power and total pressure on the electron density and energy, as well as on the deposition rate will be also discussed.

2. EXPERIMENTAL.

The configuration of the reactor used in this investigation is shown in Fig. 1. The diameter of the parallel-plate electrodes is 10 cm and they are about 5 cm apart. The r.f. power is fed to the lower electrode from a 27 MHz generator via a matching network. The upper electrode, holding the substrate (Corning glass), is grounded. The stainless steel chamber is evacuated using a combination of rotary and diffusion pumps and a base vacuum of 10^{-6} mbar is reached before each deposition run. Flow rates of SiH_4-H_2 mixture (10% SiH_4) and Ar are monitored with MKS mass flow

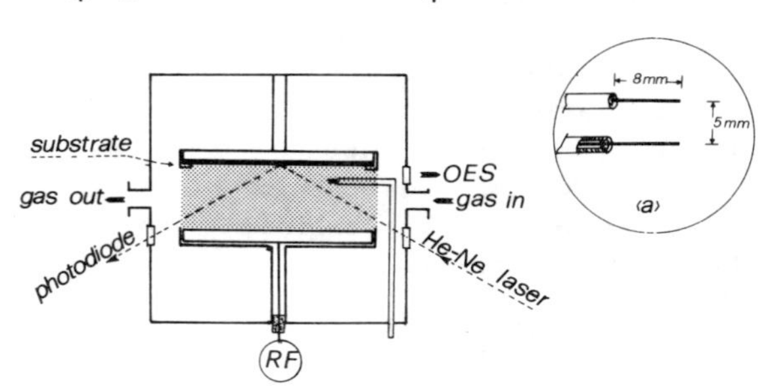

Fig. 1-Schematics of deposition chamber and electrostatic double probe (a).

controllers. The chamber pressure and r.f. power are widely varied in the ranges 0.15-1.0 torr and 5-30 watts respectively; the remaining parameters are held constant: SiH_4/H_2 = 1/9, $\Phi(SiH_4+H_2)$= 10sccm, $\Phi(Ar)$=0.2 sccm, substrate temperature= 300 °C.
The plasma emission is monitored by a Princeton Instrument OMA III attached to a Jarrell-Ash triple grating monochromator.
The I-V characteristic measurements are performed by a double probe positioned at 1 cm below the grounded electrode as shown in fig. 1. It consists of tungsten electrodes (200μm diam.) mounted in a Pyrex stem with an alumina sleeve to shield the insulating surface from the wires. Some measurements with a single probe have been also performed. In this case the bias voltage is refered to the floating potential of a second electrode.
The deposition rate and the thickness of the films are monitored during the film growth with a laser interferometer (2). The thickness of each film is also measured with the help of transmittance spectra.

3. RESULTS AND DISCUSSION.

The OES spectrum observed in SiH_4:H_2:Ar glow discharge contains the emission lines of H, H_2, Si, SiH and Ar. The emission intensity of all these species depends on the particular optical emission processes, i.e. the origin of the emitting state (upper state) and its relaxation to the lower state.
A general formulation of the emission intensity, I_X, as related to the microscopic parameters, can be expressed by the formula:

$$I_{X*} \propto k_e \cdot n_e \cdot N \quad (1)$$

where k_e is the electron impact rate constant of the electron excitation process, n_e is the electron density and N is the density of the parent molecule. This formulation includes the following hypothesis: i- the excited species X* is originated by an electron excitation process, ii- the excited species X* decay essentially radiatively. Equation 1 suggests some potentialities of the OES spectrum in giving quantitative infomation:
a) when the species involved in the emission processes is a noble gas, the trends of the intensity is related to the electron density, at least that of electrons with energy higher than that of the resonant process. In our case:

$$I_{Ar*}/k_e \cdot [Ar] \propto n_e(E \geq 13.5 \text{ eV}) \quad (2)$$

b) when the parent molecule is the same species in the ground state, the normalized emission intensity is directly related to the density of that species:

$$I_{X*}/k_e \cdot n_e \propto \cdot [X] \quad (3)$$

For this last application the normalization can be done by utilizing the emission intensity of the noble gas (actinometer) added to the gas mixture in a fixed quantity:

$$X = [Ar]\cdot I_{X*}/I_{Ar} \qquad (4)$$

This procedure, which is known as actinometric method is widely used in the diagnostics of low pressure plasma and its applicability has been demonstrated for various plasma systems (3). However, for molecular species, such as those involved in silicon deposition system, the origin of the excited state is often of dissociative type. F. J. Kampas et al.(4) have reported experimental evidence that SiH*, Si*, H* and H_2* can arise from:

$$e + SiH_4 \rightarrow e + SiH^* \text{ or } Si^* \text{ or } H^* + \text{other fragments} \qquad (5)$$

and

$$e + H_2 \rightarrow \begin{array}{l} H_2^* + e \\ H + H^* + e \end{array} \qquad (6)$$

Therefore the actinometric approach utilized on OES results can not be used in a simple way for SiH_4:H_2:Ar system. Measurements of electron density and energy by electrical probes can therefore be helpful for a better comprehension and analysis of the OES data.
Double and single probe I-V characteristics recordered under plasma conditions typical for the deposition of good quality material are reported in Fig.2. We have experimentally verified that the use of electrical probes is possible under our experimental conditions, in that we have not found any hysteresis phenomena as a consequence of the film deposition on the probe. The procedure reported by E. Eser et al.(5) has been utilized for the analysis of the I-V characteristics.

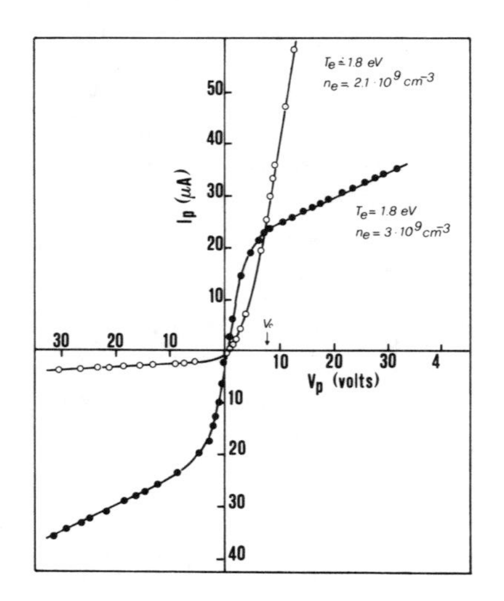

Fig. 2-Double (●) and single probe (○) I-V characteristics in SiH_4-H_2-Ar plasma.

Effect of the r.f. power. Figure 3 shows the deposition rate and the density trends of Si*, SiH*, H*, H_2* and Ar* as a function of the r.f. power. In fig. 4 the trends of n_e and T_e as obtained by the analysis of

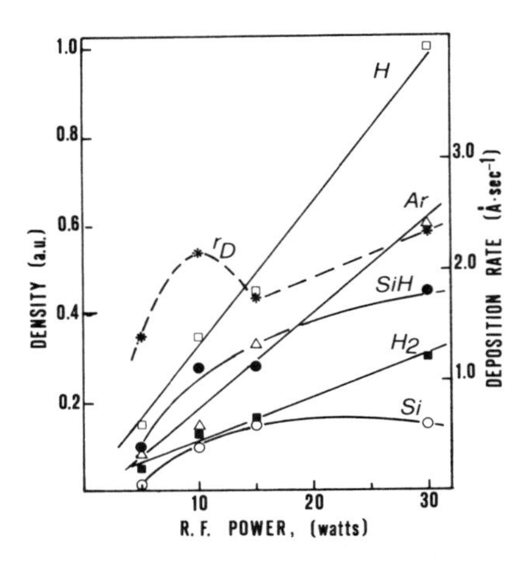

Fig. 3- Trends of excited species density and of deposition rate vs r.f. power.

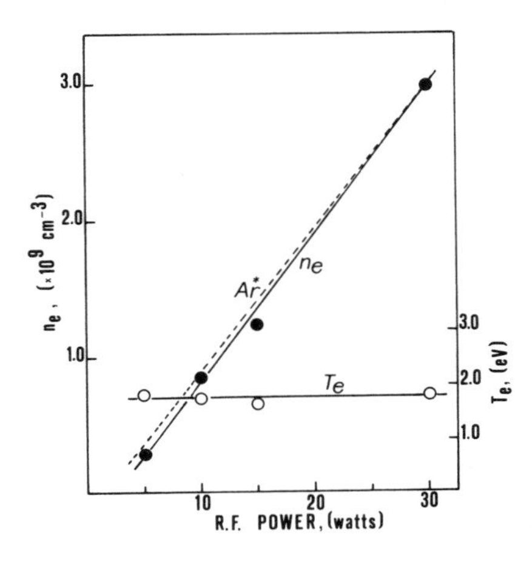

Fig. 4-Trends of electron density (n_e) and temperature (T_e) vs r.f. power.

the I-V characteristics are reported as a function of the r.f. power. In this figure the normalized profile of Ar* is also reported. The similar profile of n_e and Ar* is an indication that under these conditions the intensity of the studied Ar* line (750 nm) can be rappresentative of the electron concentration (see eq.2) This is particularly true since there is no variation of the mean electron temperature (see fig. 4). From a rough inspection of figs. 3 and 4 it comes out that there is no direct relationship between deposition rate and density of excited species as well as ions (see n_e profile). A different result is reported by A. Matsuda et al.(6), who found a linear relationship between r_D and Si* or SiH* density, by varying the SiH_4/H_2 ratio. In addition, it has been reported that also SiH_2 and/or SiH_3 radicals can be the growth precursors (1). A more complex mechanism should be invoked to explain the trends of r_D in our system.

Effect of the total pressure. Fig. 5 shows the trends of excited species density as well as the deposition rate r_D, as a function of the total pressure. Also in this case there is no exact one-to-one relationship between density of the species and the deposition rate. An important point is that, by increasing the pressure, the electron mean temperature T_e increases, while the electron density, n_e, decreases (see fig.6). The increase of T_e has to be related to an increase of the reduced electrical field, E/p, despite of the total pressure increase, which can be ascribed to the presence of high field sheath regions near

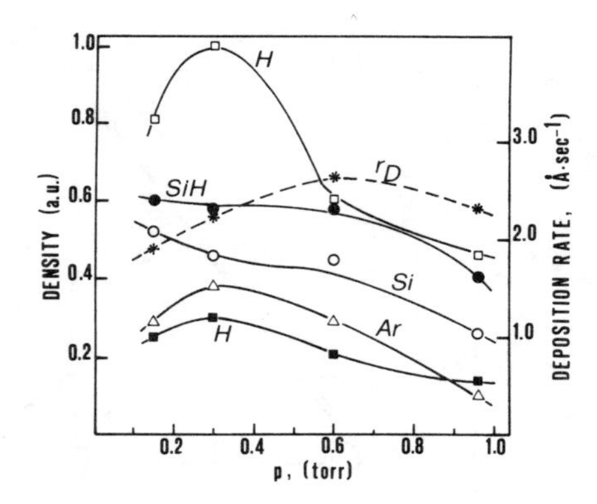

Fig. 5- Trends of excited species density and of deposition rate vs total pressure.

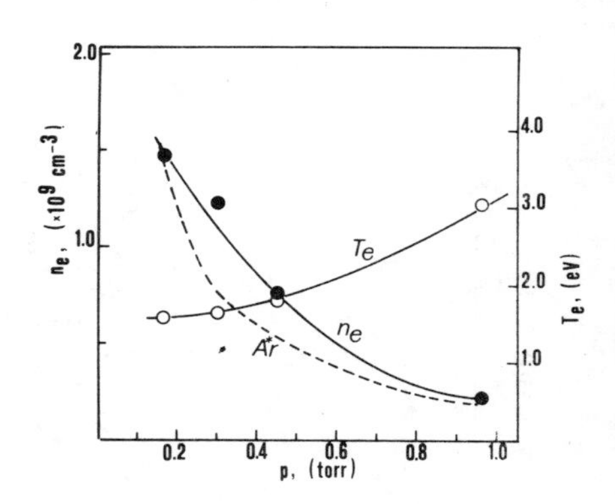

Fig. 6- Trends of electron density (n_e) and temperature (T_e) vs total pressure.

the electrodes (7) and therefore to dishomogeneity of the effective electrical field, E, in the plasma reactor. The measure of T_e can be used to evaluate such a dishomogeneity.
In fig. 6, the trend of the absolute intensity of Ar* normalized for its partial pressure is also reported. The observed different trends of Ar* and n_e and the contemporary variation of T_e indicate that in this case the normalized intensity of Ar* can not be used in the OES actinometric approach.

REFERENCES

1) P. Capezzuto and G. Bruno, Pure & Appl. Chem. 00, 00 (1987).
2) G. Bruno, P. Capezzuto, G. Cicala and F. Cramarossa, Plasma Chem. & Plasma Process., 4, 106 (1986).
3) R. d'Agostino, F. Cramarossa, S. De Benedictis, and F. Fracassi, Plasma Chem. & Plasma Process., 4,163 (1984).
4) F.J. Kampas and M.J. Kushner,IEE Trans. Plasma SCI.,PS-14(2),173 (1986)
5) E. Eser et al.,Thin Solid Films, 68, 381 (1980).
6) A. Matsuda, Proc. of 8 th Int. Symp. on Plasma Chem., ISPC-8, Tokyo, 1472 (1987).
7) M.l. Mandich, C.E. Goebe,and R.A. Gottscho, J. Chem. Phys., 83, 3349 (1985).

ELECTRO-OPTICAL PROPERTIES OF a-Si;H/μC-Si;H AND a-Si:C:H/μC-Si:C:H UNDOPED AND DOPED FILMS PRODUCED BY A TCDDC* SYSTEM

Contract Number : EN 3S-C1-602-P

Duration : 36 Months 1 July 1987- 30 June 1990

Total Budget : ESC 135,000,000 CEC Contribution : ESC 65,000,000

Head of the project : **Prof. Dr. L.J.M. Guimarães**, Uninova

Coordinator : Prof. Dr. R.F.P. Martins, Faculdade de Ciências e Tecnologia da Universidade Nova de Lisboa

Contractor : Faculdade de Ciências e Tecnologia da Universidade Nova de Lisboa/Uninova

Address : Faculdade de Ciências e Tecnologia da
Universidade Nova de Lisboa
Quinta da Torre
2825- Monte da Caparica
PORTUGAL

Summary

We have investigated the microstructure,morphology,chemical composition and electro-optical properties of a-Si:H/μC-Si:H and a-Si:C:H/μC-Si:C:H undoped and doped films that we use for glass/ITO/pin an stainless steel/pin/ITO solar cells. Film characterization was carried out using normal X-ray powder diffraction, levelling scanning electron microscopy (LSEM), Auger electron spectroscopy (AES), dark/photoconductivity and visible and IR measurements. A large variety of films were prepared in a Two Consecutive Decomposition and Deposition Chamber system- TCDDC, under various deposition conditions such as power density, dp, H_2 partial pressure, substrate temperature, T_S, and static electric fields. X-ray diffraction, LSEM and dark conductivity results show that p- and n-type μc-Si films are obtained when using dp>100mWcm^{-3}, around 20 Vcm^{-1} bias field and $p[SiH_4]/p[H_2]<0.02$. Visible and IR absorption, dark conductivity and photoconductivity data indicate that film growing is mainly ascribed with SiH precursors and the best performances of undoped a-Si(i) layers are obtained at dp<10mWcm^{-3}. These results will be presented and discussed together with ways developed for improving surface passivation, as well as for blocking interdiffusion and enhancing the open circuit voltage in these structures.

1.1 Introduction

The usefulness of a-SiH/μc-Si:H and a-Si:CH/μc-Si:C:H (undoped and doped) films for the production of solar cells depends on their capability to satisfy all cost, efficiency, and stability requirements. Nevertheless, it is difficult to reconcile all these demands as for as conventional processes, are concerned.For instance, thermal and light induced degradation depend on i layer thickness, its structural (type of SiH_n precursor, defect states and hydrogen contents C_H) transport properties and on contamination of autodiffusion impurities (1,2).Thermal diffusion from backside metal electrode also degradates cell performances by destroying the i/n junction(2).Device performances are also affected by surface states and finishing of p/i, i/n and n(p)/metal interfaces which control interface recombination processes(1).Reduction on surface roughness, transparancy and conductivity of the transparent conductive oxide layer, TCO,after a brief exposure to a hydrogen plasma, will increase absorption and ohmic losses(3).This also leaves free metal atoms which conceivably can be incorporated in the doped layer, degrading rapidly the diode junction(3).All these difficulties are not only dependent on chemical properties of constituent elements but also on defects incorporated in amorphous network, highly promoted by electron and ion bombardment, when plasma is in close contact with substrate.

Some years ago(4) we successfully developed a new deposition technique,so called "Two Consecutive Decomposition and Deposition Chamber System-TCDDC, where gas decomposition (promoted by species having high mobilities and high surface cross collisions, He,H_2)takes place far away from the deposition chamber in which a heated substrate is hold under the static electromagnetic fields(4,5).Using such system,SiH species are the main precursor during film growing and it is possible to obtain μc-Si:C(B) alloys, that, when used in the production of μc-Si:C(B):H/a-Si:H heterojunctions it enhances the open circuit voltage, V_{OC}, of single solar cells (6). In this paper some technical data on the preparation conditions and properties of a-Si:H/μc-Si:H and a-Si:C:H/μc-Si:C:H doped and undoped films are presented and their availability for the production of solar cells is discussed.

1.2 Description of apparatus and experimental measurements

a) Deposition Apparatus - a-Si:H/μc-Si:H and a-Si:C:H/μc-Si:C:H undoped and doped films were prepared in a TCDDC system, previously described(4,5). Figure 1 shows the TCDDC system which presents the following characteristics: **i)** Plasma formation can be induced by a light gas(H_2 ,He); **ii)** Carrier gas dissociation is enhanced either by using high gas flow rates or power densities; **iii)** Two grids, located above plasma region, control plasma and formed species; **iv)** Current in the plasma is controlled by a grid acting as a Langmüir probe; **v)** Substrate is placed downwards to prevent deposition of dusts and flakes which tend to cause pinhole; **vi)** Film contamination due to residual gas sorbed in-walls is reduced by using a static magnetic field; **vii)** Deposited films present uniformity within ±5% on 20x20cm² substrate area.

Before all depositions, the leak rate of the system was checked and found to be bellow 10^{-3} sccm. Prior to any deposition reactor and substrates were cleaned by a r.f. glow discharge (r.f.-g.d.) and between depositions a CF_4/O_2 r.f.-g.d. was performed in order to prevent cross contamination(7). Typical deposition conditions for the production of a-Si:H/μc-Si:H and a-Si:C:H/μc-Si:C:H films are summarized in Table-I. Amorphous undoped layers were produced at power densities, dp, bellow $10mWcm^{-3}$, while de doped ones between 5 to 130 $mWcm^{-3}$.

b) Experimental measurements - Microstructure, morphology, chemical composition, electrical and optical

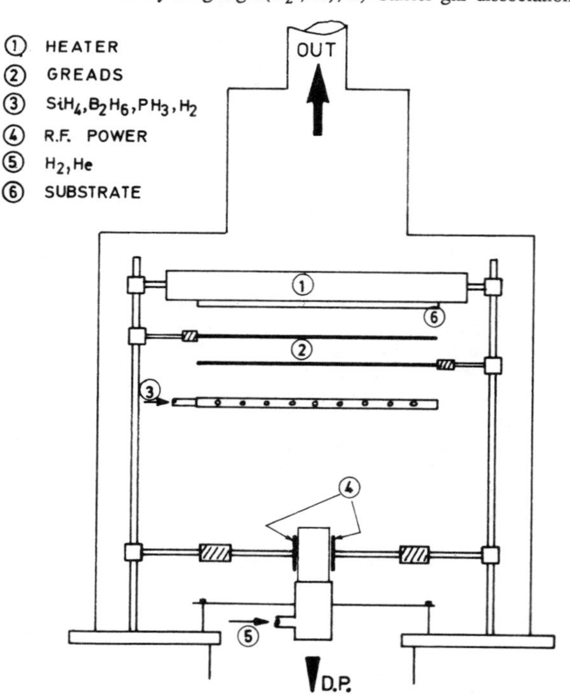

Fig.1 - Sketch of TCDDC system used

Table I - Typical depositon parameters

Gases	Gas flow (sccm)	Pressure (torr)	Substrate temp. (°C)	Power dens. ($mWcm^{-3}$)
SiH_4 CH_4 (2.5%) PH3/SiH4 (5%) B2H6/He H_2 He	1-15 1-15 1-2 1-2 50-400 10-20	0.1-0.5	100-250	3-130

properties of the produced films were investigated through normal X-ray powder diffraction, levelling scanning electron microscopy (LSEM, Auger electron spectroscopy (AES), visible and IR measurements as well as dark and photoconductivity measurements using gap cell configuration. To perform such analysis SS, alkaly- free glass and high resistivity silicon wafers were used as substrates. Glass/TCO/$p'_g i_g n$ and SS/$p^+ p i_g n n^+$/TCO solar cells (p'_g means graded μc -SiC(B):H layer; i_g means graded intrinsic layer and n^+ means microcrystalline layer highly phosphorus doped) have also been produced and characterized through spectral response of photoconductivity and I-V curves recorded under AM1 simulated sunlight (6).

1.3 Results

a) Undoped layers - a-Si:H films having dark conductivities, σ_D, lower than 10^{-9} Scm^{-1}, activation energies, ΔE, in the range of 0.75 eV-0.85 eV, pre-exponential factors, σ_0, of the order of $10^4 Scm^{-1}$, optical gaps,E_0, in the range of 1.7-1.8 eV and CH in the range of 20-10% have been produced by using dp<10$mWcm^{-3}$, $200^0C<T_s<300^0C$, and B<0.3kG. Analysis of film morphology and structure by SEM/LSEM and normal X-ray diffraction patterns reveal a smooth surface and a quasi-columnar amorphous structure(8). The dependence of photossensitivity, σ_p/σ_D, E_0 and (relative) concentration of SiH_n(n=1,2,3) species on deposition pressure, p, are shown in figures 2 and 3 (open circles). Same data for films produced by a conventional r.f.- g.d. system (dark circles) are also indicated.

Undoped amorphous silicon carbide alloys, using SiH_4 to CH_4 mixtures in the range of 1:1 to 1:2 were produced at dp<10$mWcm^{-3}$ and $T_s<150^0C$.These films have $\sigma_D<10^{-12}$ Scm^{-1}; $E_0>$ 2.5 eV; 15%< CH< 35%; $\Delta E>$ 0.9 eV; and carbon content in the range of 2% to 10%. Bond configurations and impurity contents (as observed from 770cm^{-1} and 630cm^{-1} peaks in the IR spectra (9)) depend strongly on hydrogen dilution ratio and dp used.

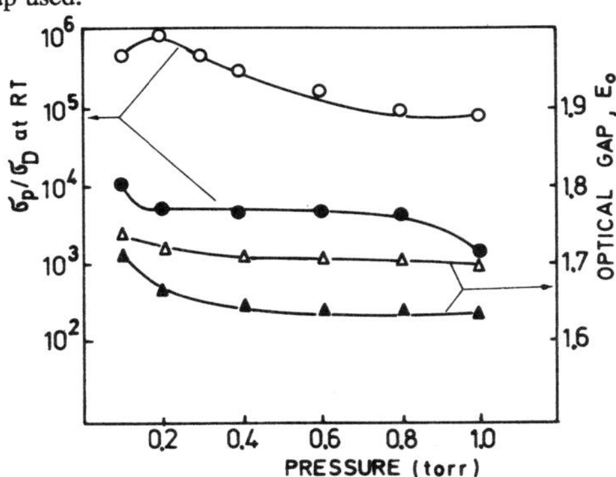

Fig. 2 - Dependence of σ_P/σ_D and E_0 on p for undoped films deposited by a TCDDC system (open circles) and by a conventional r.f.-g.d. system (dark circles)

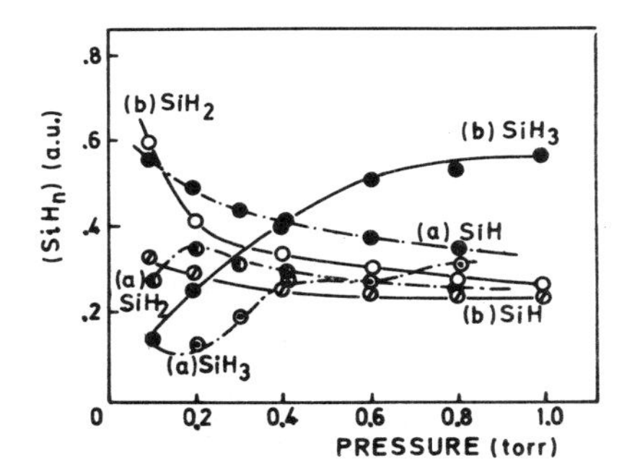

Fig. 3 - Dependence of species concentration (arbitrary units) on p. a) Films deposited by a TCDDC system. b) Films deposited by a conventional r.f.-g.d. system

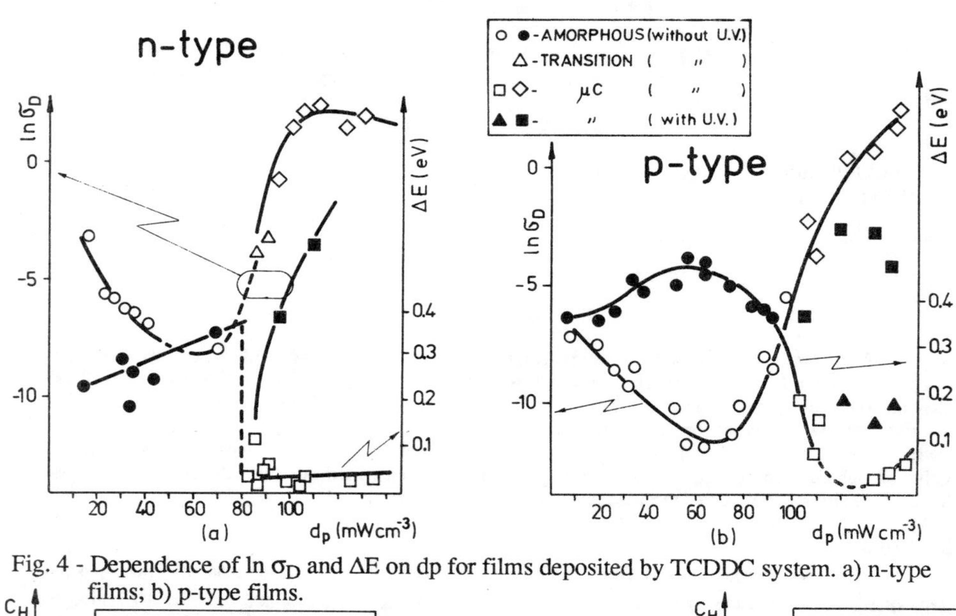

Fig. 4 - Dependence of ln σ_D and ΔE on dp for films deposited by TCDDC system. a) n-type films; b) p-type films.

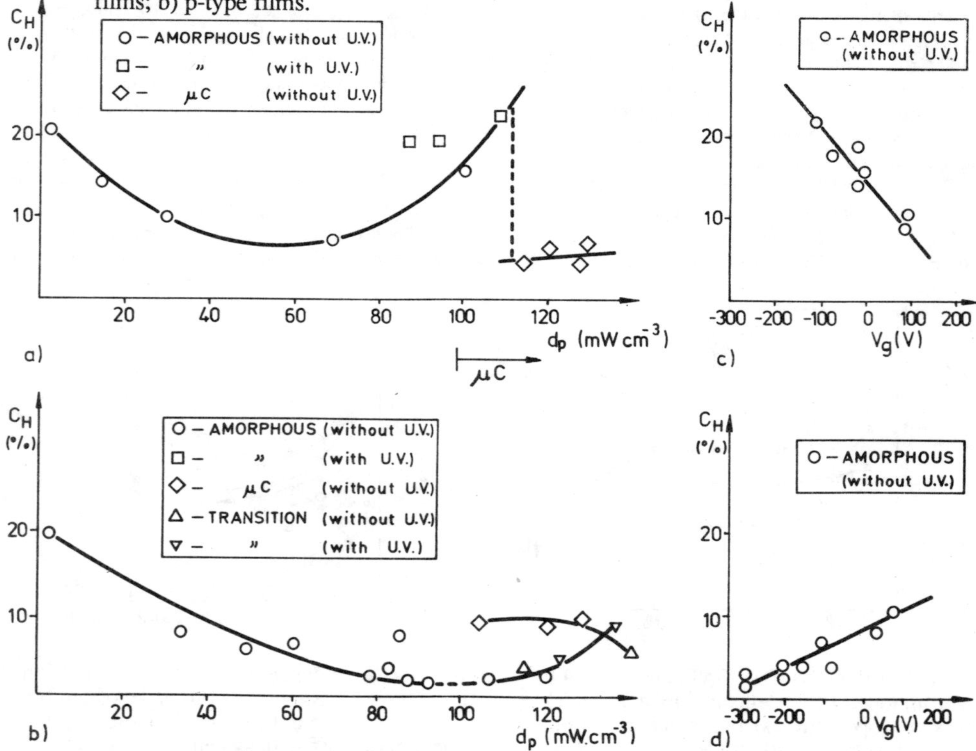

Fig. 5 - Dependence of CH(%) on dp [a) n-type films; b) p-type films] and V_g [c) n-type films; d) p-type films] for various deposited films.

b) Doped films - The dependence of σ_D and ΔE on dp for phosphorus and boron a-Si:H films is shown in figure 4. CH dependence on dp and V_g is also shown in figure 5. Transition from a-Si:H to µC-Si:H films is obtained when high dp and hydrogen dilution ratios are used. Such transition is accomplished with changes on the structure, electro-optical properties and CH , as seen in figures 4 e 5 (10).

Structure and electro-optical properties of a-Si:C(B):H films deposited at T_s=150°C using SiH_4:CH_4:B_2H_6 ratios of the order of 1:1:0.02 at different dp and hydrogen dilution ratios were investigated. As dp increases up to 120mWcm^{-3} and hydrogen dilution ratio is higher than 70%, a sharp peak appears in the IR spectra around 500-540cm^{-1}, attributed to film microscrystallization, as confirmed by normal X-ray diffraction measurements (11). σ_D and E_0 increase from 10^{-8}Scm^{-1} to 5x10^{-3}Scm^{-1} and from 1.9eV to 2.5eV, respectively, while ΔE decreases from 0.5 eV to 0.17 eV.

c) Solar Cells - Figures 6 and 7 show I-V curves and quantum efficiencies of an inverted pin and $p'_g i_g nn^+$ solar cells produced on SS substrates. The built-in voltage, V_b, was determined from the low temperature saturation of V_{OC}, and shows values of the order of 0.9V and 1.20-1.25V, respectively. The TCO layer (based on indium tin oxide) is 700 Å thick and it has a sheet resistivity, ρ_S=100 Ω/□.

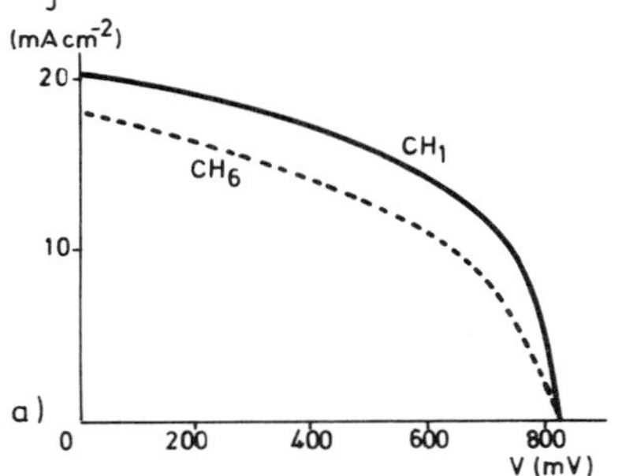

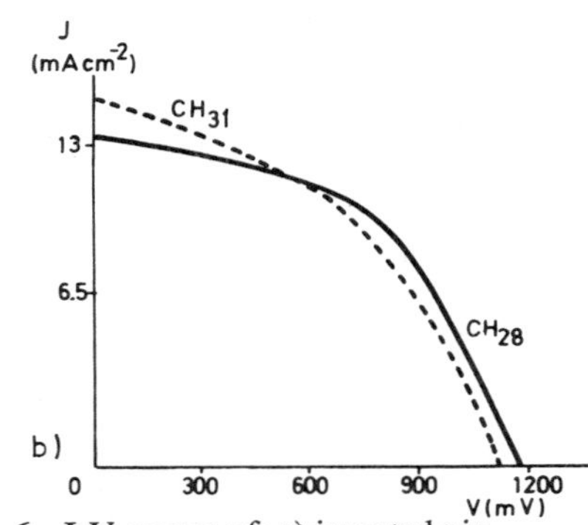

Fig. 6 - J-V curves of: a) inverted pin solar cells with 16 cm^2 in area (---) and 0.07 cm^2 in area (—); b) the same results but for a ss/n+n.i_g/p'_g/TCO silicon--carbide heterojunction solar cell.

1.4 Analysis of the results and comments.

Results shown in figures 2 and 3 indicate that structure, electrical and optical properties of films obtained by a TCDDC system are optimized in relation to those ones prepared by our conventional r.f.-g.d. system. Namely, it is observed that films produced by a TCDDC system present the following characteristics: **i)** the precursor of film growing is ascribed with SiH species which makes these films quite suitable against thermal and light induced degradation (fig. 3),(12): **ii)** $\sigma_p/\sigma_D>10^5$, E_0=1.75 eV and [SiH/[SiH_2]>10 are obtained at dp=10mWcm^{-3} and deposition pressures in the range of 0.1 to 0.3 torr, for undoped films (figs. 2 and 3); **iii)** the formation of microcrystalline films depends on the density of hydrogen radicals reaching the growing surface either for doped µc-Si:H films or µc-Si:C(B):H (see figs. 4 and 5) that is, microcrystalline films are obtained using high hydrogen dilution ratios and high dp values; **iv)** bond configuration and CH depend strongly on dp and on V_g values used; **v)** wide band gap microcrystalline doped films are ascribed with surface passivation and reduction of dangling bonds promoted by bombardment of growing surface with atomic hydrogen which also enhances doping efficiency and carbon content in silicon carbide alloys (13,14).

Solar Cells - From figure 6 we observe that SS/n+n.i_g/p'_g/TCO structures present high V_{OC}. This is due to an increase in V_b as expected from the material properties presented by i and p'_g layers. The graded p'_g layer was produced in such away that boron diffusion towards i layer and indium and tin diffusion towards the silicon carbide layer is reduced and so high V_{OC} are obtained. Nevertheless, the low conversion efficiency

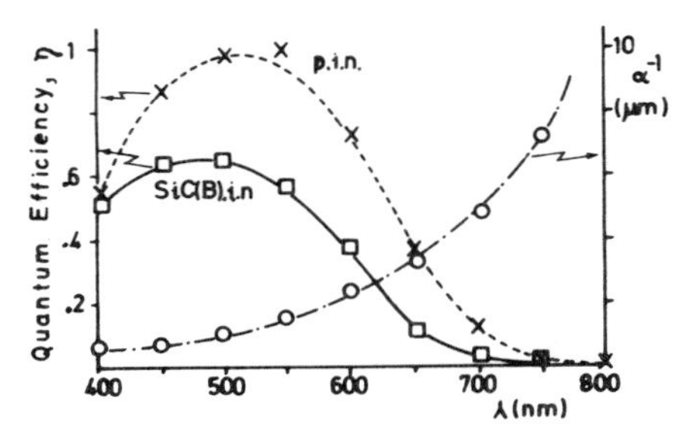

Fig. 7 - Quantum efficiency of inverted pin solar cells (---) and SS/n+n.i_g/p'_g/TCO solar cell (—) of small areas.

recorded, although dependent on ρ_S, also indicates a not yet well improved control upon thickness of p'g layer. It was also observed (fig. 7) an enhancement towards low wavelength in the photoresponse of device having a p'_g layer, when compared with the normal pin device.Such result is a clear indication that this layer works well as an effective window layer. Devices were also produced on glass/ITO substrates, and it was observed that TCO layer is not reduced when an a-Si:C:H layer is prior deposited at low dp and substrate temperatures(8).

2. Conclusions

The obtained results show : **i)** structure and transport properties of films deposited by a TCDDC system fulfil the demands for solar cells production at low costs, presenting good stability and conversion efficiencies; **ii)** high V_{OC} are obtained by using p'_g layers; **iii)** degradation of i layer and doped layers are reduced by using a-Si:C interlayers; **iv)** improving ρ_S and controlling thickness of p'g layer, will be possible to improve device efficiencies up to 10%.

Acknowledgements

We would like to thank N. Carvalho, M. Santos, M. Vieira, I. Baía, A. Maçarico, E. Fortunato, I. Ferreira, F. Soares and M. Quintela, members of the research team, for the help given during film characterization and production. Thanks are also due to M.O. Figueiredo (Centro de Investigação Tropical), Y. Shapira, J.Gordon and A. Larhea (Fac. of Eng., Tel-Aviv University) for X-ray ,LSEM and AES characterization performed in our samples.

References

1. TAWADA,Y. et al. (1986). Proc. of the Inter. Soc. for Optical Eng.-Conf. 706, Cambridge
2. YAMAGISHI, H. et al. (1986). Proc. of 7th E.C.Phot. Solar En. Conf., Sevilha
3. MAHAN, A., H. et al. (1984). App. Phys. Lett. 44, 220
4. MARTINS, R. et al.(1983). Proc. of III Simpósio Brasileiro de Microelectrónica, S. Paulo
5. MARTINS, R. et al. (1983). Proc. of 5th E. C. Phot. Solar En. - Conf. 778-783, Athens
6. MARTINS, R., GUIMARÃES, L. et al. (1986). Proc. of 2nd Inter. PVSEC, p. 445-447
7. ANDRADE, A. M., MARTINS, R. et al. (1985). Solar Cells
8. CARVALHO, N. et al. (1988). Subm.to 8th E.C. Phot. Sol.En.Conf. to be held in Florence
9. MARTINS, R. et al.(1988). Subm. to 8th E.C. Phot.Sol. En.Conf.to be held in Florence
10. MARTINS, R., CARVALHO, N. et al. (1987). Proc. of 12th I.C.A.L.S..Conf., Prague
11. SANTOS, M. et al. (1988). Subm. to M.R.S.Meeting to be held in S. Diego
12. KWANO, Y. et al. (1987). Proc. 12th I.C.A.L.S. Conf., Prague
13. VIEIRA, M., MARTINS, R. et al. (1988). Subm. to M.R.S.Meeting to be held in S.Diego
14. MARTINS,R. et al.(1988). Subm. to 8th E.C. Phot. Sol. En.Conf.,to be held in Florence.

"DEVELOPMENT OF A HIGH DEPOSITION RATE TECHNIQUE FOR DEVICE QUALITY THIN FILMS OF HYDROGENATED AMORPHOUS SILICON"

Progress Report (*)

Contract Number	:	EN3S/0039
Duration	:	36 months, 1 October 1986 - 29 September 1989
Total Budget	:	400,000 ECU CEC Contribution : 200, 000 ECU
Head of Project	:	Prof. A. Christou
Contractor	:	Research Center of Crete, Institute of Electronic Structure and Laser.
Address	:	P.O. Box 1527, Iraklio, Crete 711 10, Greece.

Summary

The basic functions of an amorphous hydrogenated silicon deposition system, based on the Ionized Cluster Beam (ICB) source have been investigated separately and the corresponding parts of the apparatus have been optimized.

Particular emphasis was given to the atomic hydrogen source. Several options were tested in an electron-gun evaporation system before the ICB facility was available. This work has led to the development of a simple and efficient hydrogen dissociator which is now in operation, together with the ICB source, in a high vacuum deposition chamber designed to take advantage of the features of both.

(*) Report prepared and presented by P. Tzanetakis

Introduction

The chemical reactions of hydrogen with silicon in the growing process of amorphous thin films determine not only the hydrogen content but, most importantly, the structure of the material and consequently its performance in any device. The presence of active, i.e. dissociated and ionized hydrogen is inherent to all the widely used techniques of silane decomposition. On the contrary, in atomic or cluster beam deposition of amorphous silicon an efficient source of activated hydrogen is indispensable.

In the case of ICB deposition, the presence of the ionizer and the high acceleration potential can cause bombardment of the growing film by relatively energetic (few KeV) protons. On the other hand we have shown that the interaction of activated hydrogen with silicon already deposited on the chamber walls produces SiH_n radicals which react with the growing film leading to increased dihydride content.

In order to avoid these problems and to be able to vary in a controlled and precise manner the hydrogen content of the films, one must use a source producing a well collimated beam of hydrogen atoms with as little molecules and ions as possible.

During the first phase of our project we have designed and assembled a high vacuum chamber for ICB deposition of hydrogenated amorphous silicon. In parallel, with the use of an electron-gun evaporation source for silicon, we have developed a simple, efficient, easy to use source of atomic hydrogen having the previously stated advantages.

Experimental

a) ICB source

In the usual mode of operation of an ICB source, the substrates are held at ground potential and the source head is raised to the high accelerating voltage (0 - 10 KV). This configuration gives a certain convenience and flexibility in the operation of the (heated) substrate holder. The power supplies of the ionizer must then float to 10 KV. This mode will probably have to be adopted in any production system.

On the other hand, in the research stage the opposite configuration, with the substrates at high voltage can be used. This greatly reduces the significant cost of the ICB power supplies.

We have purchased a commercially available ICB source head. The power supplies for grounded source operation have been built by the Electronics Service of the Research Center of Crete. The source is placed in a bell jar high vacuum system, pumped by a cryopump.

b) Hydrogen source

A tungsten filament, thermal dissociator was initially used in a small chamber at 10^{-3} torr, communicating with the deposition chamber at 10^{-6} torr, via a variable diameter orifice. This source produces a hydrogen beam with relatively poor dissociation ratio.

It quickly became evident that a reaction of hydrogen molecules at the hot tungsten surface was responsible for the dissociation in the source. This finding led to a new source design which is still under evolution. The main advantage of the new source is the elimination of the molecular component from the hydrogen beam. This allows for even lower deposition chamber pressure and reduces drastically the gas load of the cryopump.

Results and discussion

Besides bringing the ICB system to operation, the most significant result of the first phase of this project is certainly related to the understanding of hydrogen dissociation on a hot metallic surface. The scheme we are proposing includes the following steps :

1. Attachment of the H_2 molecule to the metallic surfaces.
2. Dissociation on the surface (known to have a much lower activation energy than in the gas phase).
3. Re-emission of atomic Hydrogen with little or no surface recombination.

This simple model needs further experimental evidence to be unambiguously proven. The corresponding research work is certainly not in the main line of this project and will be published separately. However the principle of operation of the hydrogen dissociator presently used in the ICB system is based on the model presented above.

The results obtained with e-gun evaporated films deposited at 250°C can be summarized as follows :

- Films were produced with hydrogen content ranging between 2 and 20%.

- The Infra-red spectra of the evaporated films indicate a high dihydride content. This was expected since thermal evaporation is known to produce non-uniform films.

- The mono- to di-hydride ratio can be improved significantly by only increasing the flux of atomic hydrogen reaching the surface of the depositing film. We believe that H-H recombinations yield extra energy to the surface layer of the growing film. This in turn, leads to a denser, higher quality material.

- The interaction of the atomic hydrogen beam with Si already deposited on the chamber walls produces SiHn radicals. We have observed the incorporation of these radicals in growing films not in direct sight of the atomic hydrogen beam, giving rise to a non negligible hydrogen content, all in di-hydride form.

A more detailed account of the results obtained with evaporated films is given in ref.(1).

Conclusion

We have developed a simple and efficient source of atomic hydrogen to be used with atomic or cluster beam deposition of a-Si:H films. The source was tested using electron-gun evaporated Si deposition.

The results obtained with the evaporated films indicate the importance of the hydrogenation method in determining the quality of the materials.

We have designed and set-up an Ionized Cluster Beam deposition system incorporating the Hydrogen dissociation source. Work is in progress to characterize the films deposited under various conditions in this system.

(1) Y. Frangiadakis, P. Tzanetakis. Hydrogen incorporation in evaporated amorphous silicon films. Submitted for publication.

PREPARATION OF SOLAR GRADE AMORPHOUS SILICON FOR PHOTOVOLTAIC CELLS BY MEANS OF AN ELECTROLYTIC PROCESS (*)

Contract number	:	EN3S-0040-I
Duration	:	17 months 1 Nov. 1985 - 31 March 1987
Total budget	:	Lit. 170.200.000
CEC contribution	:	Lit. 74.890.000
Head of project	:	Marco V. Ginatta, Ph. D.
Contractor	:	Elettrochimica Marco Ginatta
Subcontractor	:	Extramet
Address	:	Elettrochimica Marco Ginatta (EMG) Via Brofferio, 1 10121 TORINO

Summary

The objective of this work is the designing of a electrolytic process for the production of solar grade amorphous silicon, both in organic solvents and molten salts, in the form of a thin layer on a conductive substrate, suitable for the direct manufacturing of solar cells.
We have divided the R/D work in two main fields:
- silicon electrodeposition in organic solvents
- silicon electrowinning in molten salts.

Tests have been conducted in non-aqueous solvents: in acetic acid and tetraethylorthosilicate we have produced thin layers, in other electrolytes deposits were not observed.
The cell for silicon electrowinning in molten salts has been designed and is now in construction.

Our Partner Extramet has run electrodeposition test in molten salts using a cell of its design.
The results of this prefeasibility research show clearly that their simplified concept of electrorefined cell has been successful to obtain silicon in baths containing NaF, KF and K_2SiF_6 at temperature between 750-800°C without the use of lithium salts.
The electrolysis did not produce amorphous material, but tetragonal crystals.

(*) Paper presented by Pierangelo Perotti

1. INTRODUCTION AND DESCRIPTION OF THE PROJECT

The successful development of low cost solar cells requires new technology of production, cheaper and simpler.
Amorphous silicon is actively considered for photovoltaic applications, and has proved to be a promising material for the manufacture of large area solar cells of medium efficiency.

The total material costs of a-Si devices are low, in view of the fact that very low thicknesses (≈ 1 µm) are sufficient for adequate performance, and the substrates are quite inexpensive.

Amorphous silicon not containing hydrogen has a large concentration of dangling bonds, producing levels in the forbidden gap. These bonds can be saturated by hydrogen; in this way the levels disappear and doping becomes possible.

Amorphous silicon is usually made by methods like glow discharge, sputtering, thermal decomposition of silane, chemical vapour deposition and others.

The properties of a-Si films are strongly dependent on the deposition technique. Above all, the hydrogen concentration in the film is influenced by the deposition conditions.

The electrolytic process is attractive for various reasons:

- it transforms in a single step silicon containing species in a metallic silicon film on a substrate, suitable for the production of solar cells;
- it is possible to regulate the hydrogen content;
- doping may be possible adding proper dopants in the bath;
- electrochemical reaction operates a refining action, thus permitting the achievement of high purity.

Silicon cannot be electrodeposited from aqueous electrolytes because of: a) the hydrogen evolution reaction that takes place instead of Si deposition; b) the hydrolysis of most Si species, to be dissolved in the electrolyte, with water.

To remove these difficulties it is possible to work in some organic solvents or in molten salts.

The objective of this research project was the designing and testing of an electrolytic process for the production of solar grade amorphous silicon, in the form of a thin layer on a conductive substrate, suitable for the direct manufacturing of solar cells.

Critical discussion on systems already proposed employing electrolytic techniques, has allowed the choice of some baths to be investigated.

The work has been composed of these phases:

1) Bibliographic research, extending it to the papers published in the literature, patents and industrial reports, in order to complete our knowledge of the present state of the art concerning a-Si electrodeposition.
2) Design and construction of laboratory equipments used for electrolytic tests.
3) Purification of raw materials and solvents.
4) Silicon electrodeposition tests.
5) Elaboration of collected data.

We have divided the R/D work in two main fields:

- silicon electrodeposition in non aqueous solvents

- silicon electrowinning in molten salts.

2. SILICON ELECTRODEPOSITION IN NON AQUEOUS SOLVENTS

All experiments of silicon deposition in organic solvents, including preliminary tests, were done inside a dry glove box under an argon atmosphere.
The glove box has a pre-chamber, used for material exchange from outside to inside and viceversa.

A role of fundamental importance, in organic electrochemistry, is played by solvents dehydration. Infact, working with low cathodic potentials (high in absolute value), the presence of even very low quantities of water, in the order of few ppm, can dramatically influence the results.
Liquid compounds, such as propylene carbonate (PC) or tetraetylorthosilicate (TEOS), have been dehydrated over molecular sieves, used both in pellets and in powder.

The tetraalkylammonium salts, used as supporting electrolyte in order to improve electrical conductivity of the bath, were purified by recrystallization under argon atmosphere.

Tests - First group

Preliminary tests had the aim of determining electrical conductivity of the solutions.

The experiences have been conducted in a glass or teflon cell, working with two main types of bath:
- $SiHCl_3$, propylene carbonate and tetraalkylammonium salts (used as support electrolytes to improve conductivity);
- Tetraethilorthosilicate, acetic acid, methyl alcohol and tetraalkylammonium salts.

The bath investigate in the preliminary experiences was composed by PC, TEOS and tetrabutylammonium chloride (TBAC), measuring the bath electrical conductivity with increasing concentrations of TBAC.

This solution till now has not submitted to further tests and we think it's worthy of other work.

Other experiences were conducted replacing TEOS with trichlorosilane. This compound has a high volatility (bp=33°C) and is greatly reactive with humidity. This bath created many difficulties in its manipulation so we dropped it.
For this reason we are directing our efforts to TEOS and in a future we want to try other silicon compounds, with the exception of silanes and silicon halides.

Acetic acid has been used to check the behaviour of a protic solvent (instead of the generally used aprotic solvents); however there is hydrogen evolution competing with silicon reduction and so greatly decreasing current efficiency. On the other hand, for the reasons previously said, it is necessary to introduce hydrogen in the film.

Potential-time and current-potential curves have been traced.
Solution of commercial purity reagents, without purification, and solutions with accurate deydration and purification was submitted to electrolysis at constant current intensity.

With impure raw materials, the electrodeposition cannot be protracted after a certain time.

The second type of bath (TEOS in acetic acid) has produced various deposits on nickel cathodes.
SEM has showed that the film is uniform, nodular, without cracks or defects.

Further work is necessary in order to optimize layer thickness and its hydrogen content.

Tests - second group

We have then taken into account some systems of this composition:

a) organic solvent
b) silicon raw material
c) supporting electrolyte

a) We have concentrated on aprotic solvents of common use, easy to obtain and to handle.

Particularly we have chosen:
propylene carbonate (PC),
acetonitrile (AN),
dimethylformamide (DMF),
dimethylsulfoxide (DMSO).

DMSO however was rejected in the preliminary test, as it resulted immiscible with tetraethylorthosilicate.

b) In a first time we considered two possible silicon raw materials: tetraethylorthosilicate (TEOS) and potassium hexafluorosilicate.

The latter however was discarded because it is pratically insoluble in the chosen solvents, although it has a high dielectric constant.

c) The use of a supporting electrolyte is necessary because of the extremely low values of conductivity of the solutions.
We have tested both tetramethylammonium chloride (TMAC) and tetrabutylammonium chloride (TBAC).
Subsequently we directed to the latter which has higher solubility.

We have prepared the solutions, to dehydrate the solvents, and then made to flow slowly through a glass column filled with molecular sieves.
The conductivity of every solution was measured before and after the test.

Deposition tests were carried out in a glass cylindrical cell, at ambient temperature.

The electrodes were:
- a platinum foil as anode (in one test only we used silver as anode, but the result was a strong corrosion of the electrode, so we decided to use the most reliable platinum);
- a foil of silver as cathode.

3. SILICON ELECTROWINNING IN MOLTEN SALTS

The electrolysis will be performed in a controlled environment, in a cell with particular geometric design and other peculiar characteristics dictated by EMG experience of molten salts electrolysis.

The project includes the detailed design of the characteristics (dimensions and materials) of plant components (furnace, crucible, electrodes, chamber in controlled atmosphere), as well as instrumentation and control equipments.

The design of the plant consists of both the assembling and adapting of general purposes components to the requirements of the process (thermocouples, thermoresistances, pressure gauges) and the construction of more specific instrumentation (electrodes, cells, control system).

Designing and building of a lab molten salt cell

The electrodeposition of silicon to a high degree of purity, requires rigorous procedures for establishing and maintaining pure electrolytes.

One of the essential steps is the elimination of any oxygen and nitrogen.

Since it is not advisable to expose high purity deposits at temperatures higher than 50°C, the electrolytic cell was built with an expanded upper section of the cell (cold chamber) permits the running of many different tests under comparable conditions, and a specific upper closing valve, that preserves the purity of the atmosphere in contact with the surface of the bath, while changing experimental parameters, and allows the cooling of the cathode while the bath is maintained molten at a constant temperature.

The most suitable material for crucibles to hold silicon electrolytes should be graphite.

The graphite crucible used in these investigation had the following dimensions: 8 cm internal diameter, 28 cm height, and 1.5 cm wall thickness.

The outer lower shell (the hot chamber), upper closing valve and the expanded upper section (the cold chamber), were made of stainless steel 321, tungsten-inert-gas welded. Nickel tubing was used for the thermocouple protection.

The furnace used in this study was a multiple unit, 5.500 Watt, with a maximum temperature of 1.050°C; it was vertically set, had a diameter of 12.5 cm inside, and height of 60 cm.

The power supplies used in this investigation are a 50 Amp, 15 volt solid state rectifier, with a Unit-Process Assembly for periodically reversing the current, and an 8 Amp, 12 volt, for superimposing and reversing the current with dead time.

A Close Circuit Television apparatus with tape recorder is used to record the progress of the electrolytic process.

Preliminary tests

Preliminary tests has been carried out in a closed vacuum and gas tight cell in refractory alloy. It is constituted by a cylinder, closed at the bottom, of refractory steel with a diameter of about 90 mm, a height of about 500 mm and a stainless steel cover allowing the passage of gas, products and electrodes.

We have considered a separation of the winning process in two steps, electrowinning and electrorefining of the winning stage products.
We have tested the second step, using metallurgical silicon instead of an electrowon silicon.

From literature it appears that molten fluorides melts are the most suitable for electrorefining operations. Most the authors used expensive lithium fluoride eutectics as bath and a new cheap concept of silicon electrorefining should avoid if possible the use of lithium salts.
The use of copper at the anode would also be useful to favorize the dissolution of metallurgical silicon.

For our experiments we have chosen to use a KF-NaF eutectic with a melting point of 710°C. This temperature may allow to work at 750-800°C that is the range of temperature most used in the experiments described in literature.

Potassium fluosilicate has been chosen as salt assuring the electrolytic transport of silicon in the bath.
The salts used were normal products of analytical purity at least 99% purity.
Dehydration of salts was carried out by heating the salts under vacuum; K_2SiF_6 was added and the final purification of the bath was carried out by pre-electrolysis, using the crucible as cathode and a graphite rod as anode.

The deposits obtained were observed at the optical microscope, washed with water to eliminate the salts as much as possible, and examined at the electronic microscope and submissed to X ray analysis.
In view of the exploratory character of the experiments we do not have attempted, at this stage of the research, to estimate currents efficiency and refining effects of silicon deposition.

The results of this prefeasibility research show clearly that our simplified concept of electrorefined cell has been successful to obtain silicon in baths containing NaF, KF and K_2SiF_6 at temperature between 750-800°C without the use of lithium salts.

Although the purity of the deposited silicon was not particularly researched in these preliminary experiment, 99.56% silicon has been easily obtained starting from metallurgical silicon 97-98% purity.

Further research is necessary to verify the possible degree of purification and deposition efficiency and to demonstrate the feasibility of the conceived process of electrorefining.

4. CONCLUSIONS AND FUTURE OBJECTIVES

The results appear very promising.

An important feature of the process lies in performing the dissolution of silicon compounds and electrodeposition of amorphous silicon in the same cell, with a simple plant.

The objective is the direct deposition of amorphous Si in an optimal thickness film on conductive support, immediately suitable for a solar cell.

These new products will be far easier to assemble, as well as being made of less costly materials.

Silicon prepared with an electrolytic process is foreseen to be less costly because of:

- possibility of making use of cheaper raw materials (such as K_2SiF_6R or SiO_2) as compared with those utilized in other processes;
- use of a plant (electrolytic cell) of moderate cost and with the possibility of a high degree of automation.

The reduction of the cost as foreseen by the objective of this program will have a multiplying effect in reducing the cost of solar panels.

The extension of this research work is possible along these directions: after obtaining a suitable layer of silicon, the cell will be manufactured by using current technologies. Doping the film is possible by conventional techniques, but also introducing in the electrodeposition bath proper dopants, so reducing two stages of production in one only.

Afterwards the efficiency of the cell will be evaluated and all data collected will be analysed and discussed, supplying the final elements for the calculation of the process economics.

5. REFERENCES

1) Yasuo Takeda, Ryoji Kanno, and Osamu Yamamoto, "Cathodic Deposition of Amorphous Silicon from Tegraethylorthosilicate in Organic Solvents", Journal of Electrochem. Soc., Vol. 128, No. 6, 1981, p. 1221-1224.

2) T.R. Rama Mohan and F.A. Kröger, "Cathodic Deposition of Amorphous Silicon from Solutions of Silicic Acid and Tetraethyl Orthosilicate in Ethylene Glycol and Formamide Containing HF", Electrochimica Acta, Vol. 27, No. 3, 1982, p. 371-377.

3) A.K. Agrawal and A.E. Austin, "Electrodeposition of Silicon from Solutions of Silicon Halides in Aprotic Solvents", Journal of Electroche. Soc., Vol. 128, No. 11, 1981, p. 2292-2296.

4) Edward R. Bucker, Cranford, James A. Amick, Princeton, "Electrodeposition Process for Forming Amorphous Silicon", United States Patent, No. 4,192,720, 1980.

5) Alfred Ells Austin, Worthington, Ohio, "Silicon Electrodeposition", United States Patent, No. 3,990,953, 1976.

6) G.M. Rao, D. Elwell, R.S. Feigelson, "Electrowinning of silicon from K_2SiF_6-molten fluoride systems".

7) R.C. DeMattei, D. Elwell, R.S. Feigelson, "Electrodeposition of silicon onto graphite", Journal Electrochem. Soc. No. 8, Vol. 128, 1981 p. 1708

8) R.C. DeMattei, D. Elwell, R.S. Feigelson, "Electrodeposition of molten silicon", Proc. Symp. Electrocrystallization. The Electrochemical Society 1981, p. 322

9) R.C. DeMattei, D. Elwell, R.S. Feigelson, "Electrodeposition of silicon at temperature above its melting point", Journal of Electrochemical Soc., No. 8, Vol. 128, 1981, p. 1712

10) D. Elwell, G.M. Rao, "Mechanism of electrodeposition of silicon from K_2SiF_6-FLINAK", Electrochimica Acta, Vol. 27, No. 6, 1982, p. 673

11) G.M. Rao, D. Elwell, R.S. Fiegelson, "Characterization of electrodeposited silicon on graphite", Solar Energy Materials No. 7, 1982, p. 15

12) G.M. Rao, D. Elwell, "Electrolytic production of silicon", AIME Annual Meeting, Dallas 1982, p. 1107

13) D. Elwell, R.S. Feigelson, G.M. Rao, "The morphology of silicon electrodeposits on graphite substrates", Journal of Electrochemical Soc., No. 5, Vol. 130, 1983, p. 1021

14) R. Monnier, "L'obtention et le raffinage du silicium par voie electrochimique", Chimia Vol. 37, No. 4, 1983, p. 109

HYDROGENATED AMORPHOUS SILICON BY DISILANE LPCVD

Contract Number	: EN3S-0092-I (A)
Duration	: 12 months 1 September 1986-31 August 1987
Total Budget	: lit. 253 117 000 CEC contribution: ECU 50 000
Head of project	: Dr. T. A. Shamsi, TEAM S.r.l.
Contractor	: TEAM S.r.l.
Address	: TEAM S.r.l. Via G. Marconi 46/20 21027 Ispra (Varese) - Italy

Summary

Amorphous silicon films have been deposited from Si_2H_6 by using a standard LPCVD apparatus at temperatures between 370 and 475° C and at pressures in the range 1 - 10 Torr. The ranges have been characterized by IR absorption, SIMS, dark conductivity, photoconductivity, optical absorption, PDS mesurements.

The results of the characterization indicate that, at low deposition temperatures (around 400° C) and for long gas holding times, it is possible to obtain a material with hydrogen content even larger than 20%, optical energy gap around 1.7 eV, photoconductivity gains (AM1 condition) larger than 10^4, mobility-lifetime products of the order of 10^{-6} cm^2 V^{-1} and defects densities of the order of 10^{16} cm^{-3}.

All these results, which at the present stage are preliminary and must be confirmed by further and different characterization measurements, seem to indicate that our LPCVD a-Si:H films obtained with very long holding times and directly taking advantage of the continous increase of hydrogen concentration, display many features in common with glow discharge films and represent an improvement with respect to LPCVD films previously obtained.

1. INTRODUCTION

The Low Pressure Chemical Vapour Deposition (LPCVD) technique is largely used in electronic industry, because it allows to get very stable films of any kind in an economic, easy and reproducible way. LPCVD amorphous silicon however, has been obtained so far from disilane by using very small apparatuses (1,2,3), working probably far from equilibrium conditions. Nevertheless, solar cell efficiencies of 4% have been reported, with Jsc values up to 10 $mA \cdot cm^{-2}$ (4).

In the present work, we tried a quite different approach, by using a standard 5" (internal diameter) hot wall LPCVD deposition apparatus, suitably modified according to our purposes at the lowest temperatures reasonably useful and at the highest pressures still compatible with equilibrium conditions (in the range 5 ÷ 10 Torr). Moreover, we adopted very large gas holding times and took advantage of the continuous increase of hydrogen concentration along the growth tube, in order to obtain larger hydrogen concentrations in the films and to avoid polymers deposition. The films display many features in common with glow discharge films and, even if obtained at low deposition rates, represent an improvement with respect to LPCVD films previously obtained. The growth rate, in any case, does not seem to be a problem by CVD method (5).

2. EXPERIMENTAL

The depositions were carried out in a 5" fused silica ampoule, 2.20 mt long, at temperatures between 350 °C and 500 °C and at pressures between 1 and 10 Torr. The 3" substrates (silicon, quartz, sapphire and alumina) were on stand-up position. Gas supply was regulated by MFC and the normal steps used in CVD technique were adopted. Pure electronic grade 99.9% (< 5 ppm H_2O) Si_2H_6 both from "Matheson" and from "Air Liquid" was used together with electronic grade 99.999% N_2 for purging.

The flat temperature zone (within 10 °C) was generally 80 cm long. The film thickness was measured both mechanically (Taly-step) and optically, with a good overall agreement. The deposition rates are not very high (see Fig. 1) since we deliberately avoid non-equilibrium conditions occurring at relatively high pressures (above 5 Torr generally). After a series of depositions at 50 and 100 SCCM, with gas holding times of 15 s, we definitely chose the temperature of 420 °C, which apparently gave the best results, and much higher gas holding times (270 and 540 s) which were obtained at 15 SCCM, and at 5 and 10 Torr respectively, in order not to lower too much the deposition rate and also to allow slower processes than formation and deposition of polimers. In this case, we observe a variation of the deposition rate along the tube, with values between 1.7 and 0.7 Å/S. We do not have presently a real model for the growth process in these conditions : certainly the hydrogen plays an important role (6) and, likely, in radical and even in atomic form (7). By this way, total hydrogen concentrations between 14% and 22% were obtained, with partial SiH_2 concentrations as low as 0.15%. Hydrogen concentration determinations were carried out by IR absorption (8,9) while the H depth profiles were monitored by SIMS (see Figs. 2 and 3). The values of the ratio SiH/SiH2, evaluated for the stretching mode, are much higher than values quoted in literature (1). Even if a direct relationship cannot be ascertained, generally the best samples do correspond to high SiH/SiH2 ratios. Moreover, the hydrogen concentration profile, even if largely underestimated by SIMS, is practically constant. The bulk impurities concentrations are in the range 4 ÷ 10 ppm for Carbon and in the range 20 ÷ 130 ppm for Oxygen.

3. RESULTS AND DISCUSSION

Dark conductivities (σ_d), activation energies (E_σ) and pre-exponential conductivity factors (σ_o) are reported in Table I for typical samples. All these values seem to be relatively independent of deposition conditions. These values are very similar to what found in literature for GD samples. An example of dark and photo-conductivity (σ_p) behaviour as a function of $10^3/T$ is given in Fig. 4. Dark conductivity is clearly activated at room temperature.

Table II shows on the contrary, the values of the physical parameters which seem to be affected by deposition conditions. At lower deposition temperatures (Td) and longer gas holding times (t_H), the total hydrogen concentration (C_H) increases and the optical gap (E_g), as determined by Tauc's plots, exceeds 1.7 eV, a value which has never reached before by LPCVD.

A series of Tauc's plots is presented in Fig. 5, where the various samples were obtained both in different deposition conditions and with different hydrogen contents, as displayed in Table II. Also in this case, it is impossible to draw any definitive conclusion; certainly, it seems to be necessary to exceed 10% hydrogen concentrations in order to reach Eg values of 1.7 eV and these relatively large hydrogen concentrations are possible only for very large tH values. As far as the exponential Urbach's factor (Eo) is concerned, three different evaluation methods were adopted, by using Photothermal Deflection Spectroscopy (PDS), photoconductivity (PC) or optical absorption data. All the three methods present some difficulties which may account for the large discrepancies which have been observed; even if the determination has been carried out in the same energy range 1.4 ÷ 1.8 eV. In all the three methods, the exact evaluation of the sample reflectivity behaviour may present a problem. It must be noted, however, that reflectivity has been measured and not derived from other measurements or data, only on the optical absorption method, which, on the other hand, refers to the highest part of the energy interval. From our point of view, Eo is not strongly dependent on deposition conditions, even if the possibility of dropping below 60 meV cannot be excluded. It must be emphasised, concerning this point, that samples 6AP5 and 7AP5 are much thinner (around 0.2 μm) than the other ones and, consequently, may present a larger amount of disorder. Examples of PDS data and of PC normalised spectral response $\eta = \delta I/[(1-R)\cdot e\cdot Ø]$ where Ø is the incoming photon flux, R is the reflectivity and δI is the photocurrent, are given in Figs. 6 and 7 respectively.

From PDS data, and particularly from the plateau below 1.4 eV approximately, and by using a particular gap states density (10), it is possible to obtain the density of defects (Ng) (11). To a certain extent, the same determination can be carried out also by using photoconductivity spectral responses, since at lower energies they are essentially proportional to the absorption coefficients, if a suitable normalisation of data can be made in the energy region around 1.8 eV. By using the photoconductivity spectral responses one can also give an evaluation of mobility-lifetime ($\eta\,\mu\cdot\tau$) product for electrons (including the internal quantum efficiency) (11). In the present case, the evaluation has been carried out both at 632 nm and at 700 nm (with roughly coincident results),for fluxes between 10^{16} and 10^{12} photons $cm^{-2}s^{-1}$. The most representative values are reported in Table III, together with the photoconductivity σ_p in AM1 conditions (100 mW cm^{-2}) and the light-to-dark conductivity ratio in the same conditions. The results are listed for increasing hydrogen contents, which for the last samples are around 20%. Even if the values of σ_p and σ_d may depend on contacts type and geometry, and, consequently, $\eta\mu\tau$ and Ng may be considered only as order-of-magnitude estimates, it can be concluded that σ_p/σ_d ratios above 10^5, $\eta\mu\tau$ values above 10^{-6} $cm^2\,V^{-1}$ and defect densities around 10^{16} cm^{-3} may be achieved by LPCVD. These values may be favourably compared with GD data.

4. CONCLUSIONS

During a few months of deposition work after the apparatus set-up, it has been demonstrated that LPCVD, at low temperatures (400 °C), with relatively long gas holding times (270 s), and in controlled equilibrium conditions, is capable of giving samples with C_H around 20% (practically 100% Si-H bonds), with optical energy gap above 1.7 eV, logarithmic slopes of Urbach's tail around 70 meV, photoconductivity gains (AM1) above 10^5, $\eta\mu\tau$ products around 10^{-6} $cm^2\,V^{-1}$ and defects densities concentrations, as evaluated by PDS, around 10^{16} cm^{-3}. Even if the results are to be considered as very preliminary and further investigations both on deposition conditions and on parameters characterisation are certainly needed, the indication coming out from the obtained data is that the intrinsic a-Si:H films deposited by LPCVD may be considered as good candidates for p-i-n solar cells. The three orders of magnitude increase in $\eta\mu\tau$ product obtained in a relatively short time suggests clearly that further improvements in material quality, necessary in order to compete with presently available GD material, are certainly to be expected.

ACKNOWLEDGEMENT

Authors would like to thank Prof. F. Evangelisti of "Dipartimento di Fisica Università 'La Sapienza', Rome" for PDS and optical measurements, Dott. Galloni of "LAMEL Bologna" for photoconductivity measurements, and Dott. W. Gissler and J. Haupt of "CEC-JRC Ispra" for electrical and optical measurements.

SAMPLE	σ_d $(\Omega\cdot cm)^{-1}$	E_σ (eV)	σ_o $(\Omega\cdot cm)^{-1}$
10J	$4.4\cdot10^{-10}$	0.75	$2.1\cdot10^{3}$
25G	$2.7\cdot10^{-10}$	0.79	$5.3\cdot10^{3}$
26G	$1.1\cdot10^{-10}$	0.76	$3.2\cdot10^{3}$
13J	$1.6\cdot10^{-10}$	0.82	$1.1\cdot10^{4}$
6AP4	$1.3\cdot10^{-10}$	0.80	$3.2\cdot10^{3}$
7AP5	$2.5\cdot10^{-10}$	0.77	$2.2\cdot10^{3}$

TABLE I – Dark conductivity σ_d, activation energies E_σ and pre-exponential factor σ_o for differently deposited samples

SAMPLE	T_d (°C)	t_H (s)	C_H (%)	E_{opt} (eV)	E_o (meV)
16G	450	15	5.0	1.5	–
26G	420	15	5.7	1.58	–
10J	475	15	5.5	1.52	92^{A}
13J	475	15	8.4	1.53	78^{A}–80^{B}
6AP5	420	270	13.9	1.73	43^{C}–80^{B}–103^{A}
7AP5	420	540	14.9	1.68	62^{C}–105^{B}–103^{A}
27AP4	370	270	>20.	1.62	71^{A}

TABLE II – Deposition temperatures T_D, gas holding times t_H, hydrogen content C_H, optical energy gaps E_{opt} and logarithmic slopes of Urbach's tail as determined by $^{A)}$ PDS data, $^{B)}$ Photoconductivity data and $^{C)}$ optical absorption data, for samples obtained in different conditions.

SAMPLE	σ_P(AM1) $(\Omega \cdot cm)^{-1}$	σ_P(AM1)/σ_d	$\eta\mu\tau$ $(cm^2 \cdot V^{-1})$	N_E (cm^{-3})	T_d (°C)	t_H (s)
26G	$9.5 \cdot 10^{-8}$	$8.6 \cdot 10^{2}$	$1.9 \cdot 10^{-9}$	$>10^{18*}$	420	15
13J	$2.0 \cdot 10^{-7}$	$1.2 \cdot 10^{3}$	$7.9 \cdot 10^{-9}$	1.1·10^1[illegible]+	475	15
6AP4	$7.1 \cdot 10^{-7}$	$5.9 \cdot 10^{3}$	$1.0 \cdot 10^{-8}$	$3.0 \cdot 10^{16*}$	420	270
7AP4	$1.2 \cdot 10^{-6}$	$4.1 \cdot 10^{3}$	$4.0 \cdot 10^{-8}$	$4.0 \cdot 10^{16*}$	420	540
27AP4	$9.9 \cdot 10^{-5}$	$5.0 \cdot 10^{5}$	$3.0 \cdot 10^{-6}$	$1.3 \cdot 10^{16*}$	370	270

TABLE III – AM1 photoconductivities σ_P, light to dark conductivity ratios σ_P/σ_d and $\mu\tau$ products as determined at 700 nm, for different samples. Deposition temperatures T_D and gas holding times t_H are also indicated. The defect densities N_E have been evaluated both from PDS^{+} and from Photoconductivity spectral response *.

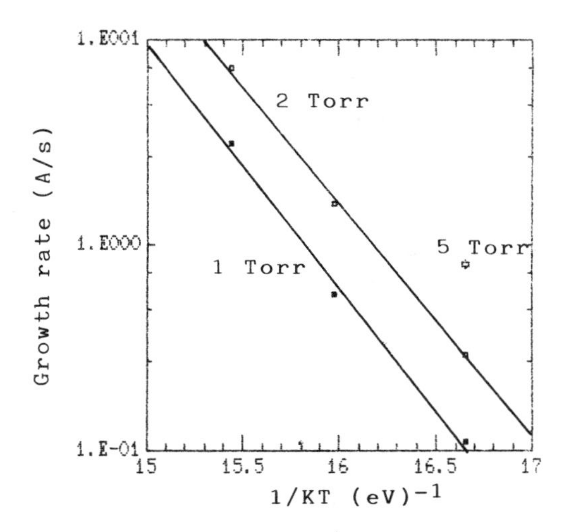

Fig. 1 - Deposition velocity vs 1/KT. At 1 Torr, 50 SCCM, the activation energy is about 2.7 eV.

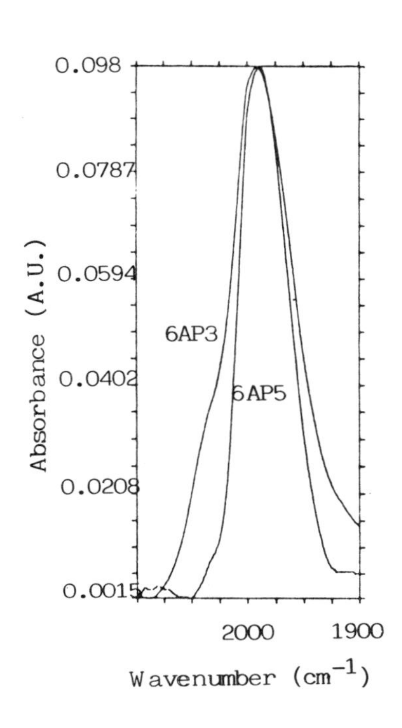

Fig. 2 - IR absorbance spectra vs wavenumber for SiH and SiH_2 stretching modes.

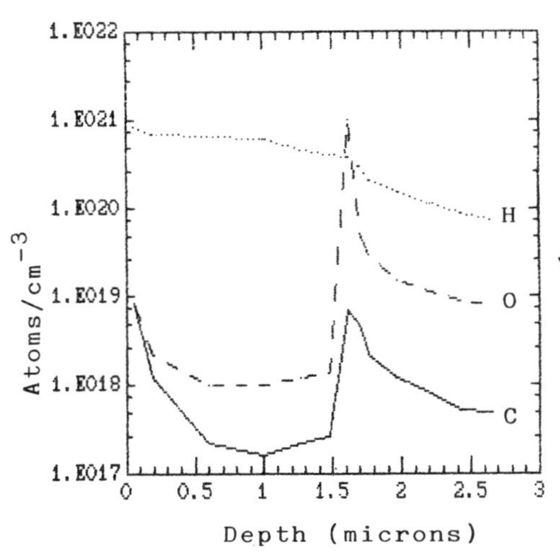

Fig. 3 - SIMS profile vs depth for a sample deposited at 475 °C, 1 Torr, 50 SCCM.

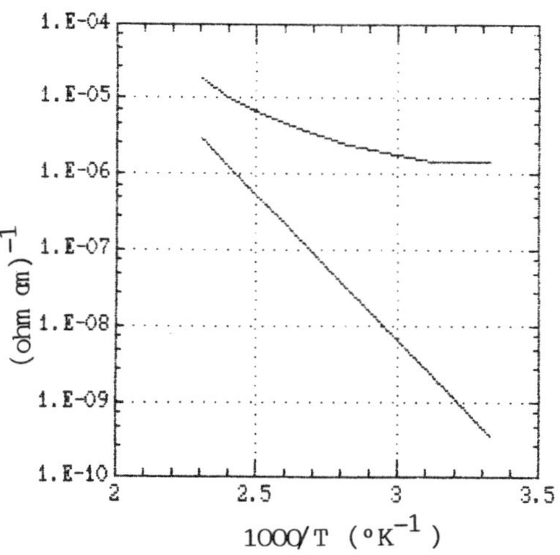

Fig. 4 - Dark conductivity and AM1 photoconductivity vs 1000/T

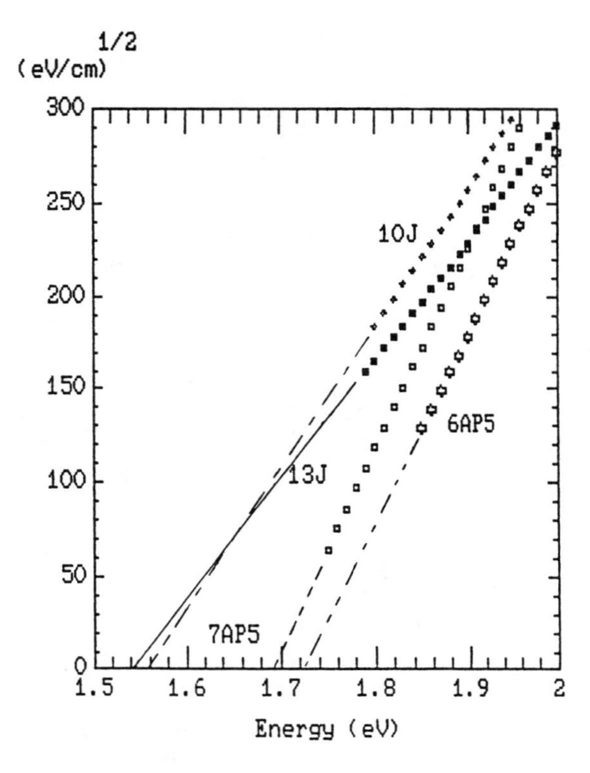

Fig. 5 - Tauc's plots for different a-Si:H samples

HIGH EFFICIENCY THIN-FILM SOLAR CELLS ON UPGRADED METALLURGICAL GRADE SUBSTRATES

Contract number : EN 3S/0074

Duration : 36 months (1/7/86 - 31/12/88)

Budget : 380 000 ECU, 190 000 ECU from CEC

Head of project : M. CAYMAX

Contractor : IMEC, MAP-division, Leuven, Belgium

Address : IMEC vzw, Kapeldreef 75, B-3030 Leuven, Belgium

Summary

A short overview is given of the aim of the project and of the tasks of the individual project partners. The activities and results of the past year are reviewed, stressing the cooperation between the different groups.
Concerning the work at IMEC, two main activities can be discerned. First, a series of solar cells has been made in epitaxial layers on various UMG-Si substrates (POLYX made in the Lab. de Marcoussis and POLYX made by Photowatt-Caen from two different feedstocks, SBB and P3). Analysis of the cells points out that the standard cell processing did not perform on the same level as before, but by comparison with SILSO-references it is possible to state that the LdM-POLYX-material has a quite good quality; the POLYX-SBB and P3 substrate did not fullfil the expectations, which could be due to contamination problems. Secondly, an improvement of the standard cell technology has been worked out and optimised : on a double epi-layer of successively grown p and n-type Si a very thin n^+-top layer is made by screenprinting. After optimisation of the n-epilayer parameters and the annealing conditions on EG-Si-SILSO slices this results in a considerably improved blue response. This has been verified on identical UMG-Si-substrates as higher mentioned.

I. General report

I.1. Introduction

In order for crystalline Si solar cells to find a more widespread use on earth, a substantial prise lowering of the completed module is essential. In view of the fact that the Si-substrate cost amounts up to 50 % of a module, looking for cheaper substrates is a straightforward way to reach this goal. In addition to this, one of the main concerns of the photovoltaic industry is the limited availability of raw materials to make substrates from, which holds for EG-Si or off-grade EG-Si as well as for the so-called SG-Si and even more for immediately useable, cheap UMG-Si. With regard to all this, the technology of epitaxial solar cells, comprising deposition of a very pure, high quality epitaxial layer on a very cheap and imperfect polysilicon substrate, made from UMG-Si without taking care of accurate selection of raw materials and of high purity arc furnaces, can offer a serious answer to these questions. The overall goal of this project is to determine whether this kind of solar cell will be an economically viable approach. Three conditions to be fulfilled are:
- the cell efficiency must be large ($\geq$ 12 %)
- sufficiently cheap substrates of an adequate quality must be developed
- a high throughput epitaxial reactor must be designed.

II.2. Cooperating teams

The partners and their tasks involved in this project are :

- IMEC, Leuven, Belgium (Contr. nr. EN3S/0074)
  - Research staff : R. Mertens, J. Nijs, M. Caymax, M. Eyckmans
  - Tasks :
    - Epitaxial growth of Si-layers on UMG-Si-substrates; optimisation of any process parameters involved.
    - Solar cell fabrication on laboratory-scale with the integral screen printing procedure; optimisation where required.
    - Investigation of gettering treatments before and after the epitaxial growth.

- CNRS, France (Contr. nr. EN3S/0075)
  - Teams involved :
    - Group M. Rodot (Lab. de Physique des Solides, Meudon)
    - Group G. Revel (Lab. Pierre Süe, Saclay)
  - Tasks :
    - Growth of UMG-Si ingots.
    - Chemical and electrical characterisation of these ingots.
    - Photo-electric characterisation of epi-layers and completed cells.
    - Study the effect of some specific impurities.

- Universita Polytecnica de Madrid (Contr. nr. EN3S/0076)
  - Research staff : A. Luque, G. Sala
  - Subcontractor : ISOFOTON (responsable J.C. Jimeno)
  - Tasks :
    - Choice and modelling of most adapted cell structure.
    - Theoretical analysis of the role of impurities and modelling of their physical and electrical characteristics during cell processing and operation.
    - Feasibility study and design of high-throughput epitaxial reactor.

- Universita di Milano (Contr. nr. EN 35/0077)
  - Research staff : S. Pizzini
  - Tasks :
    - Characterization of substrates and epi-layers
    - Study of some impurities (esp. C, O and some transition metals).

- Pragma S.p.A., Rome (Contr. nr. EN 35/0083)
  - Research staff : D. Margadonna, R. Peruzzi, R. Sposito
  - Tasks :
    - Growth of UMG-Si ingots and of deliberately contaminated EG-Si ingots.

I.3. General overview of the work carried out

From the short task description given with the project partners overview, it is clear that screenprinted epitaxial solar cells on polycrystalline UMG-Si substrates constitute a complicated structure to study. In this project several aspects of the problem are being treated at the same time.

To begin with the substrate itself, several ingots have been grown in this and previous periods by Revel (HEM) as well as by Photowatt, Caen (POLYX) as a replacement for the HEM-material (because of illness of one of Revel's coworkers) and Pragma (DS). Among these ingots are real UMG-Si crystals made from various feedstocks to test the material itself as possible substrate for solar cells, but also EG-Si has been used, deliberately contaminated with one of several impurities that are of interest because of their presence and possible harmfulness in UMG-Si. These impurities are C, B, Al, Cu (Revel, HEM) and Au, Co, Al, Fe, Cr (Pragma, DS). Analysis of this material by NAA (Revel) has revealed that the effective segregation coefficients of a.o. Cu and Co are much larger for a Bridgeman-type recrystallization than for Czochralski-solidification (Pizzini). Pizzini has also devoted an exhaustive study to the role of C and O, that are major native impurities in DS-material. From this study it is evident that the intra-grain dislocation density is intimately coupled with the C- and O-concentration, while the minority diffusion length in the substrate depends on the interaction of C and DO. The electrical activity of crystalline defects such as grain boundaries and dislocations is also influenced by the concentration of these impurities as well as by their segregation behaviour at these stressed and disturbed regions. A similar study about the role of abundant contamination by C during HEM-recrystallisation has been carried out by Revel. By means of electron microscopie analysis they found micron-size precipitates in the ingot bottom (which is a way to get rid of this impurity). In the rest of the ingot C segregates towards crystalline defects where it precipitates in a specific way together with O and metallic impurities. This finding seems to support a hypothesis by IMEC in which these precipitates act as impurity-sources during pre-epi heat treatments of the substrates in an effort to "outdiffuse" impurities. In order to study the electrical activity of microscopic defects such as stacking faults and dislocations, the CNRS-team has developed a new and powerful technique, called "STEBIC", that provides an extremely high resolution (1000 Å). A cooperation between the CNRS - lab. of Meudon and the University of Milano resulted in an original assessment of the simultaneous role of Al as an electrically active impurity as well as a dopant. BY DTLS and MCTS studies two hole trap levels have been identified, one of which is also able to trap electrons. An LBIC-analysis has the influence of the Al-concentration on the minority carrier diffusion length established. By means of fabrication of solar cells in epi-layers on the higher mentioned contaminated HEM-wafers, it has been possible for the IMEC-CNRS teams to define maximum allowable levels of Al, C, Cu and B in the substrate and, consequently, in the feedstock material. The regular UMG-Si wafers (HEM, POLYX, DS) have also been put to the test by epi-layer deposition and cell processing (IMEC). It has become clear that HEM and POLYX slices are virtually well suited candidates for epitaxial solar cells, while for the DS-substrates a new test has to be done on single-crystallised and unblended material, although this will probably also result in a good quotation. At this moment there exists a general feeling in the whole group that the crsytalline quality of the substrate itself plays a paramount role, more important probably than the contamination itself, in determining the epi-layer quality and, consequently, the cell efficiency. Nevertheless, in order to find the "best" substrate a general test will be done in a next period in which all kinds of substrates will get exactly the same treatment. Still at the level of the substrate, the work of the Luque-group has to be mentioned. They have devoted a literature survey to all aspects related to the role of impurities (behaviour as recombination centers, their diffusivity, segregation, ...) in order to come to a general view on and, eventual, a quantitative model of this phenomenon from the very beginning of the substrate on (recrystallisation) and further during epi-growth, cell processing and cell operation.

A second activity of the Luque-group is the modelling of epitaxial solar cells. An appropriate structure ($n^+ p p^+$, i.e. the BSF-cell) has been chosen and modelled, starting from three different cases : i) a strongly recombinating base; ii) an effective BSF without photogeneration in the substrate; iii) a photo-generating base. From this work it is clear that, for substrates of not-outstanding quality, the epi-cell is the best choice. In order to have an effective BSF, a substrate doping of $10^{18}/cm^3$ seems to be required. For what concerns the optimal doping of the epi-layer, modelling with the BSF-effectiveness, the doping dependent lifetime, the carriers profile and the injection level taken into account points towards a resistivity-value of 4 Ωcm, which agrees very well with an experimental assessment of the same problem by IMEC. At last, a somewhat surprising outcome of the work of the Luque-group is the idea that more attention should be paid to the high/low-junction at the substrate-epi interface, the quality of which seems to be an important parameter as well for the solar cell efficiency. More work on this is required.

A somewhat different kind of model, directed towards the IR-response of completed cells, allows the Rodot-team to evaluate the diffusion length in the epilayer on completed cells (in spite of the unfavourable geometry of this device : epi-thickness $< L_n$!) and to correlate this with the cell efficiency. Two interesting conclusions are that i) L_n (epi) decreases by the cell processing; and that ii) the wafer-position in the ingot is a more important parameter influencing the intrinsic epi-quality and the short circuit current than doping or thickness.

Spectral response-measurements by the CNRS-group of Rodot on various IMEC-solar cells have pointed towards a shortcoming of the "traditional" screen print technology, that results in a rather deep n^+-top-layer. This gives rise to a too low response in the blue-green region because of the very effective trapping of minority carriers in this region. A solution based on the already earlier proposed "two-step-junction" approach has then been worked out at IMEC, especially adapted to the epi-UMG-Si structure : on top of the thick p-epi-layer, a few microns of n-epi are grown (thus forming the junction) in the same run with adequate deposition parameters. In this n-layer a very thin but heavily doped n^+-layer is created by a screen print - and - firing step. Successive optimisation of the n-type epi-layer parameters (resistivity and thickness) and of the firing conditions of the P-paste (all this on EG-Si SILSO material) have resulted in a substantially improved blue response, which has afterwards been confirmed in a comparative experiment on various UMG-Si substrates. Spectral response measurements at IMEC as well as at CNRS, together with SIMS- and SRP-measurements will allow to make a thorough analysis of this approach in the next period. Optimisation of the metallisation pattern will also be done.

The Rodot-CNRS-team has also developed a new candidate substrate material of potentially very low cost, the so-called PLAST material, a ribbon formed by plasma torch spraying of Si powder. Very nice diffusion lengths have been measured in CVD-Si layers grown at IMEC, which is very promising for photovoltaic applications. Completion of solar cells on the other hand was not successful, due to the porosity of the samples. New and better ribbons have been made in the meantime, that will be tested further.

Last but not least is there the activity of Isofoton in studying the possibility of a high throughput epitaxial rector that does not add too much to the price of the solar cell. It is already now clear that the conventional CVD or LPCVD-machines have a too low throughput; a new design is therefore required, for which the rotary-disc reactor seems to be most promising.

In conclusion it can be stated that during this first year a tight collaboration between the different partners of this project has already delivered a number of important results. The number of publications can bear testimony to this. A positive answer on the higher raised question of economical feasibility seems to depend in the last stage on the feasibility of a high throughput epitaxial reactor.

II. Work done at IMEC in the period 01.01.87 - 30.09.87

The work done at IMEC in this period can mainly be divided in two topics :

- test of new substrate-materials with the standard technology
- optimisation and test of a new junction formation technology.

II.1. Solar cells on new substrates

II.1.1. The standard cell manufacturing technology

As described earlier (ref.1 and ref. therein) the standard cell manufacturing technology consists of an integral screen printing routine. Preparation of the as-cut substrate typically involves a wet chemical polishing step in $HNO_3/HF/CH_3COOH$-mixtures, followed by the standard RCA-cleaning. In the epitaxial reactor the native oxide-layer is stripped from the substrate by a high-temperature treatment in H_2 (typical 10 min. at 1180°C) or a HCl-etch. After this treatment a 20 µm thick p-type epitaxial layer is grown at 1120°C from SiH_2Cl_2 with B_2H_3 as doping gas at a rate of 0.5 µm/min. The resulting resistivity is 3 Ωcm. To form the metallisation and the ARC the following pastes are printed one after another : a silver paste at the frontside, an aluminium paste and a silver paste at the backside and a TiO_2 paste at the frontside. These four layers are simultaneously baked in a single operation. The cell junction is made in the epi-layer by firing a phosphorous-containing screen-printed paste in a belt-furnace at 940°C, resulting in a junction depth of 0.5-0.6 µm and a sheet resistance of about 30 Ω/□ .
Efficiencies above 9 % (for 1.8 x 1.8 cm^2-cells) are regularly obtained on good quality UMG-Si-substrates (ref. 2).

II.1.2. Description of new substrates

A number of new substrates partly provided by M. Rodot and partly by M. Fally, Lab. de Marcoussis, have been tested with the standard technology. These substrates were the following :

- POLYX-substrates made by Lab. de Marcoussis from UMG-Si without chemical treatments carried out on the feedstock before recrystallisation; the substrate used was nr. 11 (100 x 100 cm^2) (cut into 4 pieces) from ingot nr. 726. On each quarter 4 cells were made.
- POLYX-substrates (20 x 20 mm^2) made by Photowatt-Caen from UMG-Si, designated as "P3" (see paper by M. Rodot); 8 pieces.
- POLYX-substrates (20 x 20 mm^2) made by Photowatt-Caen from a special low-boron content UMG-Si, indicated as "SBB" (see M. Rodot); 8 pieces.
- PLAST-substrates (plasma-torch sprayed Si ribbons; see M. Rodot); 6 pieces.

As reference regular EG-SILSO-substrates have been added to the screenprint runs (without epi-layer).

II.1.3. Results and discussion

In table I the results of this production-run are summarised; given are only the top-values for each kind of substrate. It has to be mentioned that no ARC has been applied, so a factor of 1.3 has been used to get the corrected efficiencies. From the reference SILSO-cells (9.1 %) it must be concluded that this screen-print run did not result in very good efficiencies, compared with the normal value of 10 % on small cells. The cause of this is presently not clear. If we assume that the cell performance on other substrates is influenced by the same amount, then the efficiency of the 726/11-cells would be about 9.6 %, which is better then the previous figures on LdM-material (ref. 2 : 9 % on ingot nr. 648). The

P3-substrate on the other hand would give 8.9 %, which is too low; this could be ascribed to impurities that have contaminated the ingot during the recrystallisation process. NAA-analysis by Revel will throw light on this possibility. The SBB-material at last would result in 9.7 %, which is a rather good value; with the somewhat low fill factor taken into account, this material should be able to result in efficiencies above 10 % already with the standard technology.

Table I : Standard solar cells on new substrates.

Substrate	J_{sc} (mA/cm^2)	V_{oc} (mV)	FF (%)	η (%)	η_{cor} (%)
POLYX 726/11	18.5	530	70.1	6.7	8.7
POLYX-SBB	18.8	530	69.1	6.8	8.8
POLYX-P3	17.3	510	71.7	6.2	8.1
SILSO	18.8	535	72.3	7.0	9.1

II.2. A new junction-formation technology

II.2.1. Introduction

An old shortcoming of the integrally screenprinted solar cells, especially the cells with a simple junction, is the relatively weak blue response (as is indicated in previous reports by Rodot as well). This is due to the rather deep junction, the top-layer of which is completely composed by a heavily doped n^+-layer that constitutes a virtually dead zone - any carriers created by light absorbed here (the short wavelengths !) will recombine before they can be collected. A less severe doping of this region on the other hand would result in two severe disadvantages : it would lack the repelling action of the build-in front surface field on minority charge carriers, and it would give rise to an unacceptably high emitter resistance and consequently result in fill factor losses due to the series resistance. A solution already earlier proposed is to make this top layer in two steps : i) a slightly doped n-part, the thickness of which does not matter too much, constituting the actual junction region, and ii) a heavily doped n^+-layer, as thin as possible, acting as an FSF and for making contact with the front metallisation and for lateral conduction of the current. While this involves a rather complicated junction formation process (with its implications on the add-on cost for cell production), we have adapted this approach to our methodology, more especially to our epi-layer/UMG-Si-sandwich, without the implementation of an extra potentially expensive processing step.

II.2.2. The two-step (epitaxial) junction

A solution for the above-mentioned problem is the following. In order to make fabrication of solar cells on UMG-Si possible, growth of a very pure epitaxial layer on these substrates is essential. In our cells, this layer normally is of p-type, and is grown at 1120°C from SiH_2Cl_2 with B_2H_6 as dopant gas. If it is possible to grow a thin n-type layer immediately on the p-layer in a successive step, then the junction is formed without a dead zone. Next, by screen-printing a regular phosphorous paste and firing this in appropriate conditions, a very thin superficial n^+-layer for contacting the cell can be formed. A somewhat weak point in this was the successive growth of p and n-layers in view of the possible cross-diffusion of the dopants at the high temperatures involved; the result of this can be seen in a simulation with the SUPREM II-program (Fig. 1). This problem can be circumvented (due to the very high activation energy for diffusion of the two dopants B and P) by using a much lower temperature (e.g. 1000°C) for the growth of the n-layer. A possible problem for the growth of epi-layers at 1000°C (which is unusually low, especially for heavily defected substrates as these) is that the quality is very inferior or, even, that the resulting layer is polycrystalline. This is due to the formation of a native oxide at the substrate surface. Growth of good layers is only possible if one can work with very pure gases (especially the water-content in the H_2 has to be extremely low, preferentially below 1 ppm) in a very leak-tight system. These requirements can be relaxed somewhat if one can start with the growth of a first layer at higher temperature (1120°C), where the native oxide-removal is much easier, and continue, after temperature-adjusting, at 1000°C. This is because a fresh Si-surface has been formed at that moment ready for growth at lower temperature. Further the use of SiH_4 instead of SiH_2Cl_2 for this second step has two advantages : i) no problems with the etching action of the chlorine-components formed as by-products of the SiH_2Cl_2-decomposition, and ii) the growth rate nearly does not drop because of the lower stability of SiH_4. A complete growth cycle looks as follows :

- H_2-prebake (1180°C, 10 min.)
  purpose : removal of the native oxide (no HCl-etch !)
- Growth step 1 (1120°C, 23.3 min.)
  with 400 sccm SiH_2Cl_2 and 10^{-4} sccm B_2H_6 in 100 slm H_2 at a growth rate of 0.85 μm/min. Resulting thickness : 20 μm, resistivity : 3.8 Ωcm.
- Growth step 2 (1000°C, 4 min.)
  with 150 sccm SiH_4 and about 10^{-4} sccm PH_3 in 100 slm H_2 at a growth rate of 0.38 μm/min. Resulting thickness : 1.5 μm, resistivity : ~ 3 Ωcm.

The result of this is shown in Fig. 2, which gives a spreading resistance profile of an n-type mono-crystalline test sample. A more extensive analysis of the reproducibility and eventual reactor-memory effects is not yet completed; SRP- and SIMS-measurements are in progress at IMEC and CNRS respectively.

II.2.3. Optimisation

The optimisation of the complete junction formation involves, successively, the epi-layer parameters, the firing conditions for the top n^+-layer, and the metallisation pattern. The first two points will be described here.

II.2.3.1. Optimisation of the n-epi-layer parameters

II.2.3.1.1. Experiments

In order to prevent any problems with eventual metallic contamination we have used electronic grade SILSO material to evaluate the possiblities and to optimise the epitaxial layer parameters. Of course this technology is not interesting from an economical point of view for commercial solar cells directly made on SILSO slices. For this experiment we have grown only the n-type epi-layer without the thick p-layer. We have tried three thicknesses of the epi-layer (1, 2 and 3 µm) and two resistivities (1 and 2.2 Ω.cm). Two different sheet resistances for the heavily doped top region have been used : 30 and 50 Ω/□. Beside the SILSO slices we also added 3" ø monocrystalline monitor wafers of p-type with a resistivity of 10-15 Ω.cm. On each slice we made four cells of 1.8 x 1.8 cm^2. In order not to have problems with contact resistance we used a double number of fingers (9 in stead of 5), which gives a metallisation of about 20-21 % (in stead of 5-10 %). Also we did not make an anti-reflective coating. A correction factor for the short circuit current as well as for the efficiency of 1.4-1.5 seems to be justified. The cell parameters were measured under a solar simulator with AM1, while spectral response curves were recorded as well.

II.2.3.1.2. Results and discussion.

In table II the results are summarised for the SILSO cells with epi-layers that vary in thickness and resistivity. The first figure gives the top value, the second figure gives the mean of about 8 cells on two slices. Table III gives a comparison of cells with and without epi-layer (of 2 µm thick and 2.2 Ω.cm resistivity) and with two different resistance values for the diffused layers, i.e. 30 and 50 Ω/□. The diffusion depth for these layers is resp. about 0.5 and 0.3 µm. The best SILSO cell efficiency, after correction as indicated above, would be about 11 %. Table IV gives the results of the monocrystalline monitor slices. In Fig. 3 the spectral responses for the different epi-layer thicknesses are compared, from which it can be concluded that the thinner layers are slightly better. This was not immediately clear from table II. On the other hand, the layer-resistivity does not seem to have any influence. Fig. 4 gives the spectral response curves for a SILSO-epi-cell with a 50 Ω/□ diffusion and a "naked" SILSO-cell with a 30 Ω/□ diffusion. It is clear that the blue response is much better, but the red-infrared response is worse. This is due to a decrease in the minority carrier diffusion length in the base, caused by the thermal treatment during the epitaxial deposition, which is done at 1000 °C. This is a typical phenomenon for polycrystalline silicon, which is not seen in monocrystalline cells as is shown in Fig. 5 where a monocrystalline epi-cell (50 Ω/□)is compared to a "naked" monocrystalline cell (30 Ω/□). Here the response in the short wavelenths is much better. This reflects itself in the short circuit current (Table IV), which, for the thinner epilayer, is better than for the cell without epi-layer.

The fact that the base-diffusion lenth in SILSO-cells is negatively influenced by the epi-deposition is not important for the metallurgical cells we are envisageing, because on these substrates an epitaxial layer is built nonetheless, and a second deposition step of a few microns will not influence the quality of this layer which afterwards will form the actual active part of the complete structure.

Table II : Solar cell results for epi-SILSO-cells with various n-epi-layers.

Epi-thickness µm	Resistivity Ω.cm	V_{OC} mV	J_{SC} (mA/cm^2)	Efficiency %
1	2.2	535/530	17.0/16.1	6.4/6.1
2	2.2	535/520	17.3/16.5	6.9/6.0
3	2.2	525/520	17.0/15.7	6.3/5.7
1	1.0	530/530	17.0/16.8	6.7/6.6
2	1.0	530/530	17.3/17.2	6.8/6.6
3	1.0	525/520	17.6/17.0	6.3/6.1

Note : Sheet resistance was 50 Ω/□.

Table III : Solar cell results for SILSO-cells with and without epi-layer and with two different diffusion resistances.

Epi-layer	Sheet Resistance Ω/□	V_{OC} mV	J_{SC} mA/cm^2	Efficiency %
y	30	530/525	16.7/16.0	6.5/6.2
y	50	535/525	17.0/16.4	6.6/6.3
n	30	540/530	17.0/16.2	6.8/6.4
n	50	540/535	17.9/17.3	7.2/6.8

Table IV : Solar cell results for monocrystalline slices with different epilayers.

Epi-thickness µm	Resistivity Ω.cm	V_{OC} mV	J_{SC} (mA/cm^2)	Efficiency %
1	2.2	510/510	20.1/20.0	7.5/7.3
2	2.2	505/505	19.4/19.2	7.1/7.0
3	2.2	505/500	19.8/19.1	7.2/7.0
1	1.0	495/490	19.4/18.9	7.2/7.0
2	1.0	495/490	18.8/18.5	6.9/6.8
3	1.0	500/495	18.8/18.5	7.0/6.9
0(a)	-	510/505	19.1/19.0	7.5/7.3
0(b)	-	520/510	19.8/19.0	7.9/7.6

Notes :

(a) These slices have undergone the same temperature treatment as the others during epi-deposition, but did not get an epi-layer
(b) These slices did not get an epi-layer nor a temperature-treatment.
(c) The two series mentioned in (a) and (b) had a sheet resistance of 30 Ω/□, the others 50 Ω/□.

II.2.3.2. Optimisation of the firing conditions for the n^+-layer formation

Thirty 50 x 50 mm^2 EG-SILSO-wafers have been prepared for this by covering them with a double epi-layer : approximately 20 µm epi, p-type, ~ 2 Ωcm, followed by 1 µm epi, n-type, ~ 3 Ωcm. After screenprinting the normally used phosphorous paste, then different annealing conditions have been tried out with a continuous belt furnace on batches of 3 wafers. Variable parameters are the temperature profile (the furnace) has individually controllable temperature zones), the most important of which is the peak temperature, and the belt speed. The highest temperature used is 970°C, while the lowest one is 910°C. The belt speed varied between 12.6 and 25 cm/min. The resulting sheet resistance varied between 20 and 95 Ω/[]. It has to be mentioned that various combinations of different peak temperatures and belt speeds can give rise to the same sheet resistance, but with varying peak concentrations and layer-depth.

In this experiment as well a double metallisation pattern (coverage ≈ 21 %) was used and an anti-reflective coating omitted. The resulting (not-corrected) short circuit current densities are plotted against sheet resistance in Fig. 6. A certain tendency can be discerned for an increasing current with resistance at lower sheet resistances, while the opposite is true at higher resistances. This last feature is probably due to a decrease of the repelling effect of the front surface field with the decreasing surface doping concentration. The optimum lies around 45 Ω/□. The seeming scatter in measurements on eacht resistance value is partly due to the difference in firing peak temperature and belt speed, as is clear from fig. 7. In this figure the spectral response curves (measured by Rodot) of cells with R_s = 45 Ω/□ are compared for peak temperatures of 910°C and 940°C with resp. belt speeds of 20 and 25 cm/s (series A2 and B3). The worse performance, esp. in the red-infrared region of the series B3-cell is probably due to a shortening of the base diffusion length in the SILSO-substrate that still has an appreciable value (which means that the substrate too will contribute to the photocurrent). (so it is not impossible that later on this optimisation can shift somewhat for UMG-Si-substrates with a negligible diffusion length, but we will come back to this later on). The worse performance in the blue zone can be caused by a deeper diffusion (longer furnace residence time !), but a more detailed analysis will be made later on the base of SRP- and SIMS-measurements of diffused samples.

II.2.4. Results

In order to evaluate the performance of this new junction-formation technique, we have made a number of solar cells on substrates of the same series of UMG-Si-substrates as discussed earlier. As in previous point described, the active epi-layers consists of roughly 1 µm n-epi on top of 20 µm p-epi, while the n^+-top layer is made by firing a P-paste in the same conditions as for series A2 mentioned previously. The solar cell parameters are summarized in table V, together with the results of the classical process ("old" resp. "new"). In order to make a comparison possible, we have given also corrected values for J_{sc} and efficiency; "cor I" means a correction of 1.3 for the lacking ARC, while this figure is again multiplied by 1.15 for the abnormally high metal coverage of 21 %, resulting in "cor II". Already at a "correction I level" the difference is significant and varies between 1 and 6 % relatively (omitting the mono-crystalline cells). With the correction for metallization included, this improvement would increase to more than 20 % relatively. The top efficiency obtained in this way is 10.6 % for the LdM-POLYX as well as for the SBB-POLYX substrates. The efficiency on monocrystalline substrates would be 12.4 %, which is to be compared with the figure of 12 % for the standard process. As would be expected, V_{oc} and FF do not change noteworthy, and the integral benefit of the new technology is to be found in the short circuit current.

Table V : Comparison of the old and the new junction formation technology.

Sub-strate	J_{sc} (mA/cm^2) old unc	old corI	new unc	new corI	new corII	V_{oc} (mV) old	new	FF(%) old	new	η(%) old unc	old corI	new unc	new corI	new corII
SILSO	18.8	24.4	19.1	24.8	28.5	535	530	72.3	72.9	7.0	9.1	7.1	9.2	10.6
POLYX 726	18.5	24.1	18.2	23.7	27.3	530	540	70.1	74.2	6.7	8.7	7.1	9.2	10.6
POLYX SBB	18.8	24.4	20.1	26.0	30.0	530	510	69.1	70.9	6.8	8.8	7.1	9.2	10.6
POLYX P3	17.3	22.5	18.8	24.4	28.1	510	515	70.7	73.5	6.2	8.1	6.5	8.5	9.7
MONO X			19.1	24.8	28.5		585		75.2			8.3	10.8	12.4

II.3. Conclusion - future research

The results of the standard solar cells on new substrates, although not completely satisfying give the possibility to compare the different substrates and to do predictions about their possibilities. Taking the somewhat lower cell performance caused by the screenprint technology (as noticeable from SILSO-cells) into account, the POLYX-726 substrates are among the best with at least 10 % efficiency possible. The other materials clearly show some problems, especially the P3 ones. NAA-measurements will show whether this has to do with contamination problems. It seems also to be required to review thoroughly the status of the actual standard screen print technology in order to reach the previous level again. This will be interesting for a completion of this comparative study of different substrates, in which e.g. the DS-material still is lacking.

The new junction formation technology has already proven to result in an enhanced response in the short wavelengths although the optimisation is not yet completed. Among the points to look at in the near future are the possibilities of still thinner n^+-layers (e.g. with a short time anneal system) and an optimised metallisation pattern. The problems with the worse red-infrared response in SILSO-substrates, due to a shorter L_n, are not really fundamental for our application on UMG-Si substrates that normally do not contribute to the short circuit current.

5. BIBLIOGRAPHIE

1. R. MERTENS, P. DE PAUW, M. EYCKMANS, M. CAYMAX, J. NIJS, Q. XIANG and L. FRISSON : **Improvements in multicrystalline silicon solar cells** ; Proc. of the 6 th EPSEC, Londen, April 1985 (p. 935).

2. M. CAYMAX, M. EYCKMANS, R. MERTENS, J. NIJS, M. RODOT, J. E. BOUREE, G. REVEL, J. L. PASTOL, G. SALA, J. FALLY, S. PIZZINI ; **Solar cells made by the integral screen printing procedure on various epitaxial UMG-Si-substrates**, 7th EPSEC, Sevilla, Spain, October 1986.

3. M. RODOT, M. BARBE, J. E. BOUREE, V. PERRAKI, G. REVEL, R. KISHORE, J. L. PASTOL, R. MERTENS, M. CAYMAX, M. EYCKMANS ; **Thin film solar cells using impure polycrystalline Si**, Rev. Phys. Appl., 22, 687 (1987).

4. R. KISHORE, M. BARBE, J.E. BOUREE, V.PERRAKI, M. RODOT, J. L. PASTOL, G. REVEL, M. CAYMAX ; **Photopiles en couches minces de silicium polycrystallin : realisation, caracterisation, perspectives** ; Ann. Chim. Fr. , 12, 423 (1987).

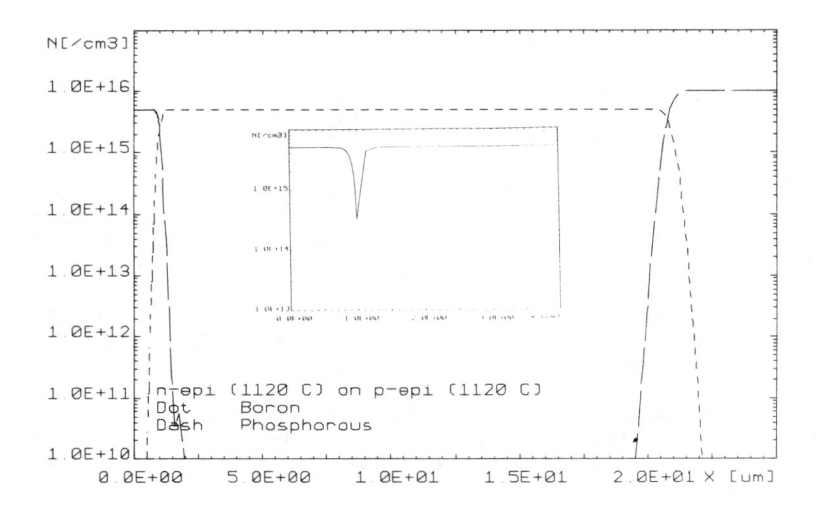

Fig. 1 SUPREM II-simulation of the successive growth of a p- and an n-layer, both at 1120 °C ; concentration profiles of P and B. Inset: netto active dopant profile at the p/n-junction.

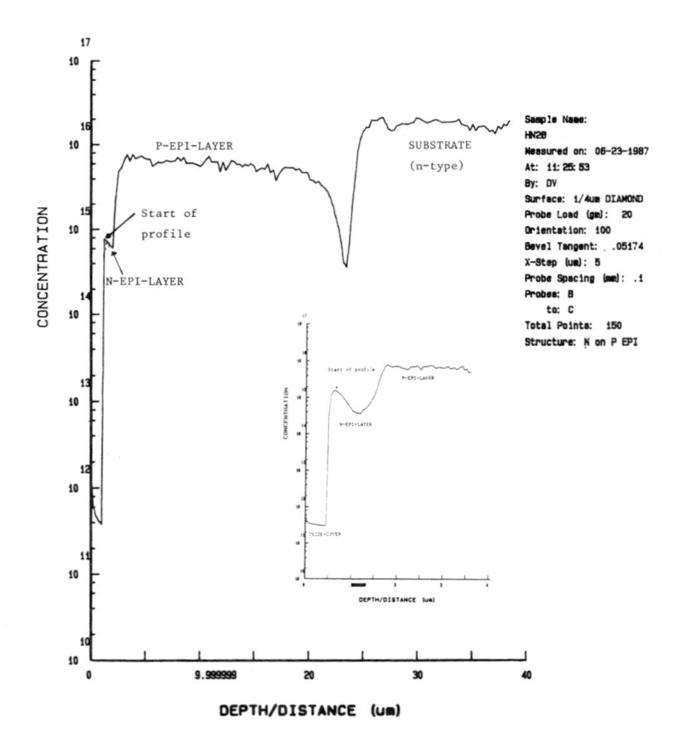

Fig. 2 Netto active dopant profile measured by SRP on test-sample HN2b with p/n-epi-layer ; inset : detailed analysis of p/n-junction. The actual profile starts anly at about 0.5 um (arrow).

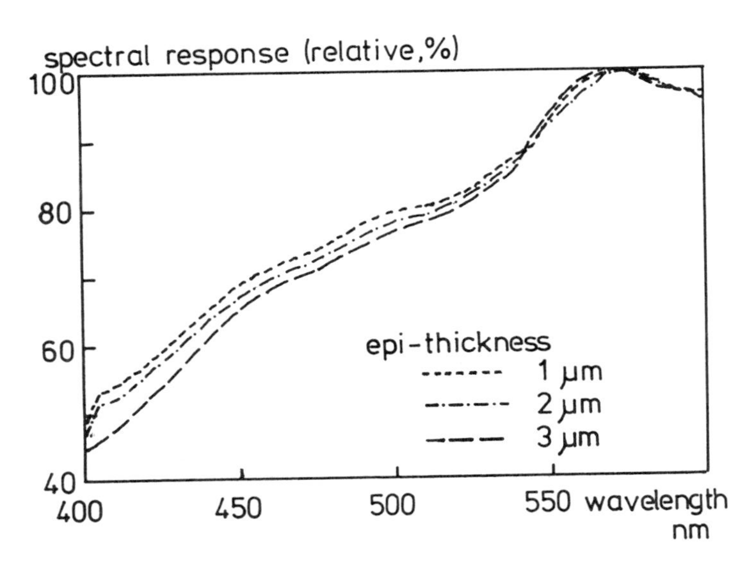

Fig. 3 Comparison of the spectral response (in relative units) of SILSO-cells with different epi-layer thicknesses.

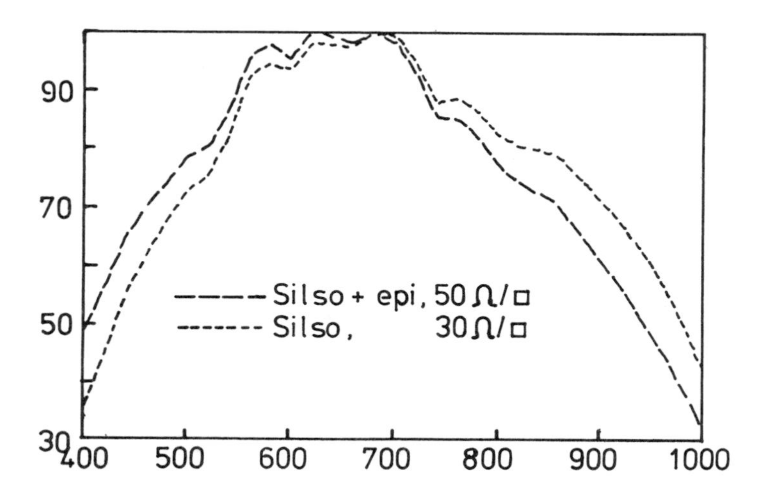

Fig. 4 Spectral response of SILSO-cells with and without epi-layers.

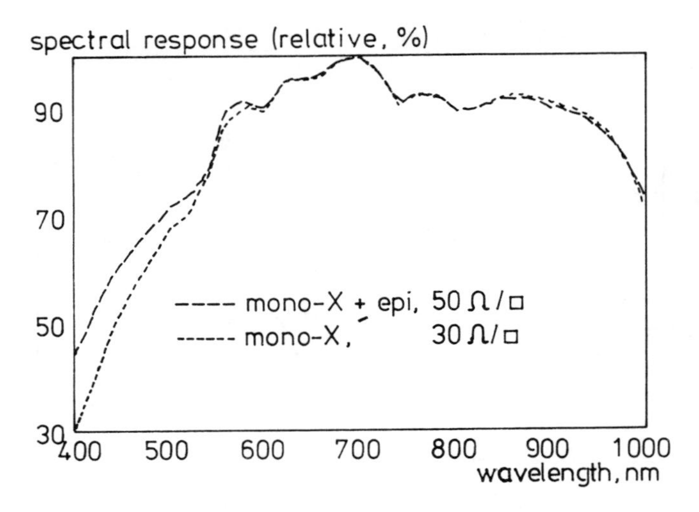

Fig. 5 Spectal response curve of monocrystalline monitor-cells with and without epi-layers.

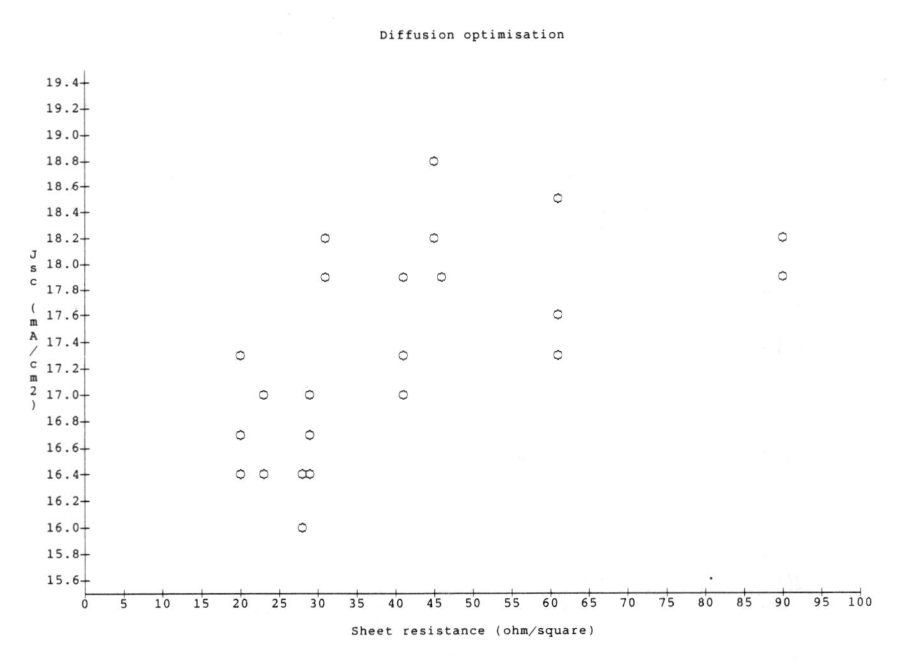

Fig. 6 Optimisation of the annealing conditions for the n_+-top layer formation from a screen-printed paste.

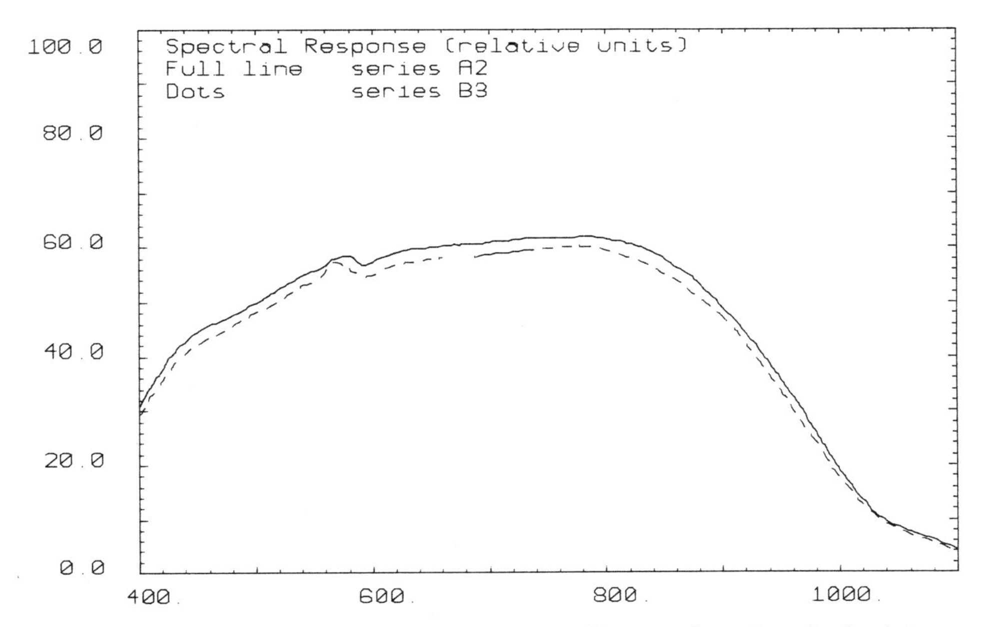

Fig. 7 Spectral response of cells from annealing series A2 and B3 with resp. peak temperatures of 910 °C and 940 °C ; belt speeds resp. 20 and 25 cm/s.

HIGH EFFICIENCY THIN-FILM SOLAR CELLS ON UPGRADED METALLURGICAL GRADE Si SUBSTRATES

Contract number: EN 3S/0075

Duration : 36 months 1 July 1986-31 dec. 1988

Total Budget : 5,567,000 FF CEC contribution: 1,000,000 FF

Head of project: M.RODOT, G.REVEL

Contractor : Centre National de la Recherche Scientifique (C.N.R.S.)

Address : CNRS/LPS 1 place A. Briand, F.92195 MEUDON

Summary

New epitaxial solar cells have been built in cooperation with R.MERTENS's team in Leuven. For this purpose, we have prepared two types of substrates: UMG-Si ingots by the POLYX process (in Photowattt's furnace, Caen) using two different Si charges,and Si ribbons by the PLAST (plasma torch) process; for comparison three other substrates were also used in this experiment. A first orientation of this work was to decrease the superficial dead layer by improving the process: varying the post-screenprinting annealing conditions led to a variation of solar cell efficiency; SIMS and spectral response measurements are in progress. A second orientation, following past work, was to correlate the cell efficiency with the epilayer electron diffusion length, the latter being determined from the infrared response using an appropriate model; this model has still to be improved. Two complementary studies were devoted to Al and C as impurities of polycrystalline Si; both implied a cooperation with S.PIZZINI. Finally the observations of polycristalline Si in electron microscopy have been completed using a new and powerful technique, "STEBIC", to determine the electronic activity of defects with a very high resolution (1000Å).

INTRODUCTION

The two teams of CNRS who take part in the EEC project on epitaxial solar cells have the following tasks:

1. grow and characterize UMG-Si substrates
2. contribute to the preparation of epilayer and complete solar cells,
3. measure the photoelectric of epilayers and cells,
4. study the effects of some specific impurities.

In April-september 1987, the following persons were involved:
M.RODOT, Dir. rech. CNRS, full time (tasks 3, 4)
G.REVEL, Dir. rech. CNRS, partial time (task 1)
C.CABANEL, Thésard, full time (task 4)
NGUYEN DINH HUYNH, Visitor, full time (tasks 1, 2)
LE HOANG THI TO, Visitor, full time (task 3).
J.E.BOUREE (LBIC measurements), R.SURYANARAYANAN (ribbons), N. DESCHAMPS, C.DARDENNE (neutron activation analysis), C.GRATTE-PAIN, M.AUCOUTURIER (SIMS) also contributed (partial time). Principal external cooperations were with M.CAYMAX (Leuven), S. PIZZINI (Milan), J.Y.LAVAL (ENSCI, Paris) and P.LAY (Photowatt-Caen).

1.1. GROWTH AND CHARACTERIZATION OF UMG-Si INGOTS.

Two new ingots have been grown by the POLYX process in the Photowatt furnace (Caen). The first charge "P3" was a rather homogeneous UMG-Si (Péchiney), as used already in the past. The second, "SBB", was a special "low-boron content" UMG-Si; this material was purer in volume, but included more impure superficial parts which are impossible to separate:

Main impurities of the charge (ppma)		Co	Cr	Cu	Fe	Mo
P3		.04	.2	.07	.11	.02
SBB	Bulk	.0002	.0004	.0007	<.01	<.0002
SBB	Surf.	.02	1.2	.11	9.	.22

	Ingot dimensions	Resistivity	Mean doping
P3	20x20x10cm	0.032 to 0.036 Ωcm	$1.2.10^{19}$ cm^{-3}
SBB	20x20x 6cm	0.28 to 0.29 Ωcm	$2.3.10^{17}$ cm^{-3}

1.2. GROWTH AND CHARACTERIZATION OF PLAST-Si RIBBONS

About 100 ribbons of UMG-Si have been deposited on different types of substrates by plasma torch projection of Si powder (from Kemanord, grains of 25 to 250µm). Using near-optimal conditions for gas flow rate (36 l/min Ar + 4l/min H_2), powder flow rate (10g/min), torch-substrate distance (12cm) and electric power (254kW), 2.5x2.5cm^2 layers, of thickness 0.02 to 0.1cm, were regularly obtained on Cu or Al_2O_3 substrates, from which they could be easily detached to give self-supported ribbons. Undoped ribbons had very little porosity and were n-type ($n \simeq 2.10^{18} cm^{-3}$, $\rho \simeq 3\ \Omega cm$). When 0.1% B_2O_3 was added to Si powder, p-type ribbons were obtained ($p \simeq 1.10^{19} cm^{-3}$ $\rho \simeq 3\text{-}4\ \Omega cm$) while porosity increased a little; more doped ribbons were too porous for direct use as substrates of epitaxial solar cells, as confirmed by first tests. Other substrates (Al, stainless steel) were found unsatisfactory. Cu substrates have the advantage of being reusable many times, but lead to smaller growth rates (0.3mm/min); Al_2O_3 substrates allowed larger growth rates (0.9 mm/min) but were reusable only 3-4 times. A series of non porous, p-type ribbons are available for future epitaxial cell fabrication.

2. CONTRIBUTION TO THE FABRICATION OF EPILAYERS AND SOLAR CELLS.

Together with A.CAYMAX ans M. EYCKMANS in Leuven, we have fabricated about 130 epilayers on different kinds of substrates, and more than 100 solar cells on these substrate-epilayer sandwiches.

A first program followed from the previous observation of a 0.4μm thick dead layer at the surface of standard epitaxial cells, which hampered the conversion of green-yellow photons. The new technology consisted of two steps: epitaxial growth of a 20μm thick p-epilayer plus a 1μm thick n (low-doped) epilayer, then screen-printing and annealing of a thin n-layer (highly P-doped). Ten different annealing regimes were tried, in an attempt to optimize the P-profile. The measured n-layer resistance and solar cell efficiency were significantly different for the different regimes. The best regime gave AM1 efficiency (corrected for reflectivity) values of :

(SILSO EG substrate) : η_{mean} = 8.6%; $\eta_{optimal}$ = 9.1%

Substrate	Cell	optimal ("standard")	optimal ("improved")
Single crystal EG	HT 7A/1		10.8%
SILSO EG	HO /3*	9.2%*	
	H35/2		9.2%
POLYX	11/B4	8.7%	
(off-grade EG)	10/B3		9.2%
POLYX "SBB"	C4	8.8%	
(low boron UMG)	C22		8.1%
POLYX "P3"	C109	8.1%	
(Péchiney UMG)	C119		8.5%
* This case only : without epilayer			

Two other programs were devoted respectively to "standard" cells and "improved" (using the above structure) cells made on different substrates. We give in the Table above efficiency of best cell for each case.

We have now in progress both SIMS measurements on epilayers (for analytical purpose) and p-n junctions (for P-profiles), as well as spectral response measurements on completed solar cells.

3. MEASUREMENTS AND INTERPRETATION OF PHOTOELECTRIC PROPERTIES.

Some LBIC measurements were made or are in progress, on different types of substrates. The main work was the measurement of spectral response curves, both on cells of previous series and cells described above (Fig. 1).

The model of epitaxial cells that we used previously (1) to deduce epilayer electron diffusion length L_{ne} from solar cell spectral response has been thoroughly reviewed and discussed. Its accurateness depends on two main assumptions:

- i) the substrate is not photoactive ($L_{ns} \ll L_{ne}$)
- ii) the recombination velocity at the substrate/epilayer interface is $s = (P_e / P_s)(D_{ns}/L_{ns})$ according to previous authors. Both assumptions are met if the substrate has a low L_{ns}: the error (by excess) on L_{ne} is only 5% if $L_{ns} = 5\mu m$ and $L_{ne} = 30\mu m$. This was the case in some of our previous measurements, but not in all. In this case, the substrate doping is also high enough for s to be low and the cell efficiency be reinforced by some BSF effect.

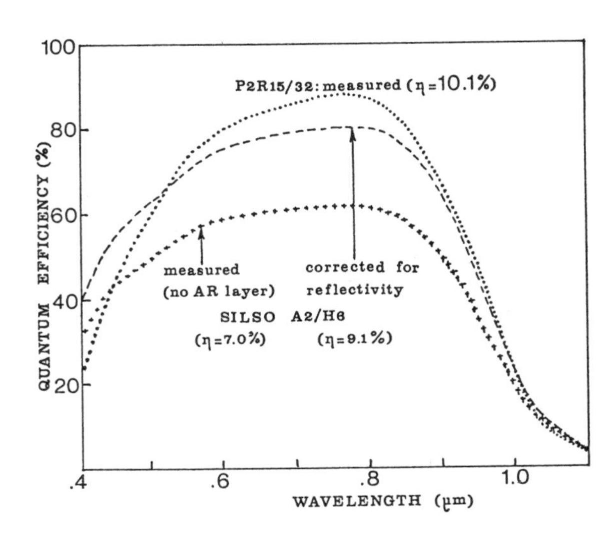

Fig.1

But another situation may also occur in epitaxial cells: the substrate is "good" enough (e.g. $L_{ns} = 20\ \mu m$) so that it is photoactive as well as the epilayer. In this case there is smaller BSF effect, but two contributions (those of substrate and epilayer) to the electron current have to be considered. This case needs to be treated theoretically, and we shall do it with the cooperation of S.N. SINGH (Delhi) who is expected to join us soon.

4. HIGH RESOLUTION ACTIVITY MEASUREMENTS: THE "STEBIC" TECHNIQUE.

The previous studies of the effects of Al(2) and C(3, 4) in polycrystalline Si, have been completed by a contribution to PIZZINI's work on C and O (5). Furthermore we were led to develop a new technique (4) to characterize the electrical activity of defects. In this "STEBIC" approach, the crystal is thinned to 0.5µm or less, then covered with a MIS junction and observed under a STEM microscope. Because the layer thickness is smaller than the electron absorption depth, the resolution reaches that of the electron beam, i.e. 1000 Å. Both profiles of the diode current and images have been obtained. They allowed to study grain boundaries (gb), stacking faults (sf) and intragrain dislocations. A number of interesting results were obtained, among which: the localization of gb's and sf's activity at dislocation emergencies, the existence of a passivated region around an active defect, the location of C-rich segregated zones on one side only of the gb's.

CONCLUSIONS.

At this contract mid-time, it is useful to concentrate the discussion on the main issue of our study, which is the epitaxial solar cell efficiency.

One point is clear from Fig. 1 : the superficial lead layer has been decreased, and the high energy efficiency increased, using the new junction technology mentioned above. This positive result may still be improved.

A second (negative) result is that, due to the red-infrared response, our total efficiency has decreased somewhat from the best value (10,3%) obtained at the beginning of our study (compare ref. (1) to the above Table). This may be assigned partly to the ingots (as analyses will show), partly to the technology (epilayer and screen printing process, as evidenced by the reached values of L_{ne} and V_{oc} respectively).

Finally, let us recall that we obtained previously detailed conclusions on the influence of Al and C in polycrystalline Si, (Ref. (2) to (5)) and on the limit of tolerance of several impurities (6).

REFERENCES.

1. M.RODOT et al., Rev. Phys. Appl. 22 (1987) 687
   M.CAYMAX et al., Comm. 2nd PVSEC Conf., Pékin, Aug. 1986
   M.CAYMAX et al., Comm. 7th PV Solar Energy Conf., Sevilla, Oct. 1986

2. M.RODOT et al., J. Appl. Phys. (to be published, Oct. 1987)

3. C.CABANEL et al., Comm. 7th PV Solar Energy Conf., Sevilla, Oct. 1986.

4. C.CABANEL, Influence du carbone sur les défauts structuraux et sur l'activité électrique du silicium polycristallin H.E.M. Thèse de Doctorat, Paris VI, Juin 1987.

5. S.PIZZINI, D.NARDUCCI, M.RODOT, Electrical activity opf extended defects in polycrystalline silicon. Rev. Phys. Appl. (to be published)

6. R.KISHORE et al., 19th IEEE PV Spec.Conf.Proc., New Orleans, Apr. 1987

(Ref. 4 and 5 are new publications of the present semester).

4TH TECHNICAL REPORT OF ACTIVITIES ON THE MODELLING OF EPITAXIAL CELLS ON UPGRADED METALURGICAL GRADE SILICON AND ON THE FEASIBILITY OF A HIGH THROUGHPUT EPITAXIAL REACTOR

Contract Name	:	"MODELLING OF EPITAXIAL SOLAR CELLS ON UMG-SILICON AND FEASIBILITY STUDY OF A HIGH THROUGHPUT EPITAXIAL REACTORS"
Contract Number	:	EN-35-0076-E(B)
Duration	:	30 months 1 July 1986 - 1 Jan. 1989
Total Budget	:	Pts. 22.500.000
Head of Project	:	Dr. A. Luque, Dr. G. Sala, Instituto de Energía Solar U.P. Madrid
Contractor	:	Fundación General de la U.P. Madrid
Address	:	I.E.S. - E.T.S.I. Telecomunicación Ciudad Universitaria E-28040 - MADRID (SPAIN) Tel. +(341) 2.44.10.60 Tx. 47430
Period of work	:	April 87 - September 87
Document Author	:	A. Luque
Report	:	UPM/IES/LS/4787

Summary

The activity of the spanish group in the period of reference is presented. Light confinement of light on the epitaxy based on the reflections at the epitaxy substrate boundary due to band gap shrinkage caused by doping is analized as well as the behaviour of the high-low junction minority carrier reflection is analized based on a non-constancy of quasi Fermi levels. A novel graded cell structure without heavily doped emitter is under analysis. Evaluation of CVD, CSVT and LPE techniques of epitaxy formation are under comparison for assessment of the more suitable technology.

1.1 Introduction

In this project the spanish group has two tasks: modelling of cells and predesign of a high throughput epitaxial reactor.

In past reports we have reported a simple model for the epitaxial cell with a synthetic presentation of important features to achieve high efficiency. These are (a) to have a high quality epitaxy (b) to have an effective minority carrier barrier at the substrate-epitaxy interface and (c) to be able to produce some light confinement in the epitaxy.

The effect of the substrate impurities on the epitaxy have been analyzed and it is fortunate to find that the most harmful impurities have small mobility and are not likely expected to pass into the epitaxy; nevertheless an out-diffussion pre-treatment is recommended. In spite of it the epitaxy is expected to be more contaminated than semiconductor grade bulk cristalline silicon. Because of it an analysis of the plausible impurities in the epitaxy suggests that higher base resistivities must be used than with cristalline cells. 4 cm is an experimental optimum found by Caymax and collaborators at IMEC and is understood on the base of these arguments.

On the side of reactor predesign a detailed analysis of the kinetics of CVD (atmospheric or low pressure) has been done jointly with an economical study of the epitaxy cost. The goal $ 0,5 per epitaxy is not achievable with present equipment but low pressure CVD reactors of novel design seem promising.

1.2. Period work on modeling

1.2.1. High Low Behaviour

Based on our previous work in bifacial cells we have evidence (1) that in some cases the high-low junction does not work following the ideal model. This model is based on the assumption of constant quasi-Fermi levels at the high-low junction boundary. This assumption is being checked based on a more accurate model (2) that predicts that in solar cells an additional injection term will appear, proportional to the cell current. Although not yet completed first estimates seem to suggest that the effect is not of importance.

1.2.2. Light confinement

Regarding the possibility of light confinement in the epitaxy an analysis of the reflection at the epitaxy substrate is being done based on the diffusion of a heavily doped layer at the interface that would cause a band gap shrinkage. The knowledge of the optical constants of the heavy doped region is being obtained from measured data completed with a kramers-kronig analysis but the scarcity of these data, particularly in the long wavelength range where they will be more interesting makes difficult the analysis. For these wavelengths, data based on theoretical-empirical relationships of absorption coefficient variation with the temperature are being used to complete the wavelength range. The overall impression at this stage is that the possibility of ligth confinement based on this scheme will be rather small.

1.2.3. Novel structures

Recent theoretical and experimental work (3) makes evident that the heavy doping of the emitter is always deletorious for the cell recombination. Although a smaller minority carrier density is present the recombination terms increase by Auger mechanisms and increase the overall recombination. This conclussion is not straightforward because of the heavy doping mechanims complexity but we consider it sufficently proven today.

Consequently a cell without a heavy doped emitter is beneficial. An analysis of a cell with an epitaxial n region over a p region is being analyzed and it is concluded that in principle it is favourable if a good surface passivation is achieved. Note that this structure is possible in epitaxial cells and not in bulk cells. This is certainly true with semiconductor grade substrates but not so clearly with epitaxial cells. We present, in Fig. 1 for the case of high quality bases the expected efficiency of a one-dimensional cell vs the emitter sheet resistance (which determines the losses due to series resistance) for a good diffused emitter and an epitaxial emitter cell. It can be appreciated that the performance of the epitaxial emitter cell is higher.

However, in the epitaxial cells the base recombination is higher than the emitter recombination so that its ellimination is not very practical at this stage.

Nevertheless the problem can be different if the epitaxy is of graded doping so that the minority carrier transport is done by electric field and the residence time of the minority carriers is greatly reduced. The band diagram of the suggested cell is shown in Fig. 2. It is unfortunate that the structure leads to lower voltages than the flat band ones due to the higher minority carrier densities at the p-n junction. A model of this structure is being developped and some expectancies of noticeable improvement are expected from it.

1.3. Epitaxial reactor predesign

1.3.1. Novel epitaxy approaches

Before to start with the properly speaking predesign work it has been decided to examine two different novel approaches these are Close-Spaced Vapor Transport (CSVT) and Liquid phase epitaxy (LPE).

The first method has been principally used for the growth of III-V and II-VI compound semiconductors, in the last case at low temperatures (≈600°C) and high growth rates (7-8 μ m/min). Silicon requires a higher temperature (≈900°) to achieve a reasonabale growth speed. A major problem is the large area on which the growth is to be done. At a first glance it seems not to be a reliable solution but a definitive assessment is not yet available.

LPE is a technique that is receiving extensive attention. Its application to silicon is recent and is based on the use of Si solutions in Sn or Pb at low temperatures. The technique is indeed promising from a cost point of view but certainly it is less flexible than CVD not allowing easily graded structures and other improvements. Also the quality of the obtained epitaxy is much less proven.

1.3.2. Experimental set-up

In our Institute we have an ASM epitaxial reactor that for many years has been devoted to deposit SiO_2 and Si_3N_4 layers. The reconditionement of it for epitaxy has required some modifications and some safety measures. The following modifications have been done

Installation of an outdoor housing for gases

Hydrogen perifier

Modification of the reactor for SiH_2Cl_2 operation

Air extraction and breathing equipment (for safety)

The reactor is expected to be in operation room.

Conclusions

Although not yet finished it is expected that final assessment about the possibility of the light confinement in the epitaxy will be given soon.

More insight on the high-low operation is being achieved. A novel structure for epitaxial cells with graded epitaxy will be presented with evaluation of its potential. This structure seem to be, at least theoretically the optimal for epitaxial cells.

A final assessment of CVD, CSVT and LPE will be soon available to start with the predesign of a fast reactor.

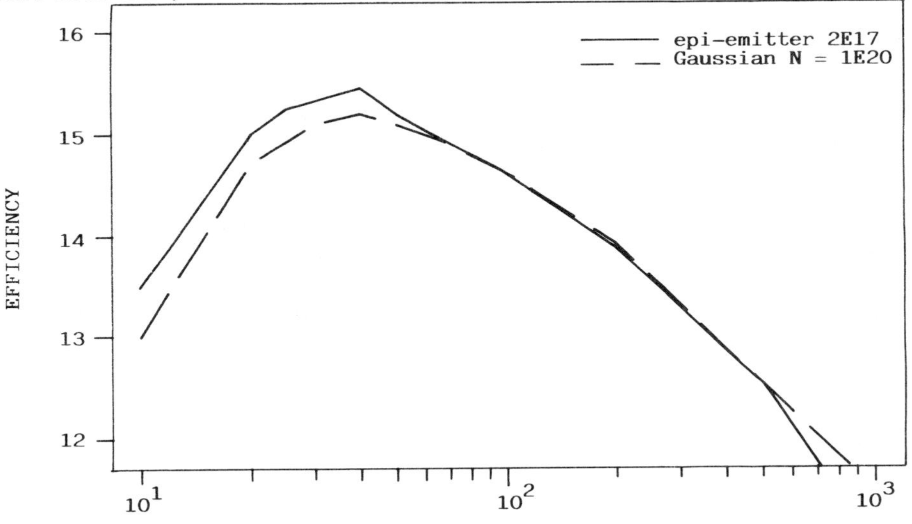

Fig.1 -Efficiency vs. emitter sheet resistance for a one dimensional diffused emitter cell and epitaxial emitter all with good recombination parameters.

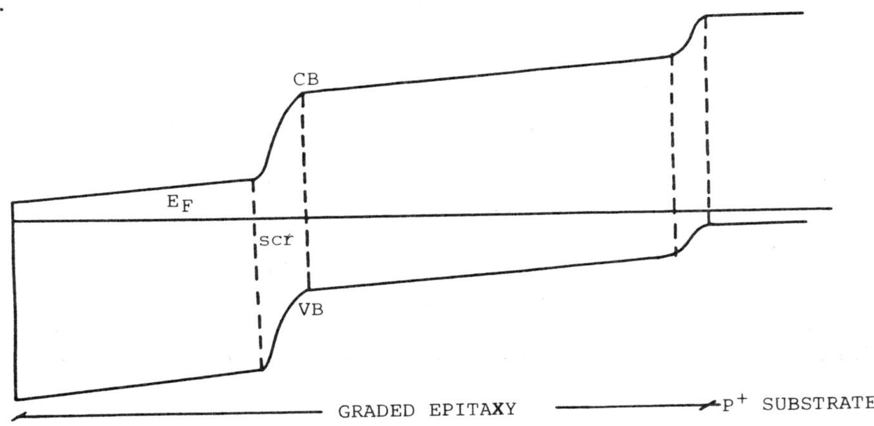

Fig.2 -Band diagram of a graded epitaxy solar cell.

HIGH EFFICIENCY THIN FILM SOLAR CELLS ON UPGRADED METALLURGICAL GRADE SILICON SUBSTRATES

Authors :D.MARGADONNA, R.PERUZZI, R.SPOSITO
Contract number :EN3S-0077-I (A)
Duration :24 months 1 September 1986-31 August 1988
Total budget :LIT 206.000.000 CEC contr.:LIT 103.000.000
Head of project :Dr. D. MARGADONNA, ITALSOLAR S.p.A.
Contractors :ITALSOLAR S.p.A.
Address :ITALSOLAR S.p.A.,Via A. D'Andrea 6
00048 NETTUNO (RM) ITALY

Summary

Some results related to silicon wafers obtained from second directionally re-crystallization of metallurgical ingots are reported. The MG feedstock, delivered by SAMATEC was obtained from a new run of smelting experiments. The upgraded feedstock has a boron and phosphorus concentration of 2.5 ppmw and 5.5 ppmw, respectively. The aluminum concentration is around 60 ppmw. From this feedstock ingots of first directional crystallization have been obtained with a resistivity of about 0.06 Ohm x cm. Some of these ingots have been re-crystallized again obtaining 2 ingots of second crystallization with a resistivity of 0.2 Ohm x cm. Selected wafers cut in different positions along the growth axis from the first and second crystallization ingots are now available for epitaxial growth.

1. INTRODUCTION

In our previous report we presented photovoltaic characteristic referring to wafers obtained from upgraded metallurgical silicon prepared by SAMATEC. The concentration of boron and phosphorus in such feedstock was 6 ppmw for both impurities. Silicon ingots were then prepared by double directional solidification starting from the upgraded metallurgical silicon blended with 50 % of electronic grade silicon. This operation was necessary in order to reduce the resistivity of the resulting wafers with the aim to achieve the photovoltaic measurements. The same material has been employed as substrate for epitaxial growth, obtaining encouraging results in term of P.V. performance, as reported in ref. 1. An interesting results, in term of reached efficiency, was obtained growing epitaxial solar cells on this metallurgical sustrate. Obviously, these substrates, coming from a double directional solification and blended with electronic grade silicon, are not competitive from the economical point of view for realizing cheap solar cells. Meanwhile, a new set of MG silicon smelting experiments have been achieved by SAMATEC and the resulting material exhibits a lower impurities concentration than the previous one, in particular with only 2.5 ppmw of boron (Table n.1). With this material ITALSOLAR can now growth ingots of first and second crystallization with low impurities concentration without blended the feedstock with EG silicon, reducing in this way the costs of the wafers.
We report here some results obtained on wafers prepared from such feedstock together with the most important experimental conditions. The wafers are now available for the epitaxial regrowth and other measuraments.

2. EXPERIMENTAL

2.1 Preparation of upgraded metallurgical silicon (*).

The metallurgical "solar grade" feedstock was prepared according to the process described in ref. 2. In order to reduce the overall contaminations a number of improvements were introduced in some steps of the process:

- the three-phase arc furnace was changed in a single-phase D.C.,thus improving the smelting process yield and decreasing consequently the impurity concentration in the liquid.

(*) This part of the work was carried out by SAMATEC by the financial support of E.N.E.A.

- the arc furnace operation and the handling of the raw materials were performed under filtered air, avoiding any pollution coming from the external environment.
- the mechanical mixture of the starting materials (quartzites carbon black and SiC) has been agglomerated as briquettes, using sucrose as a binder and employing a new equipment, particularly suited for clean operation (i.e. low iron contamination).

The impurities concentrations of the starting materials and their evolution for different steps of the process are shown in Table n. 1, together with the analysis of the smelted silicon, indicated as run MNP/9. The analysis of the silicon prepared in previous experiments (HPS 10/2) is also reported for comparison. It is evident that the improvements introduced in the process lead to a reduction of the boron content by a factor 2, aluminium by a factor 7 and the deep impurities (Fe+Ti) are also reduced by a factor 20. On the other hand, the P content remains equal.

2.2 Purification of upgraded metallurgical silicon by directional solidification

50 Kg of upgraded metallurgical feedstock, labeled in table n.1 as MNP/9 were employed to growth 4 silicon ingots following the directional solidification method described in ref.3. Such ingots (MG 1) are listed in table n.2 together with the electrical resistivities measured as an average across a vertical section. A vertical cross section of the ingot TW 112 is shown in fig.1: the breakage line of the solid-liquid interface as a consequence of the impurities (mostly SiC) segregation towards the top of the ingot appears as a boundary line between the columnar and dendritic growth of the crystal grains.

The chemical analysis of the ingot TW 111 and TW 112 subjected to a single solidification process (MG 1) was made as follows: a slice, 1 cm thick, was cut off from the middle part of the ingots and then, by further slicing, 120 cubes of one cubic centimeter were obtained and analysed for impurity content by ICP-ES technique. By this way the impurity distribution can be easily obtained for both length and height of the ingot. Some results of the chemical analysis on 8 cubes forming a vertical row in the middle part of the original slice are reported in table n.3. Although the data here reported must be considered as preliminary (boron results still aren't available), the segregation effect of impurities is remarkable for Al, Fe and Ti.

In reference to the crystalline structure, MG 1-type ingots can be divided in two regions: the columnar region, which

extends from the first solidified part (i.e. the bottom of the ingot) up to 2/3 of the total height, and the dendritic region. This was cut off and the remainder of all ingots MG 1-type was remelted for the growth of the double solidified ingots to which we refer as MG 2-type (TW 117 and TW 118). MG 2-type ingots differ from MG 1 for a higher resistivity (Table n.2) and for the columnar structure extending across the total height of the ingot. In order to determine the impurity distribution in the MG 2-type ingots, the same procedure was adopted as for the MG 1. In Table n.3 the analytical results are shown for the central row (row n.8) of the ingot TW 117. The purification induced by the second solidification is very effective in the case of Fe and Ti; whereas the behaviour of Ca and Al is harder to explain, their concentration in TW 117 being of the same level as TW 111.
Finally, a number of wafers (10x10 cm x 0.04 cm) were cut from ingot TW 112 (MG 1) as well as from TW 117 and TW 118 (MG 2) and will be subjected to resistivity measurements before to be processed for solar cells or epitaxial regrowth.

3. CONCLUSIONS

Preliminary measurements carried out on "solar grade" silicon wafers prepared as described above, seems to indicate that our process (carbothermic reduction + directional solidification) is able to meet some fundamental requirements for the production of low cost solar cells. In particular, we presume that MG 1-type wafers should be suitable as low cost substrates for epitaxial regrowth.
In order to confirm these indications we need more detailed measurements which will carried out during the next 6 months as follows:
- To complete chemical analysis, mainly for B and Al.
- To study the resistivity behaviour of the wafers (both MG 1 and MG 2) to know if some resistivity variation exhists depending on the position of the wafer in the ingot.
- To achieve the photovoltaic characteristic.

REFERENCES

1 -M. Caymax et al., Proc. 7th E.C.P.S.E.C., Sevilla, 1986 (D. Reidel, Dordrecht 1987), p. 800.
2 -M.Rustioni et al., Proc. of the Flat-plate solar array workshop on low cost polysilicon for terrestrial photovoltaic solar-cell applications, Las Vegas, October 1985 (JPL publication 85-11), p. 297
3 -S. Pizzini et al.,It. Pat. 203501 (April 1984).

TABLE n. 1 - Impurities content in raw materials and in the silicon smelted in the run MNP/9

Element / ppmw	SiO_2	Carbon black	Briquette SiO_2 + C	SiC	Briquette SiO_2+SiC	Si RUN MNP/9	Si RUN HPS/10-2
Al	60	50	30	20	50	50$\pm$20	350
B	< 0.5	< 0.5	< 0.5	< 0.5	< 0.5	2.5$\pm$1	6
Ca	-	8	10	-	20	130	150
Fe	30	25	40	40	-	50	980
Mg	< 0.5	10	-	-	-	7	30
P	2	0.8	6	1	1	5$\pm$2	6
Ti	11	3	3	-	15	30	380
Mn	< 0.5	-	-	-	1	1	-

TABLE n. 2 - List of ingots grown from Upgraded MG Silicon

Ingot n.	Type	Weight Kg	Resistivity (p) ohm x cm
TW 111	MG 1	12	0.06 $\pm$ 0.005
TW 112	MG 1	12	0.065$\pm$ 0.005
TW 113	MG 1	12	0.06 $\pm$ 0.005
TW 115	MG 1	10	0.04 $\pm$ 0.005
TW 117 111 + 112	MG 2	10	0.2 $\pm$ 0.05
TW 118 113 + 115	MG 2	9	0.15 $\pm$ 0.05

TABLE n. 3 - Impurities distribution measured on the ingot TW 111.

Element / ppmw	MG-Si MNP/9 C_o	CUBE n.								$\bar{C}_s / C_o$
		A8	B8	C8	D8	E8	F8	G8	H8	
Al	50± 20	7	3	3	3	4	3	3	350	0.06
B	2.5± 1	not yet measured								
Ca	130	11	5	7	3	4	4	3	14	0.03
Fe	50	5	1	1	2	2	-	4	9	0.03
Ti	30	0.4	0.3	0.3	0.3	0.3	0.7	0.7	25	0.01
P	5± 2	3	5	6	6	5	6	6	14	1

Solidification proceeds from A to H. $\bar{C}_s = \left(\sum_{B}^{F} C_s \right) / 5$

TABLE n. 4 - Impurities distribution in the ingot TW 117

Element / ppmw	CUBE n.						
	A8	B8	C8	D8	F8	G8	H8
Al	n.d.	1	2	3	n.d.	55	10
B	2	1	1	2	1	4	4
Ca	n.d.	5	13	15	6	10	18
Fe	n.d.	n.d.	n.d.	n.d.	n.d.	9	11
Ti	n.d.	n.d.	n.d.	n.d.	n.d.	n.d.	n.d.
P	5	5	3	4	5	5	9

Solidification proceesd from A to H. n.d. = not detected

Fig. 1 - Vertical cross section of the ingot TW 112

HIGH EFFICIENCY THIN FILM SOLAR CELLS ON UMG-SUBSTRATES

Contract number :EN3/35-0083-I(S)

Duration :36 mo nths 1 May 1986- 30 April 1988

Total budget :Lit.81.473.OOO CEC contribution: ECU 30.000

Head of project :Sergio Pizzini

Contractor :University of Milano

Address :Dipartimento di Chimica Fisica ed Elettrochimica, Via Golgi, 19, Milano (Italy)

Authors :S.Pizzini- A.Sandrinelli- D.Narducci- M.Beghi

Summary

The role of impurities on the recombination losses in polycrystalline silicon is not yet fully understood, mostly because the specific role of interactions between deep level impurities, native impurities (oxygen and carbon), shallow impurities and microstructure is unknown.

This knowledge is also necessary for understanding the role of impurities in materials used as substrates for thin film epitaxial solar cells, as impurities segregated at extended defects not necessarily are electrically inactive, thus contributing to the overall efficiency losses, if the substrate contributes to the generation and collection of photogenerated carriers.

The work carried out in the III semester, after the significative achievements of the last semester with respect to the influence of oxygen and carbon content on the microstructure of polycrystalline silicon has been dedicated to the set-up of an LBIC (Light Beam Induced Current), in view of its future application for the detection of impurities microsegregation at extended defects.

INTRODUCTION

We have reported in the II semiannual Progress Report that oxygen and carbon are or could be very harmful impurities in polycrystalline silicon.
In the absence of carbon-oxygen compensation, infact, their presence results in an enhancement of the electrical activity of dislocations (case of oxygen excess) or in the increase of dislocation density (case of carbon excess).
We have also shown that the segregation features of oxygen and carbon depend on the presence of impurities presenting high reactivity with oxygen (case of titanium-doped silicon).
As the microstructure of the substrate has certainly a significant influence on the electrical properties of an epi-layer deposited on it, a thorough knowledge of the influence of oxygen and carbon impurities on the structural and electrical properties of the substrate is compulsory.
Still in the absence of new MG silicon samples to be characterized, the activity of this semester has been therefore enterely dedicated to the set-up of our LBIC system, in view of its future application for the detection of microsegregation of impurities at extended defects.

EXPERIMENTAL

The experimental set-up is illustrated in fig.1. It consists of an optical bench, on which the beams of three lasers (λ=635,780,1150 nm) are suitably collimated by means of dichroic filters and focused on the sample by means of an optical microscope objective.
The sample itself ($25\times25mm^2$ in size) is mounted on a x-y traslation stage, driven by two stepping motor (0.1 μm steps).
The light intensity is monitored by a silicon photodiode.
The data aquisition system consists of a EG&G mod.5207 lock-in data amplifier , wich serves for the measurement of the photogenerated current and of a stepping motor control unit, which serves for the monitoring of the traslations.
Both are suitably interfaced to an IBM mod. 6151 microcomputer.

The samples used for preliminary experiments were policrystalline silicon solar cells manufactured at the Pragma factory using the conventional phosphorous diffusion and aluminum metallization substrate. As the evaluation of the LBIC profiles requires an accurate knowledge of the beam size, the determination of the beam size ahs been preliminary carried out using two different metodologies.

a) The beam is scanned over the edge of a metallization line. The current collected increases in a way which is characteristic of the beam width, being the integral over the intensity profile except for the region covered by the metal. According to Marek (1), as the coordinate x is changed, the collected fraction of the light results in

$$v(x) = 1/2 + 1/2\mathrm{erf}(x/\sqrt{2\sigma})$$

were σ is the beam width.
The beam width is then determined as the smaller value of which could be experimentally observed by changing the distance between the objective and the sample at steps 1 to 5 μm.
We report in Fig. 2 a typical set of experimental results obtained with the He - Ne laser (λ = 635 nm).
By this procedure a spot width of 3.8 μm, a beam width of about 2 μm and a depth of focus of less then 10 μm has been determined.

b) The beam is focused on a photographic plate and the beam width is then determined at a convenient magnification of the photographic image. Results relative to this procedure are not yet avaible as the work has not been yet completed.

Being the width of the Cottrell atmosphere (e.g. of the impurity cloud) around an extended defect of the order of 50 - 100 μm (2), this resolution is certainly large enough to detect any significant deviation of the experimental EBIC profile from the theoretical one.
We expect, therefore that on a good polycrystalline sample the half - width of the EBIC profile should be totally determined by the beam width and by the diffusion length L_d of the minority carriers in the grains. Preliminary experiments carried out on a polycrystalline cell presenting

an average diffusion length of 100 μm and an efficiency larger than 11% showed that experimental half - width (15 μm, see Fig. 3) closely corresponds with the calculated one (see Fig. 1) for a beam width of 2 μm.

CONCLUSIONS

Being the results of this semester largely preliminary in their nature, any conclusions about the influence of impurity segregation on the electrical activity of extended defects in polycrystalline silicon could be extracted. As the influence of the carbon / oxygen ratio is however expected to be very important on oxygen - carbon compensation effects and the reliability of oxygen and carbon determinations via FTIR is scarce, in the absence of a primary standard, it is our intention to propose to dr. Revel to arrange for us some standard oxygen samples, to be used for absolute FTIR calibration curve.

REFERENCES

(1) J.Marek - J.Appl.Phys. 55 (1984) 318

(2) S.Pizzini,P.Cagnoni,A.Sandrinelli,A.Anderle and R.Canteri - Appl.Phys.Lett. Aug.1987

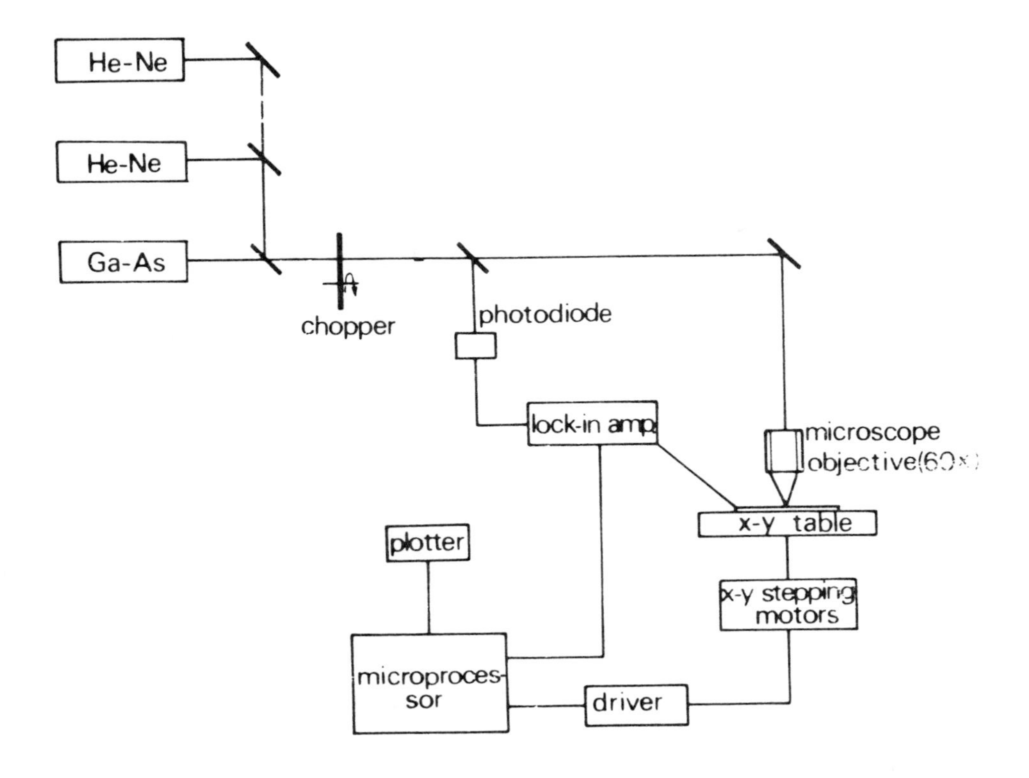

Figure 1 - Schematic diagram of the experimental arrangement.

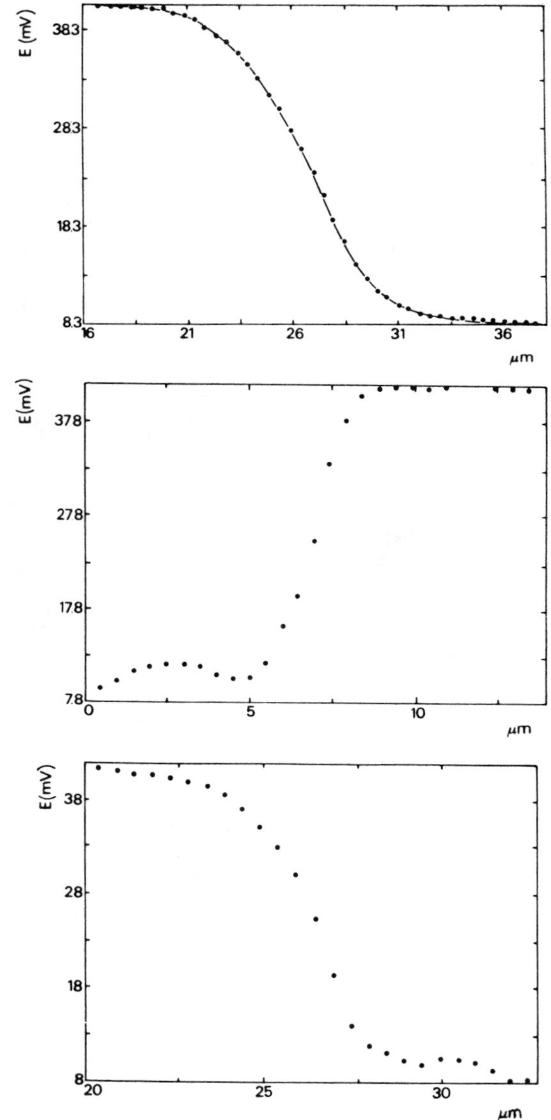

Figure 2 - LBIC profiles obtained by scanning the beam over the edge of the metallization line:

a) + 10 μm out of focus

b) on focus

c) - 5 μm out of focus

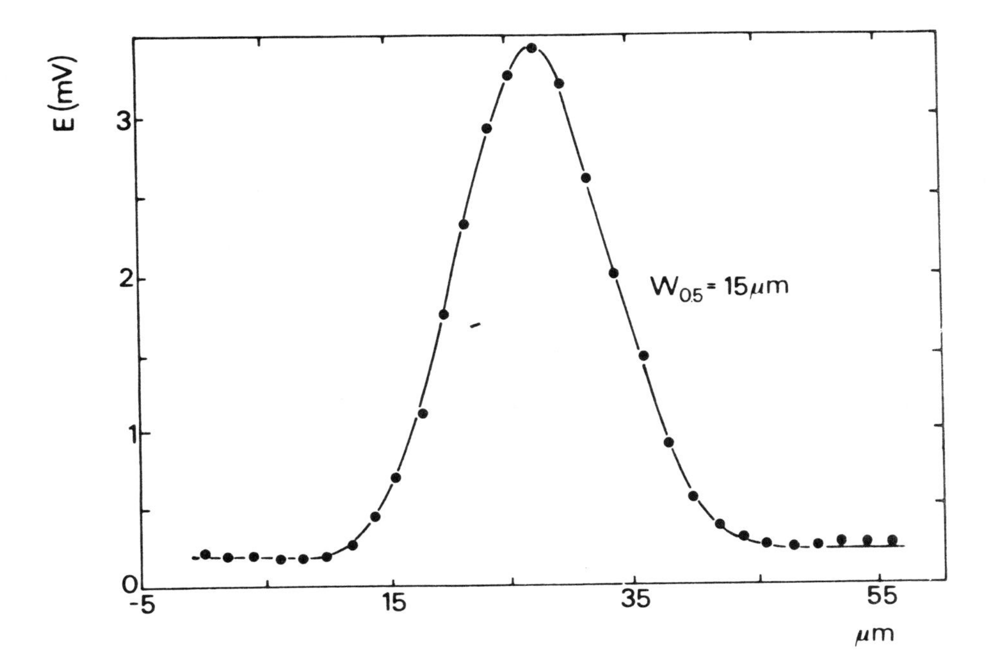

Figure 3 - Example of an LBIC profile when the beam is scanned over a grain boundary. The half width is 15 μm.

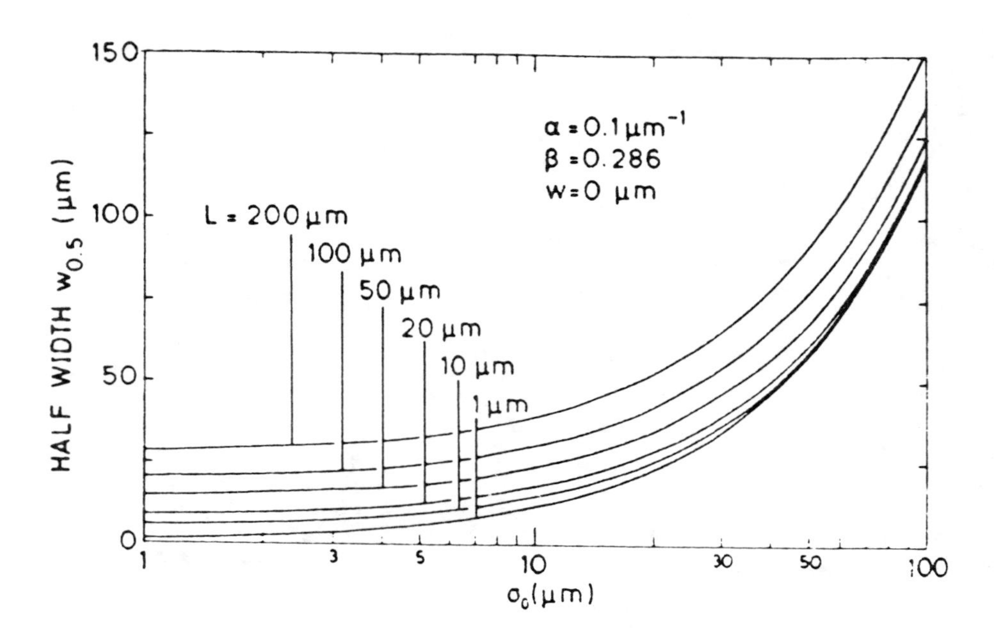

Figure 4 - Dependence of the half - width on the value of the beam width. The diffusion length is the parameter used in the calculation.

"THIN FILM SOLAR CELLS BASED ON THE CHALCOPYRITE SEMICONDUCTOR $CU(GA,IN)SE_2$"

Contract Number: EN 35-0067-D

Duration : 36 month, 1 Jan. 1986 - 31 Dec. 1988

Total Budget : DM 885 000.- CEC Contribution 885 000.-

Head of Project: Dr.-Ing. H.W. Schock, Inst. f. Phys. Elektronik

Contractor : Universitaet Stuttgart

Address : Keplerstr. 7
D-7000 Stuttgart 1

SUMMARY

$Cu(In,Ga)Se_2$ based heterojunction cells are investigated in a joint effort of groups from France, Italy, Great Britain and Germany. Each group uses a different approach for cell fabrication such as spraying, sputtering, evaporation and selenization. Furthermore electrochemical methods for the modification and characterization of the films are applied. The comparison of the results gives important informations on the specific properties of $CuInSe_2$ as a photovoltaic material. At the University of Stuttgart the quaternary alloy $Cu(In,Ga)Se_2$ is studied. Various heterojunctions based on the quaternary alloy have been fabricated and efficiencies up to 6% have been achieved. The experiments show that (Zn,Cd)S can be replaced by a ZnO/ZnSe bilayer window. For the selenization process, the reaction kinetics and formation of phases are studied in detail. First experimental heterojunctions with selenized $CuInSe_2$ and $CuGaSe_2$ show promising performance.

JOINT RESEARCH ON SOLAR CELLS BASED ON THE CHALCOPYRITE SEMICONDUCTOR $CU(GA,IN)SE_2$

The research work within this group progressed as planned in the proposals. Major achievements of the single groups are:

USTL Montpellier:

Fabrication of a 6.4% efficient all sprayed CdS-$CuInSe_2$ backwall cell. Deposition of n-type $CuInSe_2$ films for $CuInSe_2$ pn junctions.

University of Parma:

Single layer $CuInSe_2$ all sputtered $CuInSe_2$-CdS cell with 5% efficiency. Low resistive CdS films by sputter deposition.

ENSCP Paris:

Measuring semiconductor parameters by electrochemical methods. Method for the determination of composition evaluated.

Newcastle Polytechnic:

Achievement of large area $CuInSe_2$ films by strip sources. $CuInSe_2$ band edge observed in laser processed films.

University of Stuttgart:

Several heterojunctions with quaternary absorbers reach 5-6%. First hetrojunctions with selenized $CuInSe_2$ and $CuGaSe_2$ realized. New windows (ZnO/ZnSe) yield high I_{SC}.

Besides the meetings of the contractors at the EC Photovoltaic Solar Energy Conference, one day seminars are hold semiannually at the single labs. These seminars serve to get to know the facilities of the different labs and to get familiar with the different technologies and specific problems. Furthermore an efficient exchange of experience is possible. Researcher working on selected fields without EC contract are invited to these meetings and to contribute to special scientific problems. Meetings were hold in Stuttgart April 1986, Paris December 1986 and Montpellier June 1987. The next meeting will take place in Parma. An in depth discussion of the role of bilayer $CuInSe_2$ in the different technologies is planned in order to improve the basic understanding of the material. This is of paramount importance for optimization of the devices.

1 Introduction

The quaternary chalcopyrite semiconductor alloy $Cu(Ga,In)Se_2$ covers the bandgap range from 1.02 to 1.7 eV. The purpose of the work is to demonstrate the possibilities to realize heterojunctions for any bandgap in this range and to find optimum absorber/window constructions. Besides vacuum evaporation and the methods applied in the other groups in the joint program selenization of metal films are investigated as a potential low cost method.

2 Evaporated films

In the first phase of research the deposition and properties of quaternary $(Cu_2Se)_{1-x}[(Ga_yIn_{1-y})_2Se_3]_x$ films have been studied. For these experiments an arrangement of four effusion sources with feedforward control (thermocouples and microprocessor control) was applied (1). Films with defined compositional gradients on the substrate allow to study the influence of deviations from stoichiometry and variations of the Ga/In ratio. Detailed studies of the film morphology and the dependence on growth conditions have been carried out (2). From the experiments optical and electrical properties of the quaternary films are obtained. Fig. 1 shows the variation of the optical bandgap with the Ga content. The data were obtained by extrapolation of the linear part of the normalized absorption coefficient $(\alpha h\nu)^2$ to the energy axis. Furthermore the absorption data allow an estimation of the change of the reduced effective mass in the quaternary alloy system. The effective electron mass may be higher by a factor of two for the Ga-rich films compared to pure

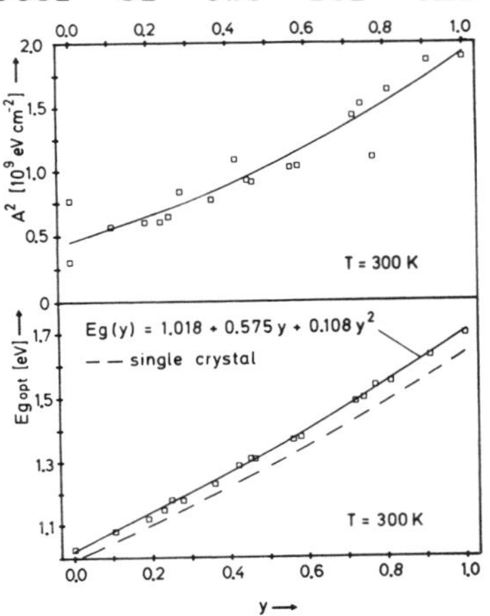

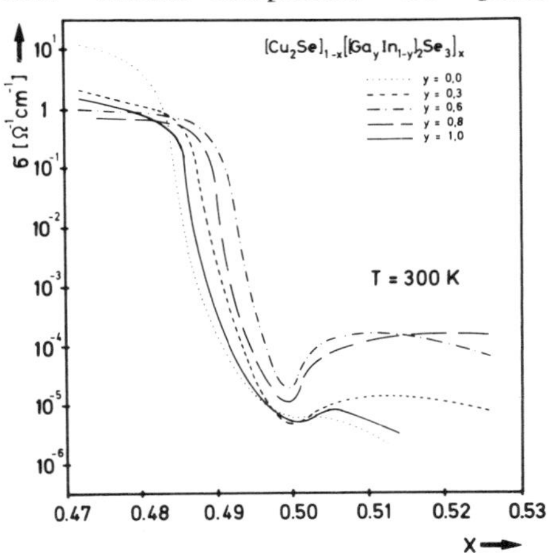

Fig. 1: Variation of optical bandgap with Ga-content

Fig. 2: Conductivity as a function of deviations from stoichiometry

$CuInSe_2$, so that a somewhat lower mobility has to be taken into account (3). In fig. 2 the conductivity as a function of deviations from stoichiometry does not show significant differences for the various Ga-contents. On the Cu-rich side, the conductivity is dominated by an additional $Cu_{2-x}Se$-phase as it has been detected by X-ray diffraction. A minimum of conductivity is found at the stoichiometric composition. First heterojunctions with (Zn,Cd)S windows have been prepared with single layer quaternary films. Improved performance was found with bilayer films with a lower Cu-concentration towards the surface. Results are discussed in section 4. Triaxial sources are beeing designed for the evaporation of quaternary films.

3 Selenization

Experimental research on selenization reactions was extended from the Cu-Ga-Se systems to Cu-In-Se and basic results were obtained. First solar cell devices with selenized chalcopyrite thin films were realized.

3.1 Deposition of metal films

Cu-In metal films were prepared by thermal vacuum evaporation as described for Cu-Ga films in previous reports. Careful handling of evaporation parameters for the In-deposition is necessary because of the formation of a liquid intermetallic Cu-In phase which disturbs the homogeneity of the films during recrystallization. To achieve more homogeneous metal films the deposition rate of In has to be decreased. Metal thin films deposited with high In rates show also extremely orientational ordering of the crystallites as measured by X-ray powder diffraction.

3.2 Selenization in sealed quartz ampoules

Choosing a closed system for the experiments based on the easy estimation and control of Se-vapor pressure by measuring temperature along the ampoule. The experiments are carried out in sealed quartz ampoules. Temperature gradient in the two furnace zone was measured with a NiCr-Ni thermocouple and Se-vapor pressure was calculated by equation (3):

$$\log p\ [\mathrm{mmHg}] \approx -5010.7/T + 8.1105$$

Se-vapor pressure, substrate temperature, reaction time and dynamics of heating and cooling down turned out to be the most important selenization parameters because of influencing more or less the selenization reaction kinetics. The most common desturbing effect is segregation of Cu-Se phases on the surface of the films due to a high reaction velocity of these binary phases and loss of (In,Ga) on the surface by reevaporation of $(In,Ga)_2Se$ (see fig.3a)). For optimized substrate temperature and reaction time single phase chalcopyrite thin films are obtainable (see fig. 3b)). Se-vapor pressure and substrate heating ramp influences the stoichiometry of the Cu-Se phases as shown in table I.

Fig. 3: Morphologies of selenized $CuGaSe_2$ thin films.
a) with large XX of $Cu_{2-x}Se$ as a secondary phase
b) single phase chalcopyrite thin film

	Cu [At.%]	Ga [At.%]	Se [At.%]	phase
ramp: 50·C/min p_{Se}: 2000 Pa	62,11	0,78	37,11	$Cu_{2-x}Se$
ramp: 5·C/min p_{Se}: 2000 Pa	48,53	1,35	50,12	CuSe

Table I: Influence of substrate heating ramp and Se-vapor pressure on binary reaction kinetics. Cu-Se segregations measured by EDS.

4 Heterojunctions

Besides the experiments with evaporated (Zn,Cd)S windows new window materials and double layer windows have been investigated. The results for heterojunctions based on various evaporated and selenized absorbers are listed in table II. The best results have been obtained with (Zn,Cd)S windows, however ZnO turns out to be a very promising wide gap window. The blue response of the heterojunction is enhanced according to the wide band gap. The open circuit voltage increases according to the bandgap of the quaternary absorber. A set of IV characteristics for quaternary based cells is shown in fig. 4. IV curve 2 refers to a single layer absorber. The main limitations are due to the low fill factors. Optimization of doping gradients and the thickness of the layer are necessary for the improvement of the layers. The very first results of heterojunctions with selenized films demonstrate the feasibility of this method.

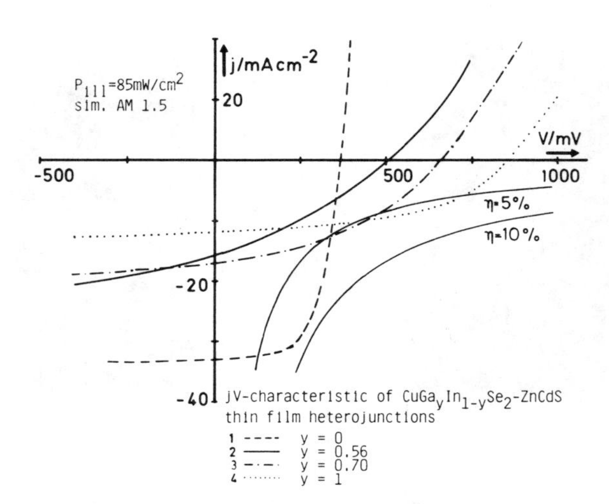

Fig. 4: IV characteristics of various Cu(Ga,In)Se_2 based heterojunctions

P_{ill} = 85 mW/cm^2 sim. AM 1.5	V_{oc} [mV]	j_{sc_2} [mA/cm^2]	ff [%]	η [%]
$CuInSe_2$-CdS	366	33	64	9.3
$CuGaSe_2$-$Zn_{0.3}Cd_{0.7}S$	845	11.6	50	5.8
$CuGaSe_2$-ZnO	218	14.5	26	-
$CuGaSe_2$-ZnSe-ZnO	466	15.4	34	2.8
$CuGaSe_2$-$Zn_{0.5}Cd_{0.5}S$-ZnO	470	10	35	2.0
$CuGa_{0.7}In_{0.3}Se_2$-$Zn_{0.3}Cd_{0.7}S$	670	18	40	5.7
$CuGa_{0.56}In_{0.44}Se_2$-$Zn_{0.25}Cd_{0.75}S$	522	15.6	30	2.9
$CuInSe_2$-CdS (selenized)	300	13.2	33	1.7
$CuGaSe_2$-(Zn,Cd)S (selenized)	730	2.9	-	-

Table II: Performance of various Cu(Ga,In)Se_2-based heterojunctions

5 Conclusions

The work has progressed as it has been planned in the proposal. Quaternary Cu(Ga,In)Se_2 films show promising performance (n = 5,85% for $CuGa_{0.7}In_{0.3}Se_2$ absorber films). The development of new window materials for the heterojunctions show the potential for improved short circuit current. First heterojunctions with selenized films have been successfully realized.

6 References

(1) Dimmler, B., Dittrich, H., Menner, R. and Schock, H.W.: Proc. Int. Symp. on Trends and New Appl. in Thin Films, Strasbourg (1987), pp. 103-107.

(2) Dimmler, B., Dittrich, H., Menner, R. and Schock, H.W.: Conf. Rec. 19th IEEE Photov. Spec. Conf., New Orleans (1987), IEEE New York (to be published).

(3) Zingaro, R.A. and Cooper, W.C. (Editors): Selenium, Van Nostrand Reinhold Company, New York (1974).

RESEARCH AND DEVELOPMENT ON SPRAYED $CdS-CuInSe_2$ THIN FILM SOLAR CELLS

Proposal contract n° : EN3S/C2/102/F

Duration : 36 months 1/01/86 - 31/12/88

Budget CCE : 1 373 600 FF

Head of Project : Dr. M. Savelli - Professeur
: Dr. J. Bougnot - Chargé de recherche CNRS

Co-workers : Dr. S. Duchemin - Maître de Conférences
: V. Chen - J.C. Yoyotte (students)

Address : Centre d'Electronique de Montpellier
Université des Sciences et Techniques
du Languedoc - Place Eugène Bataillon
34060 Montpellier Cédex - France

Summary

This work concerning the $CdS-CuInSe_2$ solar cells entirely sprayed is a part of an European research programm on chalcopyrite semiconductors.

In this paper we present the last results obtained on this type of solar cells.

The optimization of the $CuInSe_2$ layers which was described in our precedent contract reports, in particular the use of two $CuInSe_2$ layers in place of only one, had lead to an efficiency at about 4.8% under 100 mW/cm^2 (AM 1).

In the course of this study we are working on the amelioration of the CdS layers in particular by the use of an intermediary CdS layer doped with In in place of Al. Moreover the fact that, actually, all the cell is elaborated in the same apparatus with optimized fabrication parameters improves the efficiency up to 6.4%.

I. First generation of photocells

As mentioned in our precedent reports the device is constituted by a stack of layers as below

I.T.O./doped CdS/CdS/$CuInSe_2$ high ρ — $CuInSe_2$ low ρ.

I- a Fabrication

The standard fabrication parameters of the sprayed layers constituting the solar cell were summarized in our precedent report (1).

In this first generation of solar cells the transparent electrode (I.T.O.) and the two $CuInSe_2$ layers are sprayed with the same apparatus (named A) using a nozzle sweeping along the two axis direction X and Y. The two CdS(Al) and CdS layers are fabricated in an other apparatus (named B) used in past times for the CdS-Cu_2S photocells with an elliptical sweeping for the nozzle. The diameter size of the nozzle is smaller (0.3 mm) than used for $CuInSe_2$ films.

The resistivity is 15 Ω cm for the active CdS layer, 10^3 Ω cm for the Al doped CdS layer in the standard fabrication conditions (1). The optimization of the two $CuInSe_2$ layers was carried out in precedent studies (2).

I- b Characterization

After an optimization of the two CdS thicknesses (th_1 = 1.5 µm for CdS(Al) - th_2 = 2.5 µm for CdS) the solar cell efficiency was 4.8% for an area of 0.125 cm^2 with a good short circuit current (36 mA/cm^2) and an open circuit voltage near 235 mV. This weak V_{oc} was related to the poor quality of the junction surface and also to a disorder growth of the CdS layer.

By replacing the Al doped CdS layer by a layer doped with In we can improve the crystallinity growth of the CdS layers as shown in Photo 1 a-b, by reducing the defects induced by the CdS(Al) buffer layer. Moreover the resistivity of the doped CdS layer is decreased for some doping concentrations (Fig.1) and a minimum near 2 Ω cm is obtained for a In/Cd concentration ratio near 1% at.

By this way the V_{oc} is slight increased (250 mV) and the I_{SC} is up to 37 mA/cm^2. By decreasing the thicknesses of the two CdS layers (doped and undoped) near 1 - 1.3 µm the fill factor and the efficiency are improved (FF = 0.47 ; η = 5.5%).

However compared to the values obtained by another fabrication process our V_{oc} is still weak ($V_{oc} > 350$ mV by thermal evaporation (3)). So we have think that an improvement of the V_{oc} coulb be obtained if all the layers were sprayed in the same apparatus so as to obtain a strict control of the fabrication parameters, especially the substrate temperature.

II. Second generation of photocells

II- a Fabrication

At that time all the layers are sprayed in the same apparatus (named A) the growth of the CdS layers is three time faster than in the B apparatus, so a readjustement of some fabrication parameters is needed; in particular for the CdS layers the substrate temperature must be taken lower (330°C in place of 450°C) and so, for the same molar concentrations in the CdS solutions, the resistivity of the active CdS layer is 3 Ω cm (Fig. 2).

II- b Characterization

The better photovoltaic characteristic obtained after different studies related to the In doping concentration of the buffer layer and to the two CdS thicknesses are shown in Fig. 3. An efficiency near 6.4% is obtained with a $V_{oc} \simeq$ 310 mV when the thickness of the active CdS layer is thin.

Conclusion

By still moving the substrate temperatures of the active and absorber layers and adjusting the different thicknesses of the layers a strong improvement of the open circuit voltage and efficiency must be expected.

However the part hold by the intermediary doped CdS layer is not well understand. So we hope to eliminate this difficulty by using $CuInSe_2$ homojunctions entirely sprayed in place of CdS-$CuInSe_2$ heterojunctions. In this goal we begin the study on the feasability of n-type $CuInSe_2$ layers by chemical spray pyrolysis. First n-type layers are actually obtained by decreasing the Cu/In ratio (< 1) and the Se content in the initial solutions.

However, before the homojunction realization, numerous studies will be undertaken.

References

(1) J. Bougnot, S. Duchemin, M. Savelli, Report n°2, Contract n° EN3S/C2/102 F, October 86-April 87

(2) S. Duchemin, V. Chen, J.C. Yoyotte, C. Llinares, J. Bougnot, M. Savelli, 7th EC Phot. Sol. En. Conf., october 86, Sevilla (Spain)

(3) R. Noufi, R. Axton, D. Cahen, S.K. Deb, 17th IEEE Phot. Spec. Conf., Orlando (may 1984)

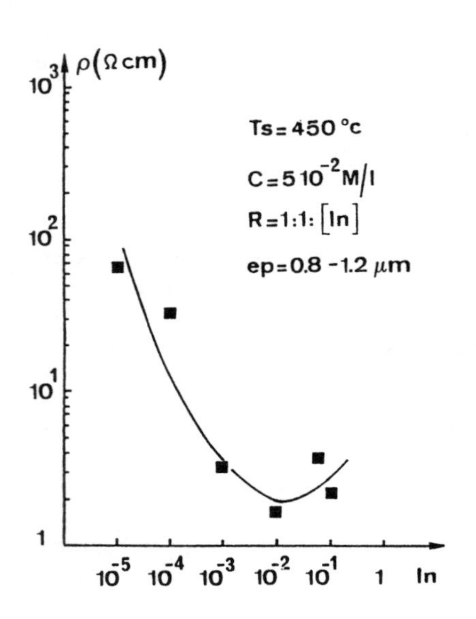

Fig. 1 - Resistivity variation of the In doped CdS layer with the In atomic concentration

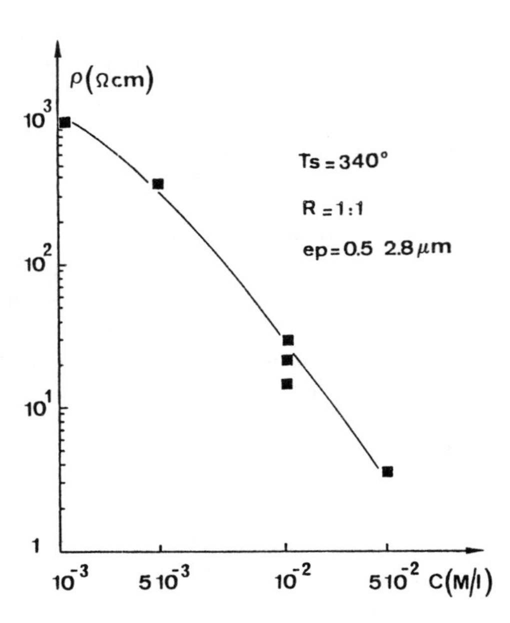

Fig. 2 - Resistivity variation of the active CdS layer with the molar concentration in the spraying solution

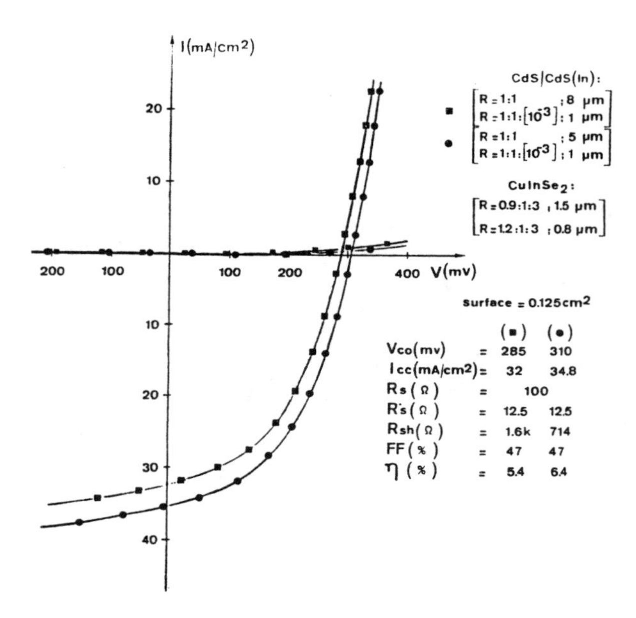

Fig. 3 - I-V characteristic of a cell entirely sprayed in the same apparatus (n° A)

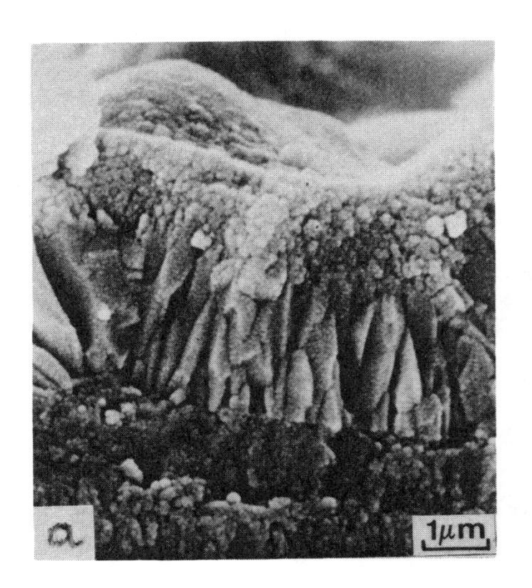

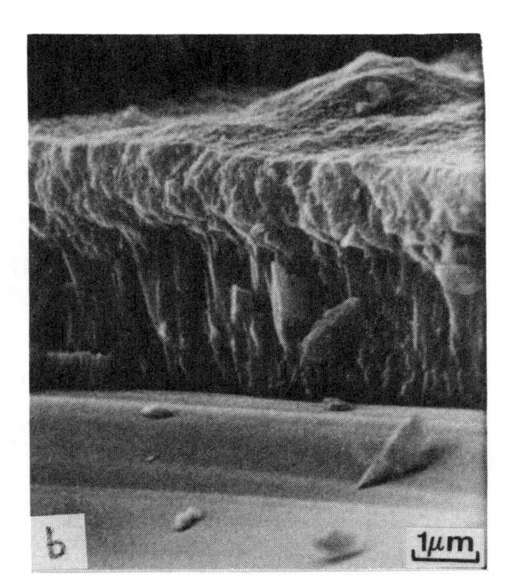

Photo 1 - Transversal cross section of a solar cell

a - the buffer layer is Al doped CdS
b - the buffer layer is In doped CdS

$CuInSe_2$ /CdS Thin Film Solar Cells by R.F. Sputtering

Contract number : EN3S–0069–I(S)

Duration : 36 months

Total budget : Lit.1.000.000.000
CEE Contribution Lit 500.000.000

Head of Project : Professor Nicola Romeo
Department of Physics,
University of Parma,Italy.

Contractor : University of Parma-Italy

Address : Department of Physics,
via M.D'Azeglio 85,
43100 PARMA-ITALY

Summary

The maximum efficiency obtained so far in our[1] or other laboratories for the $CuInSe_2$ /CdS thin film solar cells prepared by R.F. Sputtering is around 5%. In order to further increase the efficiency it could be useful to prepare the $CuInSe_2$ film by the subsequent deposition of an In-rich layer on the top of a Cu-rich layer as it is done when more than 10% efficient $CuInSe_2$ /CdS thin film solar cells are prepared by evaporation[2,3,4,5]. Several targets with different stoichiometries ranging from a Cu/In ratio of 1.2 to 0.66 have been prepared with a technique developed in our laboratory. Three main parameters have shown to be important for the preparation of a double-layer $CuInSe_2$ film: target composition, substrate temperature, and relative thickness of the two layers. First results show that the quality of the bottom layer can be improved by the deposition of the top layer. However, much work has to be done in order to establish the exact values of the parameters involved in the preparation by R.F. sputtering of a good quality $CuInSe_2$ double layer.

1 Introduction

The aim of our research is the achievement of an all sputtered $CuInSe_2$ /CdS thin film solar cell on low cost substrates with an efficiency larger than 10%. In a preceding paper[1] we described the preparation of an all thin film $CuInSe_2$ /CdS solar cell by R.F. Sputtering with an efficiency of about 5%. Here a single layer $CuInSe_2$ film has been used as the absorber. In order to further improve the efficiency we thought to prepare the $CuInSe_2$ film by the subsequent deposition of two $CuInSe_2$ layers with different stoichiometries as it is done when high efficiency $CuInSe_2$ /CdS solar cells are prepared by multi-source evaporation [2,3,4,5]. In this case, the bottom layer should exhibit a stoichiometry in which the Cu/In ratio is more than or close to 1, while in the top layer the Cu/In ratio should be less than 1. The reason why such a double-layer is more suitable than a single-layer for the fabrication of high efficiency solar cells is not clearly understood. It seems that the deposition in sequence of a Cu-rich and an In-rich film one on each other favours the formation of a single phase $CuInSe_2$ film with the chalcopyrite structure. However, when the films are grown by R.F.Sputtering, it is not exactly established what must be both the composition of bottom and top layers and their relative thicknesses. Here we describe the first experiments that we have done in order to prepare a high efficiency $CuInSe_2$ /CdS solar cell by using as the absorber a double-layer $CuInSe_2$ film.

2 Experimental System and Device Preparation

For the deposition of both $CuInSe_2$ and CdS thin films we used the R.F. Sputtering system Z 400 supplied by Leybold-Heraeus. This system is provided with three magnetron targets, a rotatable substrate-holder/heating station, the temperature of which can be controlled up to 650°C and an Inficon XTM thickness monitor and deposition rate meter. Mo, CdS and $CuInSe_2$ targets were mounted in the same sputtering system so that a complete cell could be prepared without interrupting the vacuum.

As substrates, 1 inch-square 7059 Corning Glass, covered by a 4 μm thick crystalline Al film, are used. An 0.2–1 μm thick layer of Mo is deposited on top of the Al film. Then, without interrupting the vacuum, a single or double-layer $CuInSe_2$ film is deposited on Mo. The cell is completed by an 0.6 μm thick CdS film sputtered in an $Ar+H_2$ atmosphere [6]. As transparent high conducting top layers both an In-doped CdS film sputtered from a 99% CdS + 1% In_2Se_3 target supplied by Cerac or a sputtered ITO thin film have been successfully used. The outline of the cell is shown in fig.1

Excluding the Al layer, which is deposited by Electron-Gun, all the films that compose the cell are deposited by R.F. sputtering.

3 $CuInSe_2$ Target Preparation

The $CuInSe_2$ targets have been prepared with a tecnique entirely developed in our laboratory [6]. Cu, In and Se elements are put together in a quartz crucible having a flat bottom whose diameter (3 inches) is equal to that of the target holder. Then the quartz

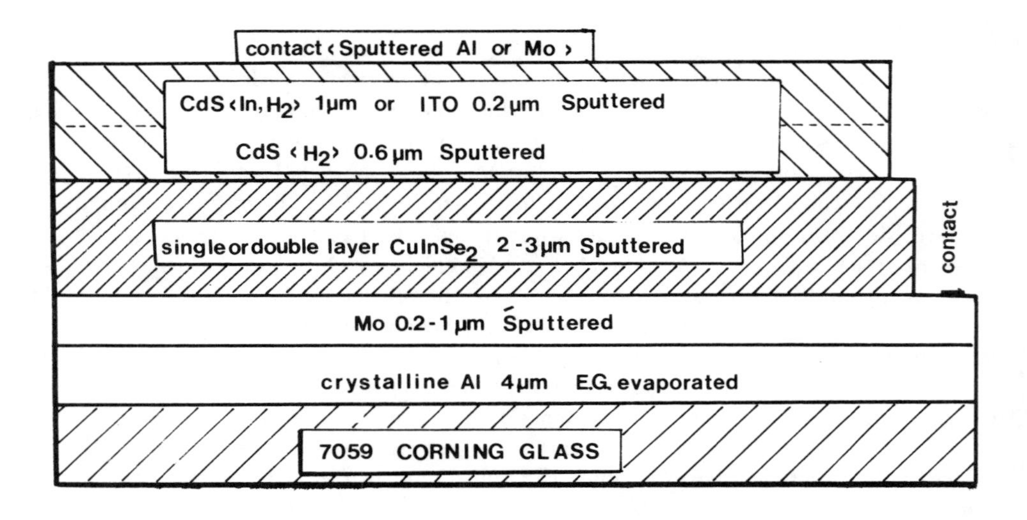

Figure 1: Outline of the $CuInSe_2$ /CdS solar cell.

crucible is mounted inside a high pressure furnace. In order to avoid the evaporation of the elements, the liquid encapsulation tecnique with B_2O_3 as an encapsulant has been used. The reaction of the elements and the melting were carried out under a nitrogen pressure of 50 Bar. In this way, compact, homogeneous and uniform policrystalline $CuInSe_2$ targets with almost any composition could be prepared. For the deposition of the bottom Cu-rich $CuInSe_2$ layer four targets with a Cu/In ratio of 1.2, 1, 0.98 and 0.96 respectively have been checked, while for the top In-rich layer two targets with a Cu/In ratio of 0.8 and 0.66 have been used.

4 Experimental Results

First of all, we observed that the $CuInSe_2$ films generally contain a Cu amount larger than that contained in the target. This could be due to a different scattering of sputtered Cu and In atoms from the argon gas. For this reason, Cu-rich targets seems to be not suitable for the preparation of bottom $CuInSe_2$ layers. In fact Cu-segregations are visible at the optical microscope in films prepared by targets with a Cu/In ratio of 1.2 or 1. Even though the Cu-segregations are not visible at the optical microscope, they come out when an In-rich layer is deposited on top of a Cu-rich layer. So far, bottom layers prepared by using targets with a Cu/In ratio of 0.98 or 0.96 have been demonstrated to be more suitable if deposited at high substrate temperatures (350–450°C). Probably targets with less Cu-content could give better results. In any case, an improvement in a double-layer in respect to a single-layer $CuInSe_2$ thin film solar cell both in the photovoltage from 250 to 300 mV and in the photocurrent from 30 to 35 mA has been observed by depositing an 0.5 μm thick layer from a target with a Cu/In ratio of 0.66 at 500°C substrate temperature on top of a 2 μm thick layer obtained from a target with a Cu/In ratio of 0.98 at 400°C substrate temperature. From these experiments, we have found out that three parameters are relevant for the preparation of a high quality double-layer $CuInSe_2$ film: the target composition, which determines the Cu/In ratio in the film, the substrate temperature which can vary the Se-content in the film and the relative thickness of the two layers which determines the final stoichiometry and the electro-optical properties of the composed layer. Experiments are in progress in order to establish what are to be these three parameters for both layers for the achievement of a high efficiency $CuInSe_2$/CdS solar cell.

References

[1] N. Romeo, A. Bosio, V. Canevari, 7th European Photovoltaic Solar Energy Conference, Sevilla, Spain, D.Reidel Publish. Company, P.O.Box 17, 3300 AA Dordrecht, Holland, 1986.

[2] R.A.Mickelsen, W.S. Chen, Y.R. Hsiao and V.E. Lowe, IEEE Trans. Electron Devices, 31 (1984), 542.

[3] R.Noufi, R.J. Matson, R.C. Powel and C. Herrington, Solar Cells, 16 (1986), 479.

[4] R.W. Birkmire, L.C. Dinetta, P.G. Lasswell, J.D. Meakin and J.E. Phillips, Solar Cells, 16 (1986), 419.

[5] R.R. Potter, Solar Cells, 16 (1986), 521.

[6] N. Romeo, V. Canevari, G. Sberveglieri, A. Bosio, and L. Zanotti, Solar Cells, 16 (1986), 155.

AN ELECTROCHEMICAL STUDY OF CHALCOPYRITE SEMICONDUCTORS

Contract Number : EN3 S/C2/132/F
Duration : 36 months
Budget : CEC Contribution (marginal costs) : FF545000
Head of Project : Dr J. VEDEL, CNRS
Contractor : Ecole Nationale Supérieure de Chimie de Paris
Address : Laboratoire d'Electrochimie Analytique et Appliquée
ENSCP
11, rue Pierre et Marie Curie 75005 PARIS France.

Summary

Electrochemical techniques are used to analyze the composition of Copper Indium Diselenide (CIDS) or Copper Gallium Diselenide (CGDS) and their semiconductor properties (spectral response, bandgap, doping level). The samples were prepared by vacuum coevaporation on Mo covered glass sheets at the Institut für Physikalische Elektronik (Stuttgart). Chemical analysis was performed on CIDS by fractional anodic oxidation followed by pulse polarographic analysis of the dissolved ions. It was found that the ratio Cu/In can be obtained, and in case of an excess of Cuprous Selenide that this material can be selectively oxidized and analyzed. Precise analysis needs further improvements (change of substrate, study of the influence of diffusion). It was also shown that optical and electronical properties of CGDS (Cu/Ga varying from 0.5 to 2) can be obtained by use of electrochemical techniques without alteration of the materials. All samples are p type, for any Cu/Ga ratio, the bandgap varies with Cu/Ga, together with the doping level. A first attempt of classificcation is proposed, in relation with chemical composition.

INTRODUCTION.

The aim of the research work is to establish a correlation between the composition of $CuMSe_x$ materials (M = Ga, In, Ga + In) and their electrical and optical properties, in relation with their use in photovoltaic devices. Two directions are simultaneously investigated : (*i*) the determination of the composition, and its controlled modification ; (*ii*) the characterization of some semiconductor properties of the layers. Both use electrochemical techniques for they allow, by a convenient choice of the electrolyte, either the controlled decomposition of the solid (for analysis), or the realization of a non destructive contact, sometimes called a liquid diode (1).

In the previous report, the basis of compositional analysis was presented, using $CuInSe_2$ layers sprayed on ITO, and also photoelectrochemical characterization of bulk $Cu\ Ga_x\ In_{1-x}\ Se_2$, showing the suitable properties of the selected electrolytes. In this work, we present results obtained with layers prepared in Stuttgart (IPE) by the coevaporation technique, on Mo substrates.

2 - EXPERIMENTAL.

1 - Composition analysis is performed by controlled anodic oxidation in a solution of MERCK Suprapur 0.1 M Chlorhydric Acid + 0.1 M MERCK Suprapur Potassium Chloride. A sample electrode is obtained by taking an electrical contact on the conductive substrate with conductive epoxy, and by keeping free an area of about 0.2 cm^2 by means of an isolating coating. The electrode potential is imposed using a three electrode Tacussel potentiostat (reference electrode Ag/AgCl/KCl 1M) and the current crossing the cell is integrated. It is either fixed at a constant value, or slowly driven towards positive values using a ramp generator. Dissolved species (Cu^{2+} and In^{3+} ions) are analyzed *in situ* by pulse polarographic analysis.

2 - The non destructive photoelectrochemical characterizations were performed in a 1M HCl solution containing the V(II)/V(III) redox couple : under oxidizing conditions, V(II) is oxidized instead of the SC, which is thus protected. The electrode is made as before. The electrochemical set up allows to record current/voltage curves in darkness or under illumination, capacitance/voltage curves and spectral responses.

RESULTS

1 - Détermination of the composition.

Analysis of Cu^{2+} and In^{3+} by pulse polarography (P.P.). PP uses a dropping mercury electrode. The polarogram of an electroactive species shows a peak with the position specific of the species and the height proportional to its concentration. Figure 1 curve 1 presents a pulse polarogram obtained after oxidation of a Mo substrate. It shows that Mo is dissolved and disturbs the analysis of copper.Curve 2 was obtained after oxidation of a CIDS layer. It shows that dissolved copper and indium are analyzed and the disturbing action of dissolved Mo.

Controlled partial oxidation. In figure 2 are shown various current/voltage curves corresponding to various samples deposited on Mo by coevaporation. Curve 1 was obtained by oxidation of bare Mo. It is oxidized, as confirmed by P.P. analysis. Curve 2 was obtained with a CIDS sample with an excess of cuprous selenide. It presents a peak at + 0.25 V which was shown to correspond to copper dissolution and was attributed to the reaction :

$$Cu_2Se - 4e \rightarrow 2\,Cu^{2+} + Se \qquad (1)$$

There is no apparition of the Mo peak, showing that the substrate is well protected by the layer. Curve 3 was obtained with a sample with an excess of indium.The first peak disappeared. Again no Mo peak appears,but Mo is observed in the solution at the end of the observed peak, together with In and Cu.The second peak corresponds to reaction :

$$CuInSe_2 - 5e \rightarrow Cu^{2+} + In^{3+} + 2Se \qquad (2)$$

but the presence of Mo disturbs the precise determination of Cu.

In case of a copper excess, oxidation at + 0.25 V allows the elimination of the cuprous selenide phase and its determination and then, at 0.6 V, the determination of the Cu/In ratio in the purified solid. In case of indium selenide excess, no selective dissolution of the excess compound was observed.

Precise and quantitative results cannot be given yet : the dissolution of Mo is a drawback for Cu analysis (causing also a peeling of the layer if one attemps to oxidize the remaining Se) and, in case of cuprous selenide excess, there is a problem to determine the end of its oxidation probably related to copper diffusion in the solid phase. These questions will be considered in the next step of our work.

Finally, it should be noted that CGDS gives similar current/voltage curve than CIDS. However, we are not able to analyse the dissolved Ga(III) by P.P. yet.

2 - Photoelectrochemical studies of CGDS layers.

The effect of the Cu/Ga ratio was studied for it has been observed that the photoelectrochemical studies were easier for these compounds than for CIDS.The compositions used here were obtained from EDS determinations at IPE. They were characterized by two quantities,the deviation from stoichiometry, Δs, and the deviation from molecularity, Δm. Δs is a measure of the excess or defect of Se, as Δm is related to the ratio of cuprous selenide and indium selenide. If CIDS is written as $Cu_aIn_bSe_c$, $\Delta m=(a/b)-1$ and $\Delta s=(2c-3b-a)/2b$.

Current voltage curves. Figure 3 presents several I-V curves obtained for various values of Δm (d = dark, i = illumination). All the samples are of the p-type (easy dark oxidation and cathodic photocurrent), in agreement with thermoelectric power measurements (2). The dark curves shows that the increase of the Cu or Ga content results in decrease of the shunt resistance, more pronounced on the $\Delta m < 0$ (gallium selenide rich) side.The higher shunt resistances were obtained for samples with Δm close to -0.15. A crossover between i and d curves is observed indicating some photoconductive effet in the layers as for CdS-Cu_2S cells.

*Spectral responses and quantum efficiencies.*The variation of QE with the wavelength for the various studied samples is given in figure 4. According to their shapes, they are classified into three families. The curves were characterized by their gap and the integral of QE from 500 to 900 nm. The results are given in figure 5, where it is seen that , with a few exceptions (K97,K12, K102), the order of magnitude of these values respects the classification by shapes.

The changes in observed properties may be tentatively related to the phase distribution. No data was published on CGDS phase diagram. By similarity with CIDS (3), a solid solution may be supposed to extend towards the gallium selenide rich side ($\Delta m < 0$) and no solid solution towards cuprous selenide side. The shift of the gap towards lower values when Δm becomes negative is attributed to the increase of transitions involving band tails and due to a decrease in the crystalline quality : it is known that samples with an excess of cuprous selenide present a better crystallinity than others (4). This is also in agreement with the gap value obtained with a bulk CGDS sample prepared by CVD at the University of Parma (Eg = 1,66 eV). Explanations for the various shapes of QE curves are under study : the shape of type I may be attributed either to a dead layer or to a nonohmic back contact. The shape of type III is attributed to the presence of two phases which also explains the low value of its maximum, and the shape of type II to a monophase system. Acceptor concentrations were also obtained from capacitance voltage curves (figure 6), but they are difficult to interpret without knowledge of the phase status of each specimen.

CONCLUSION.

The aim of the work is (1) to determine the sample composition, (2) to characterize the properties of the solids, and (3) to correlate them with composition.

The following points can be summarized :

1. The Cu/In ratio may be obtained, by fractioned anodic oxidation and analysis of the solutes by pulse polarography.
2. In case of cuprous selenide excess, it is possible to oxidize it selectively, and to determine its quantity.
3. In case of indium selenide excess, no selective oxidation has been found yet.
4. The oxidation of substrate (Mo), and the difficulties to determine the end of reactions disturbs the obtaining of precise results.
5. Photoelectrochemical characterization of thin layers samples has been made possible.
6. For CGDS, all samples were of p type for any value of the Cu/Ga ratio.
7. The bandgap and the quantum efficiency change with the sample composition.
8. These variations are tentatively correlated with the presence of various phases in the solids.

In the next future our program will be :

1. Improvement of sample analysis and extension to CGDS.
2. Determination of quantities suitable to correctly define the solid composition.
3. Interpretation of the observed electronic properties.

REFERENCES.

(1) E. Fabre, M. Mautref, Acta Electronica, 18 (4), 331 (1975).

(2) W. Arndt,H. Dittrich,F. Pfisterer,H.W Schock, Proc. 6th Photov. Solar Energy Conf., London, Apr. 87,(D. Reidel, Dordrecht,86),p. 260.

(3) L.S. Palatnik,E.J. Rogacheva,Sov. Phys. Dokl.,12,(1967),503.
C. Rincon, C. Bellabarba,J. Gonzalez, G. Sanchez Perez, Solar Cells, 16,(1986),335.

(4) H.W. Schock, B. Dimmler,H. Dittrich, J. Kimmerle, R. Menner, Proc. 7th Photov. Solar Energy Conf., Sevilla, Oct. 86,(D. Reidel, Dordrecht,86),pp. 446-469.

LEGENDS OF THE FIGURES.

Figure 1. Pulse polarograms : curve 1 = dissolved Mo,; curve 2 = Cu^{2+} + In^{3+} + Dissolved Mo.

Figure 2. Anodic oxidation of CIDS: curve 1 = Mo substrate ; curve 2 = excess of copper, curve 3 : excess of indium.

Figure 3. Current-voltage curves : d : dark,i : illumination. The number of crosses is proportional to illumination.

Figure 4. Variation of QE with the wavelength.

Figure 5. Schematic representation of Eg and QE variations with the composition.
---- QE : arbitrary units ; - - - - $E_g = 1.58 + \Delta E_g$

Figure 6. Schematic representation of acceptor concentration with the composition.

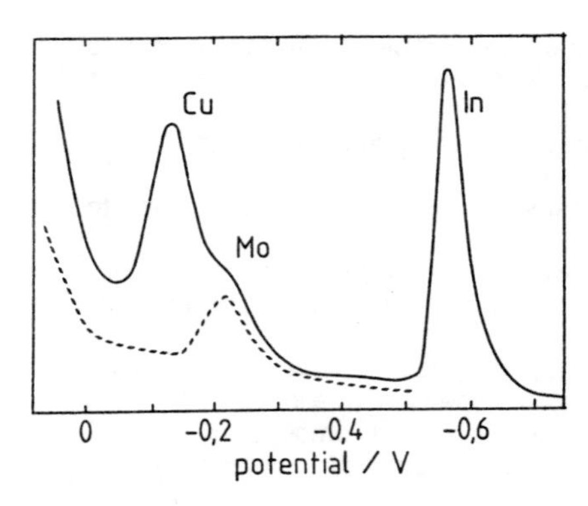

Figure 1.

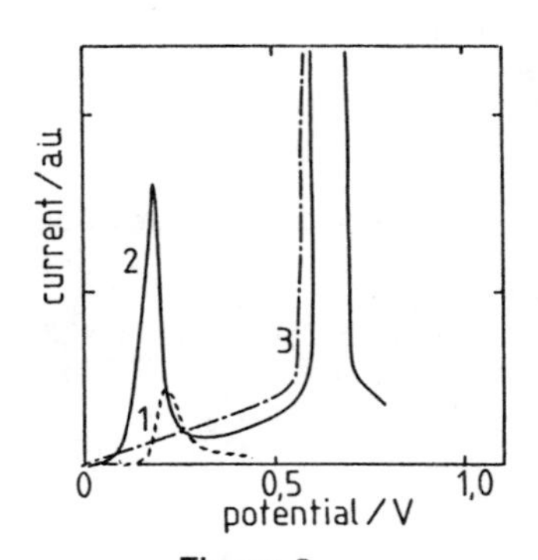

Figure 2.

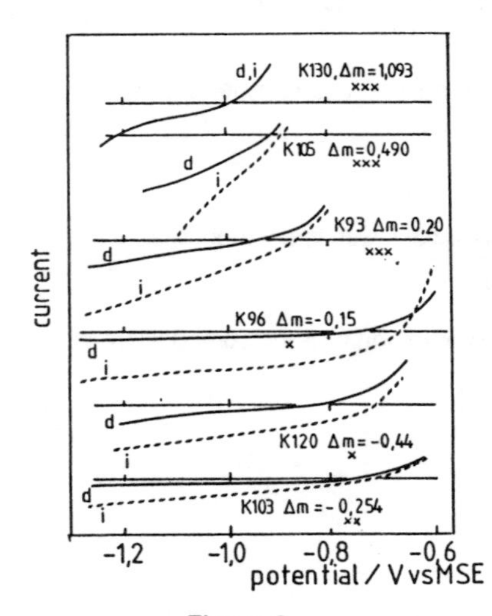

Figure 3.

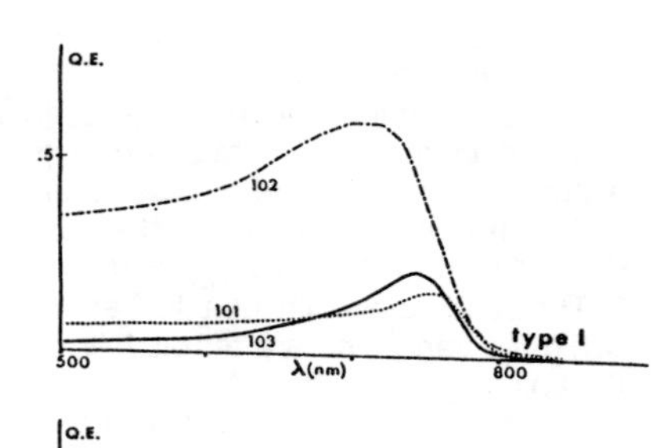

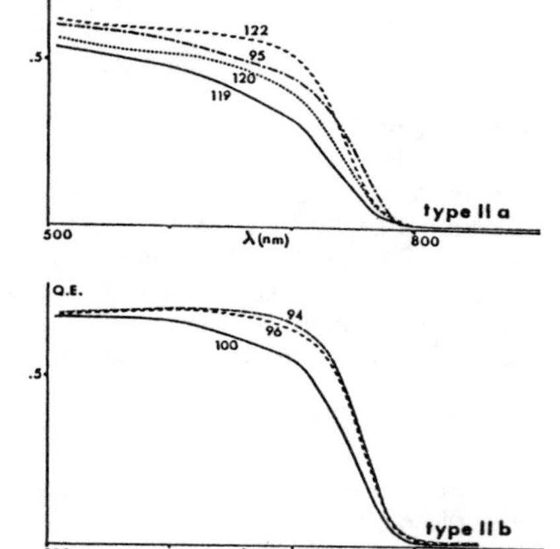

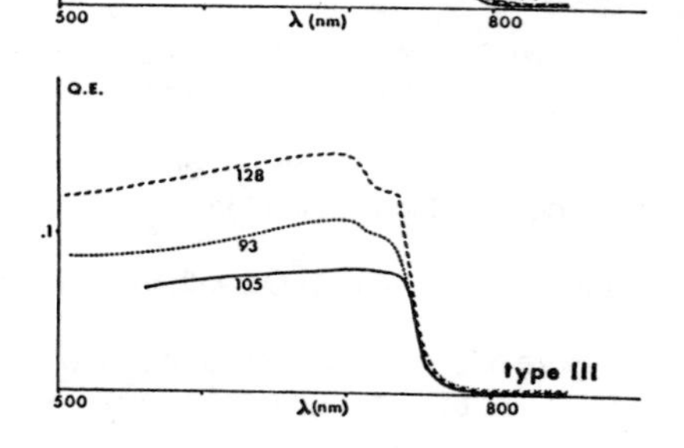

Figure 4.

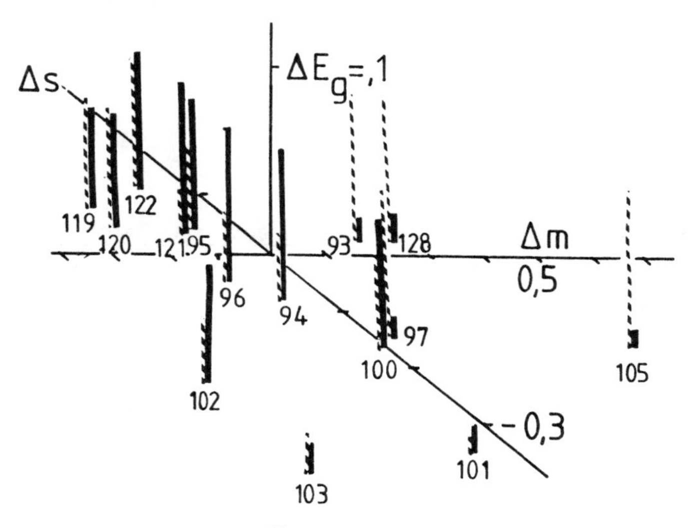

Figure 5.

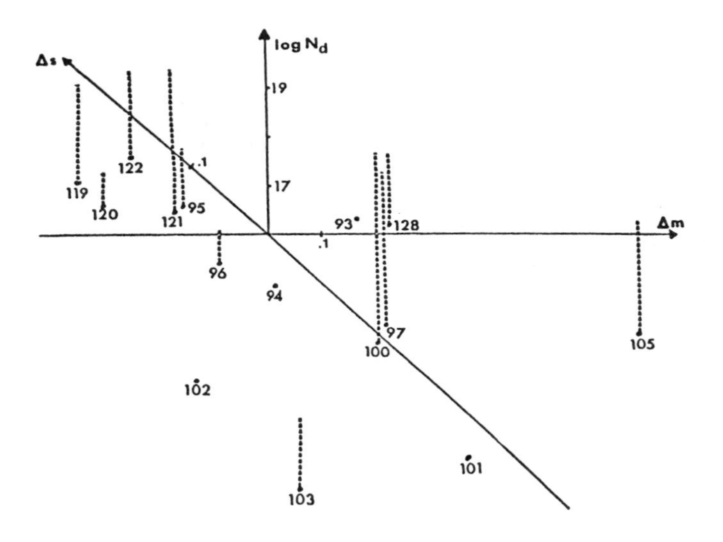

Figure 6.

THIN FILMS OF COPPER INDIUM DISELENIDE FOR PHOTOVOLTAIC DEVICES

Contract Number	:	EN3S-0071-UK	
Duration	:	36 months	1 Apr 1986 - 31 Mar 1989
Total Budget	:	£106,000	CEC contribution : 100%
Head of Project	:	Professor R. Hill, Newcastle Polytechnic	
Contractor	:	Newcastle upon Tyne Polytechnic	
Address	:	Ellison Place, Newcastle upon Tyne, NE1 8ST, UK	

Summary

In order to be competitive for large area modules, the good efficiencies obtained for small area copper indium diselenide cells must be achieved on large area susbtrates. The work discussed here involves studies on techniques for achieving reproducible, high quality large area films of copper indium diselenide carried out at Newcastle Polytechnic. The studies include the investigation of control techniques for deposition of copper indium diselenide by thermal evaporation. Improvements in film thickness and optical transmission uniformity have been obtained using plane strip sources and a feed forward control mechanism. In addition, the laser processing of elemental layer structures has led to the formation of copper indium diselenide and the laser irradiation of co-evaporated copper indium diselenide layers leads to modification of the electrical and optical properties. The extent and nature of the modification depends on the as-deposited stoichiometry of the film. This work forms part of a research programme, being undertaken being undertaken in collaboration with other European laboratories. The laser processing studies to date have provided information which will be used in further work, including the laser processing of cells produced at the collaborating institutions.

1. Introduction

Thin film solar cells based on copper indium diselenide have shown good efficiencies (around 12% AM1.5) for small area devices of approximately 1 cm^2. To be competitive for power modules, this efficiency must be achieved on large area substrates. However, the reproducibility of co-evaporated coatings is often poor and there is tendency for the coatings to exhibit a large variation in properties over large area depositions. It is possible for the coating to change conductivity type across the sample, because of the configuration of the sources and changes in evaporation rates of the co-evaporated elements. Since the ratio of the elements in the deposited film defines the electrical, optical and structural properties, strict control of the evaporation is of paramount importance. The work at Newcastle Polytechnic includes studies of control techniques for large area depositions by thermal evaporation and laser processing of elemental layer structures, copper indium diselenide films and copper sulphide/copper indium diselenide cells.

2. Progress of Work

2.1 Large Area Depositions

To date, Electron Impact Emission Spectroscopy (EIES) has been the most commonly used feedback control system in order to produce coevaporated copper indium diselenide of the required stoichiometry. However, this system has limitations which make it unsuitable for critical large area depositions. It is hoped that another feedback control system, using a crossed beam ion source, will be adequate and this type of system is in the process of being designed and built. The capability of production and critical control of coevaporated copper indium diselenide will then be evaluated.

For large area depositions, a feed forward control system is often the most convenient and this is under study. Ring sources are capable of producing uniform deposition but these tend to be very expensive Radak type furnaces made from high purity materials and using high precision manufacturing technologies. A low cost, low technology feed forward system using a long thin plane strip source allows uniform deposition over large areas. In practice, an open boat about 10 cm long and 1.4 cm wide, with a twisted wire lying in it, allows copper or indium to wet the wire, giving good even distribution of the evaporant. A tungsten boat is used for the evaporation of copper and a molybdenum boat for indium. The selenium source used is a high capacity Knudsen cell, made in-house, originally from graphite. This exhibited slight cracking due to long term heat cycling, allowing selenium to permeate the cell, and so a replacement source has been fabricated from stainless steel.

Results to date indicate an optimum source-substrate distance of 30cm, for a separation of 3 cm between the copper and indium strip sources. This gave the most uniform distribution over an unheated substrate area of about 75 cm^2. Evaporation rates are controlled by supplying constant power to the evaporation boats, together with a conventional quartz crystal monitor to indicate coating thickness.

Initially, separate depositions of copper, indium and selenium were carried out, showing thickness variations within 5% across the substrate area for a total coating thickness of about 500 nm. Coevaporations of copper/indium, have also been carried out, giving variations within 10% for a coating thickness of approximately 200 nm. Three element co-evaporation was also carried out and, again, a variation within 10% for a coating thickness of 1000 nm was found. It should be noted that the substrate holder/heater needs good temperature control (within +/- 0.5^oC) over the whole of its surface if high quality uniform coatings are to be produced. These results are a great improvement when compared to those from samples produced in our coevaporation system using the EIES feedback system. This system gives a variation of 5% for an area of 4 cm^2.

At present, several strip sources of different designs, including some having alumina coatings, are being evaluated. The alumina coating is of particular importance, since it restricts the tendency of the molten copper to migrate due to surface tension effects. This can have the effect of producing a non-uniform spread, affecting the distribution of the evaporable material and, hence, the reproducibility of the results. This basic investigation has allowed the area of high quality material in the coating to be increased greatly.

Preliminary optical and electrical measurements also exhibit an improvement over the EIES controlled system. Changes in optical transmission of less than 5% over an area of 75 cm^2 can be obtained. It is expected that further improvements will result from modification of this process, together with the development of a more sensitive feedback control system.

2.2 Laser Processing

The production of copper indium diselenide films over large areas by laser irradiation of elemental layer structures is being investigated. In the case of laser induced synthesis of the compound from elemental layers, the stoichiometry depends on the composition of the initial structure (assuming all the material reacts). The high vacuum evaporation of single element layers is more straightforward than coevaporation of the compound and can be controlled with a high degree of accuracy and repeatability. Therefore, the production of multilayer structures of the required composition over large areas should be more easily accomplished than with the coevaporation techniques already discussed.

The compound, copper indium diselenide, has been produced by laser synthesis (1). In the work discussed here, structures of between three and nine layers were irradiated with an argon ion laser. The layer thicknesses were chosen so as to give the correct stoichiometry. The laser operates on all lines (4579A–5745A), producing a maximum output of 4.4 Watts. The sample mount is a rotating disk, which allows the sample to be scanned at a range of speeds and, hence, total energy inputs. The layer structures are always terminated with copper, since indium terminated structures showed a tendency to blister.

The best results to date were obtained with a nine layer structure, consisting of Cu/Se/In/Cu/Se/In/Cu/Se/In. A slow scan speed of about 0.05 cm/sec for a spot size of 1.25 mm diameter and laser output power of about 3 Watts gave the most pronounced absorption edge in the correct region for copper indium diselenide. In initial experiments, the structure required several scans in order to increase the transmission in the infra red region and to develop the absorption edge at 1250 nm. By using a very slow scan speed, more energy per scan can be coupled into the sample and, hence, the number of scans required can be greatly reduced. However, an attempt to use lenses for beam concentration, in order to get the same energy input at a faster scan speed, resulted in partial vapourisation of the film. Although the first experiments were carried out on coatings on 7059 glass substrates, more recent work has involved the use of quartz substrates. These are capable of withstanding higher energy dissipations without mechanical failure. The laser processed films exhibit a second absorption edge at around 500 nm, attributable to the formation of another compound, and the transmission at wavelengths above the 1250 nm bandedge remains low (approximately 10%). Figure 1 shows transmission curves for laser processed nine layer structures, where the layer thicknesses have been varied to make the material either rich or deficient in copper or indium. It can be seen that the shape of the transmission curve is strongly dependent on the initial ratios of the elemental layer thicknesses. The work in progress includes the study of variations in process parameters, layer thicknesses an layer sequence, in order to increase transmission and inhibit the formation of compounds other than copper indium diselenide.

Another aspect of laser usage has been the annealing of coevaporated copper indium diselenide films. Changes in resistivity have been observed following exposure to the laser, with the direction of change dependent upon the initial stoichiometry of the film. Slightly copper rich films showed up to a four fold increase in resistivity, whilst slightly indium rich films showed a decrease in resistivity by factors of up to ten. This annealing process will now be extended to cadmium sulphide/copper indium diselenide cells produced in the other laboratories involved in this project. It may be possible to 'tailor' the electrical properties of the cell and improve the window/absorber interface, so improving cell efficiency.

3. Conclusion

The work undertaken at Newcastle Polytechnic in the last contract period has been a continuation of studies on control techniques for large area depositions of copper indium diselenide by thermal evaporation, together with studies on laser processing in order to form copper indium diselenide or modify the properties of evaporated material. The development of strip sources for the evaporation of the metal components has led to a much improved uniformity of the film over a substrate area of 75 cm^2. A significant reduction in the variation across the substrate of film thickness and optical transmission when compared to samples produced using the EIES control system has been observed.

Copper indium diselenide has been produced by the laser processing of evaporated elemental layers, although there is a tendency for additional compounds to form during the laser processing step. It is believed that some material also remains unreacted. A comprehensive investigation into the influence of the initial layer structure in terms of layer compositions, thicknesses etc., and of the laser processing parameters is in progress. Several factors affect the percentage formation of copper indium diselenide with the most influential observed to date being the laser scan speed. The processing of evaporated copper indium diselenide has also been investigated, with changes in film resistivity observed, and this study will be extended to copper indium diselenide/cadmium sulphide cells during the next phase of the work.

Reference

1. M.J.Carter, I.I'Anson, A. Knowles, H.Oumous and R.Hill, Proc. of the 19th IEEE Photovoltaic Specialists Conference, New Orleans, May 1987, to be published.

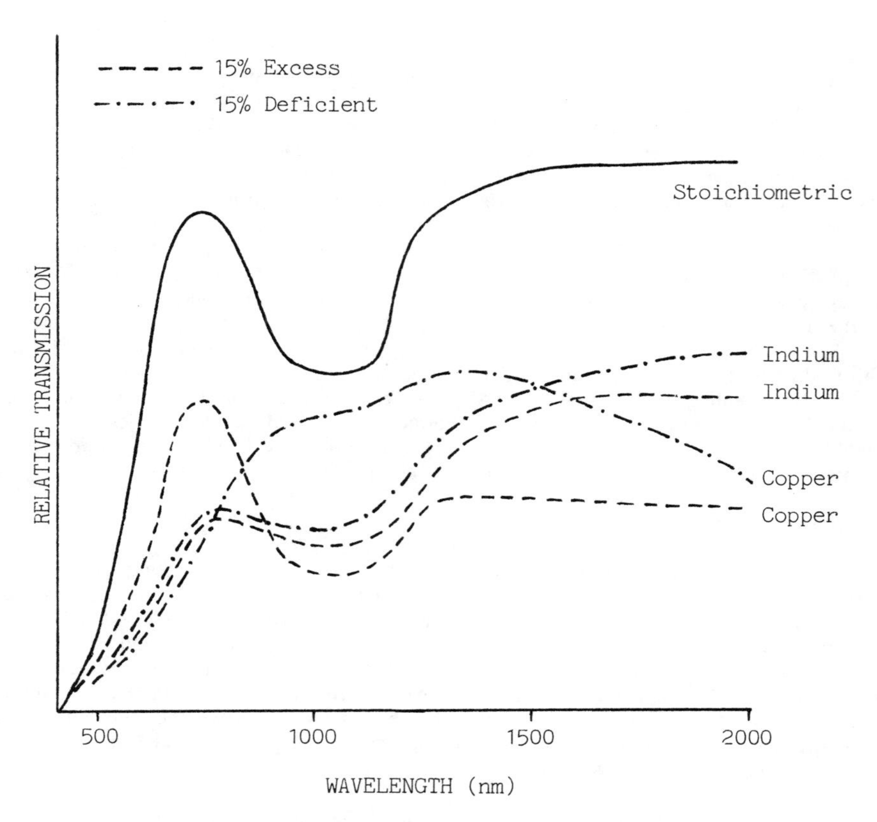

Figure 1 Transmission curves for laser processed nine layer structures of differing stoichiometry.

DEVELOPMENT OF CdSe THIN FILM SOLAR CELLS WITH REGARD TO THE REQUIREMENTS FOR THE USE IN TANDEM STRUCTURES

Contract Number : EN3S/0041

Duration : 48 months
1 July 1985 - 30 June 1989

Total Budget : DM 5,027,050.00
CEC Contribution: DM 1,320,000.00

Head of Project : Dr. H. Richter

Contractor : Battelle-Institut e.V.

Address : Battelle-Institut e.V.
Am Roemerhof 35
D-6000 Frankfurt 90

Summary

Polycrystalline n-CdSe MIS solar cells have been produced and investigated by I-U measurements and admittance spectroscopy. A high density of interface states ($\geq 10^{13}$ $cm^{-2}(ev)^{-1}$) has been estimated, possibly due to disorder phenomena at the n-CdSe surface.

INTRODUCTION

Thin film solar cells do have the potential of extreme cost reduction and therefore attract much more interest compared to the situation 5 to 10 years ago. The amorphous Si thin film solar cell is already in production /1/ or at least in a pilot-line operation /2/. Stability and yield, that means mean efficiency of cells produced in a production line, are still a problem and have to be considered in more detail.

With respect to II-VI compound thin film solar cells, some pilot-line productions are running /3/. One approach, the Cu_2S-CdS thin film solar cell, has been produced successfully in a pilot-line production /4/, but has been abandoned, finally, due to some reasons. Two of those are:

- the CdS layer still had to be of a thickness of 30 µm, at least about 20 µm: that is too thick,
- the mean efficiency achieved in a pilot-line production was 6%, the expected maximum value of the mean efficiency of this type of cell was estimated to be 8%: that was considered to be too low.

The CdSe thin film solar cell developed at Battelle /5, 6/ is a real thin film solar cell: the thickness of the cell (electrode plus semiconductors plus electrode) is below 5 μm (typically around 3 μm), efficiencies above 10% should be obtainable and the band gap of the absorber matches well to the requirements of a high band gap cell in a tandem solar cell system.

The highest efficiency obtained up to now with the CdSe thin film solar cell is still 7% /6/. The efficiency up to now is limited primarily by the low open circuit voltage achieved. According to the band gap of the CdSe of 1.7 eV, an open circuit voltage of up to 1 volt should be possible, but only values below 700 mV have been observed.

To overcome this decisive barrier, a fundamental understanding of the limiting factors as

- Fermi energy level pinning
- Density and energetic distribution of surface states
- Diffusion barrier heihgt
- Current transport mechanisms through the junction

is required.

This paper summarizes the results obtained up to now on the above mentioned issues.

EXPERIMENTAL RESULTS

The produced CdSe solar cells have been investigated by admittance spectroscopy and measurements of I vs. bias at different temperatures. In addition to a former paper /7/ the frequency range of the admittance spectroscopy has been extended from 50 kHz up to 1 MHz. Fig. 1 shows the large frequency dependence of the measured a.c. capacitance C_m of a typical CdSe cell (with antireflective coating). The simulataneously measured a.c. conductance G_m is depicted in Fig. 2 for various frequencies. The dramatic increase in conductance for bias voltages > 0.5 V restricts the admittance measurements to the depletion region. Fig. 3 shows the typical log (I) vs. bias dependence at different temperatures indicating strong tunneling processes (excess current at low bias voltages).

DISCUSSION

For the high-temperature (> 260 K) log(I) vs. bias curves, after a correction for series resistance /8, 9/, the data points fit a straight line except for the low bias region (tunneling). The results of this correction procedure are given in the table 1.

The attempt of series resistance correction for the low temperature (> 260 K) log(I) vs. bias curves undoubtly reveals a Schottky barrier height distribution of the CdSe MIS structure indicated by "N-shaped" log(I) vs. bias curves /8/. Effects of Schottky barrier height distribution are not considered in the analysis of the high-temperature curves. The variations of I_s and ϕ_B might be due to this effect.

TABLE 1 Results of I vs. bias analysis

Temperature [K]	Saturation current I_s [A/cm²]	Schottky barrier height, ϕ_B [V]	Ideality n	Series resistance R_o [Ω*cm²
325.7	1.9×10^{-9}	0.97	1.4	4.2
308.6	6.4×10^{-10}	0.94	1.4	4.6
296.0	4.1×10^{-10}	0.91	1.5	4.8
277.7	1.5×10^{-10}	0.88	1.6	5.0
259.1	8.3×10^{-11}	0.83	1.7	5.8

The admittance data are treated applying the simplified /10, 11/ equivalent circuit (Fig. 4) of Lehovec and Slobodskoy (1964) /12/. The following assumptions are made /7/:

1) The series resistance R_o of the CdSe solar cell is negligible.
2) The insulator capacitance $C_i \gg C_{sc} + C_{ss}$ because the ZnSe layer thickness is about 5 nm with C_{sc} being the depletion layer capacitance and C_{ss} the interface states capacitance.
3) The frequency dependence of C_{ss} follows the continuum model, i.e. for $\omega\tau \ll 1$ $C_{ss} = q^2 \cdot N_{ss}$. N_{ss} denotes the interface state density.

Therefore, a) $\omega\tau \ll 1 : C_m^{LF} = C_{sc} + C_{ss}$ and

b) $\omega\tau \gg 1 : C_m^{LF} = C_{sc}$

with τ denoting the interface state relaxation time.

To evaluate the interface state density from C_m, first the dependence of the surface potential of the sample from the applied bias V_B has to be determined. Applying the procedure of Kar and Dahlke /10/ this relation may be extracted from the C_m^{HF} data. The impurity concentration N_d acts as a parameter to fit the barrier height derived from the capacitance data to the barrier height derived from the analysis of the I vs. bias data.

The Fermi energy in the bulk CdSe locates ≈ 0.14 eV below the conduction band corresponding to a impurity concentration around 10^{16} cm^{-3}. The Fermi energy in the ZnSe arranges due to Cu acceptors 0.65 eV above the ZnSe valence band /13/. The Schottky barrier height ϕ_B = 0.9-0.95 eV and the build-in voltage V_{bi} = 0.8-0.85 V. In accordance with the surface potential vs. bias relation a strongly increasing interface state density ($N_{ss} \approx 10^{13}$ $cm^{-2}(eV)^{-1}$) around 0.6 eV below the CdSe conduction band is evaluated from the capacitance data.

The frequence dependence of the conduction well /17/ (Fig. 2) increasing from 1.7 eV at 0.1 kHz to ≈ 3 eV at 100 kHz suggests even surface state densities fairly above 10^{13} $cm^{-2}(eV)^{-1}$.

To summarize the analysis, the following assumptions are made:

1) A maximum of interface states correlated with chemical and structural disorder /14, 15/ exists in the energy range of $E_c > E > E_c - 0.6$ eV.
2) The dramatic increase of the interface state density around 0.6 eV below the CdSe conduction band pins the CdSe Fermi level and severely limits the open-circuit voltage of the CdSe cell.

CONCLUSION

From temperature dependent I-U measurements and admittance spectroscopy data a high interface state density of the order of 10^{13} $cm^{-2}(eV)^{-1}$ has been derived. The interface states are located in the energy range of $E_c > E > E_c - 0.6$ eV of the energy gap of the n-CdSe. These states are suggested to originate from disorder phenomena in the n-CdSe surface and to limit the open-circuit voltage of the solar cell. A drastic reduction of these surface states is expected from the improvement of the CdSe deposition process. Generally, for thin film solar cells the "... greatest gain is to come in the open-circuit voltage and via a reduction of surface states" /16/.

ACKNOWLEDGEMENT

The authors greatly acknowledge for technical assistance by K.H. Jaeger, U. Luke and E. Seipp. This work is sponsored by the Ministry of Research and Technology, Federal Republic of Germany, and the Commission of the European Communities, Brussels.

REFERENCES

/1/ For example: ARCO Solar Inc., Cronar Corp., USA; Sanyo Electric Corp., Japan
/2/ Winterling, G.; in Photovolt. Power Generation, Vol 1, D. Reidel Publishing Company, 1986, p. 15
/3/ ARCO Solar Inc. ($CuInSe_2$), AMETEK Inc. (CdTe), USA; Matsushita Battery Industry Corp. (CdS/CdTe), Japan
/4/ van Campe, H.; Hewig, G.H.; Hoffmann, W.; Huschka, H.; Schurich, B.; Wörner, J.: 6th E.C. Photovolt. Sol. Energy Conf., London, 1985, p. 778
/5/ Bonnet, D.: Proc. 2nd E.C. Photovolt. Solar Energy Conf., Berlin 1979, p. 387
/6/ Rickus, E.: 5th E.C. Photovolt. Solar Energy Conf., Athens 1983, p. 746
/7/ Richter, H.: Proc. Conf. Opt. Mat. Techn. f. Energy Efficiency & Sol. Energy Conversion VI, San Diego, 1987
/8/ Chekir, F.; Lu, G.N.; Barret, C.: Sol. State Electronics 29 (1986) 519
/9/ Sze, S.M.: "Physics of Semiconductor Devices", Wiley, New York, 1969
/10/ Kar, S.; Dahlke, W.E.: Sol. State Electronics 15 (1972) 221
/11/ Nicollian, E.H.; Goetzberger, A.: Bell Syst. techn. J. 46 (1967) 1055
/12/ Lehovec, K.; Slobodskoy, A.: Sol. State Electronics 7 (1964) 59
/13/ Kaldis, E. (ed.): Current Topics in Materials Science, Vol. 9, North Holland Publishing Comp., Amsterdam, 1982
/14/ Frese jr., K.W.: J. Appl. Phys. 53(3) (1982) 1571
/15/ Haak, R.; Tench, D.; Russak, M.: J. Electrochem. Soc., Nov. 1984, 2709
/16/ Wagner, S.: 7th E.C. Photovolt. Solar Energy Conf., Sevilla, 1986, p. 452
/17/ Shewchun, J.; Waxman, A.; Warfield, G.: Sol. State Electronics 10 (1967) 1165

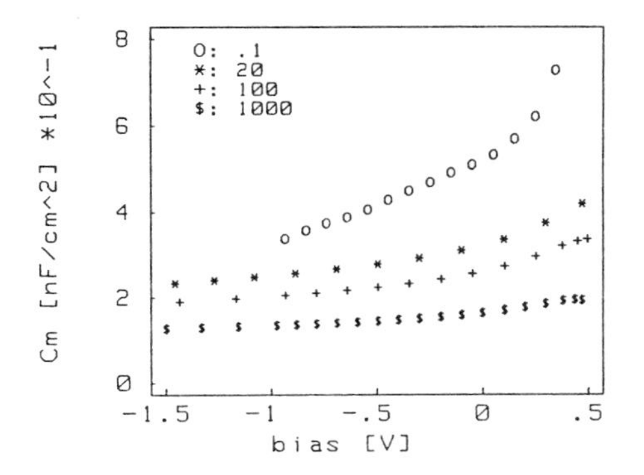

Fig. 1. A.c. capacitance of a typical CdSe MIS structure. Frequency as a parameter in kHz.

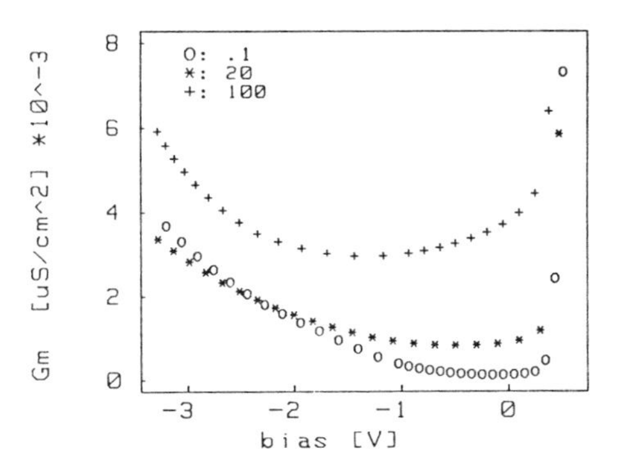

Fig. 2. A.c. conductance of a typical CdSe MIS structure. Frequency as a parameter in kHz.

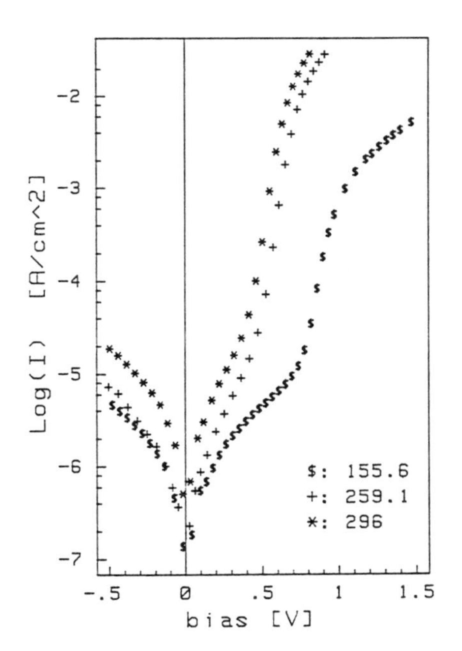

Fig. 3. I vs. bias dependence of a typical CdSe MIS structure. Temperature as a parameter in K.

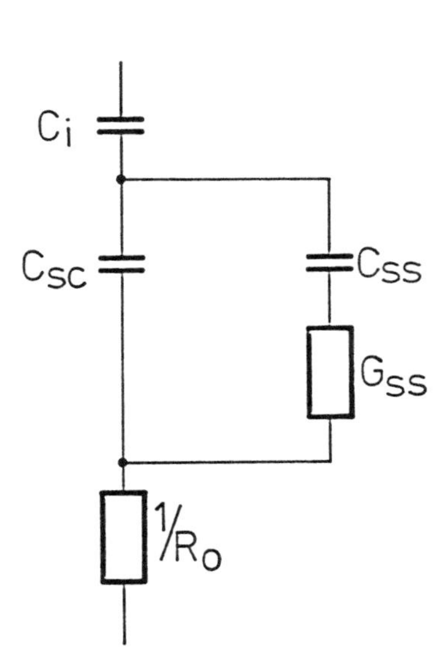

Fig. 4. Equivalent circuit of the CdSe MIS structure.

THE FABRICATION OF MEROCYANINE DYES FOR USE AS PHOTOVOLTAIC MATERIALS

Contract Number: EN35/0043

Duration: 24 months 1.1.86 - 31.12.87

Total Budget: £4500 CEC Contribution £4500

Head of Project: Dr. R. W. Buckley
Twyford Church of England High School

Contractor: Humberside County Council

Address: Twyford Church of England High School
Twyford Crescent
Acton
London
W3 9PP

SUMMARY

In the search for cheap, efficient and reliable solar cells, the potential for organic semiconductors is currently being assessed. The use of appropriate, molecularly engineered and chemically doped systems has led to the development of metal-insulator-organic semiconductor cells with efficiencies approaching 1% (1). This work seeks to develop a photovoltaic p-n junction between a variety of merocyanine dyes and cadmium sulphide. Optimum conditions have been identified for the spin-coating of merocyanine dyes on top of CdS and work on the fabrication of a number of dyes has been undertaken in order to optimise their photovoltaic potential.

1. Introduction

Two merocyanine dyes have been fabricated in the school laboratories, viz. Quinolylidene - Rhodanine (Q-Rh) and Benzothiazole - Rhodanine (BTh-Rh). The chemical structures of these compounds are shown in Figure 1.

In general terms a merocyanine dye consists of an electron donating group linked by a bridge to an electron accepting group (Figure 2). In a normal inorganic solar cell the carrier generation process involves the absorption of a photon of energy greater than the band-gap. This leads to a direct generation of an electron-hole pair. The transport of these carriers results in the photovoltaic current in these cells. In organic cells the photocurrent is somewhat different. The absorption of a photon creates a high energy excited state of the dye called an exciton. The dye then reacts with a halogen present in its layer to form a negatively charged halogen ion and a positively charged dye derivative which are the free carriers. Only those excitons that reach the merocyanine - CdS junction by diffusion dissociate into carriers. These then separate due to the field and are able to diffuse through the junction, thus constituting a current with the majority carriers originating from the dye layer. This mechanism is illustrated in Figure 3.

2. Experimental details

The merocyanine dyes were prepared using the method described by Brooker et al (2). Equal amounts by weight of 4-Phenylmercapto-pyridine and methyl p-toluenesulphonate were heated together for 90 minutes at 100°C. The viscous mass of methyl-4-phenylmercapto-pyridine p-toluenesulphonate (QSI) produced was washed with ether and used without further purification. 3.9 grm of QSI were then dissolved in 30ml of ethanol along with 1.6 grm of 3-phenyl-5 (4H) - isoxazolone. The mixture was then refluxed for 30 minutes. The addition of a further 1.6 grm of 3 Ethylrhodanine and further reflux produced the Q-Rh format of the dye.

3. Experimental results

The dye was dissolved in dichloromethane and spun coated onto CdS layers. The optimum conditions for this process have been reported previously (3). In sum they are one drop of boiling dye solution (in the ratio 10mg dye to 1ml of solvent) applied from a height of 2cm centrally onto a disc at 60°C spinning at 3,000 rpm. Figures 4, 5 and 6 illustrate the experimental observations to support this conclusion.

4. Future work

The dyes have been fabricated and spin coated by sixth form students at Twyford Church of England High School. Another school in the

North of England is involved with the work and will evaluate the photovoltaic properties of the resultant cells. We are now looking to develop other merocyanine dyes and study the depletion layer of our cells using capacitance and photocurrent measurements.

5. References

1. G. A. Chamberlain J. App. Phys. 53 6262 1982
2. L. G. Brooker et al. J. Amer. Chem. Soc. 73 5332 1951
3. R. W. Buckley and L. Murray. Proc. 7th E-C. P-V. Conference. Seville. P.651

Fig.1. Dye Structures

Fig.2. Donor and Acceptor Groups

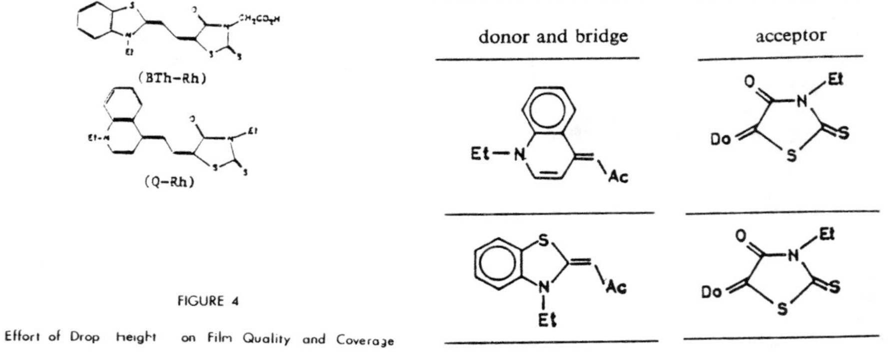

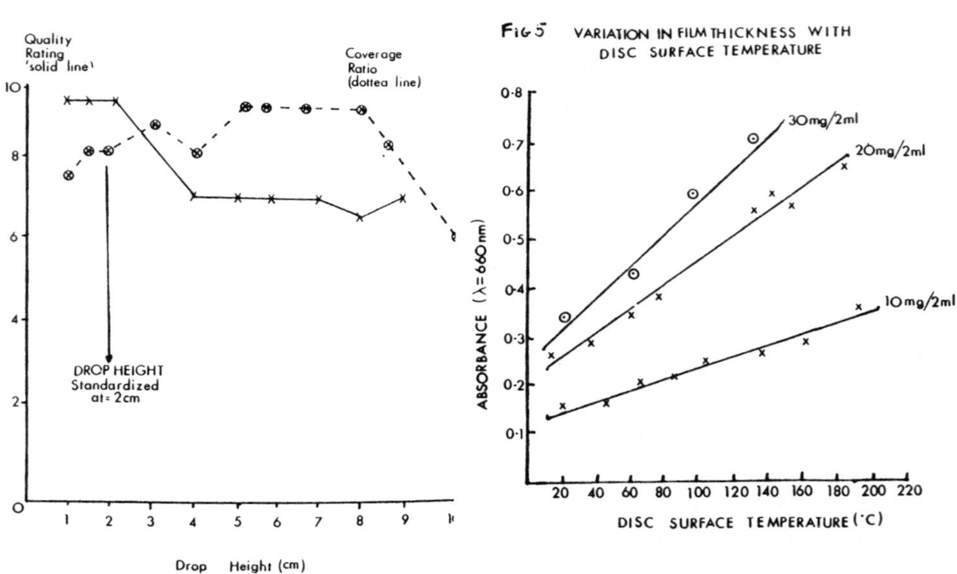

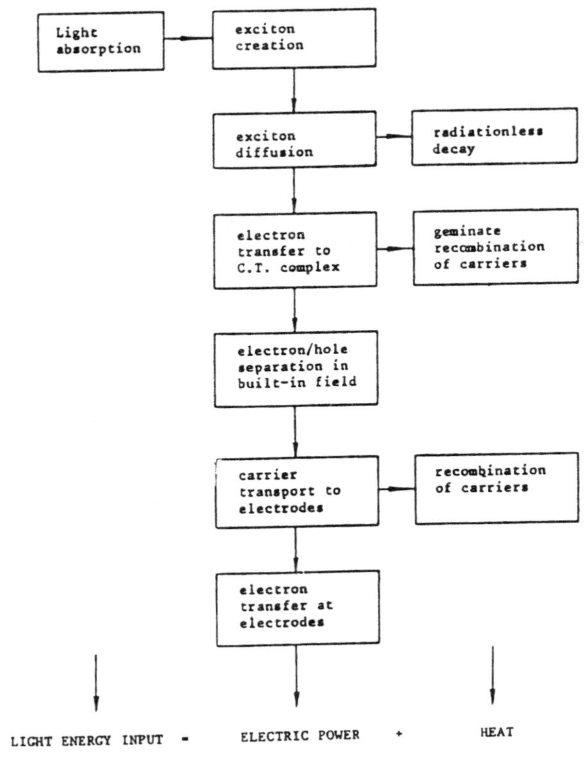

Fig. 3. Mechanism of photovoltaic behaviour.

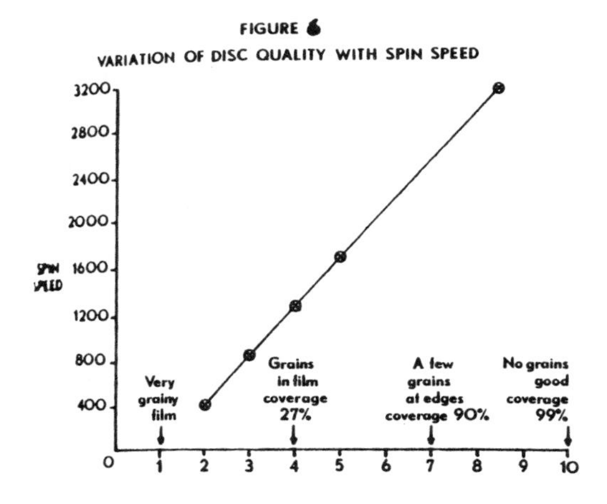

OPTIMIZATION OF HIGH EFFICIENCY MULTILAYER SOLAR CELLS BASED ON III-V COMPOUNDS.

Authors : M.H.J.M. DE CROON, L.J. GILING

Contract number : EN 3S/C2/135/NL

Duration : 36 months 1 March 1986 - 28 February 1989

Total budget : Hfl 1.143.717
CEC contribution Hfl 1.143.717

Head of project : Prof. dr. L.J. Giling, Dep. of Solid State Physics III

Contractor : Faculty of Science, University of Nijmegen

Address : Faculty of Science, University of Nijmegen, Toernooiveld, 6525 ED Nijmegen, The Netherlands

Co-contractors : IMEC, Leuven, Belgium
Maspec, Parma, Italy

Half-year report, September 1987

1. Summary

In this report we will discuss the progress made in the MOCVD growth - and doping behaviour of GaAs and AlGaAs epilayers and of MOCVD grown solar cells. We also will discuss the possibility of making cells by an ion implantation technique. The type of cell made with this method will be compared to specimens made via other techniques in the near future. Next a description will be given of PL studies on interfaces and epilayers and of EBIC measurements on defects in GaAs substrates. Further we will discuss the feasibility of hardening GaAs substrates by In alloying without the generation of misfits in epilayers grown on top of these. The final section is devoted to a short discussion of an investigation on return flows in horizontal MOCVD reactors, a study we are performing to reduce concentration gradients at interfaces of multilayer structures.
In the very near future we are going to extend our investigations to MBE and LPE grown samples. These samples will be grown at IMEC, Belgium and at CISE, Italy, respectively.

2. Preparation of GaAs and AlGaAs epilayers and of solar cells

2.1 Si-doping of MOCVD-grown GaAs.

Under MOCVD growth conditions, doping with Si will give n-type GaAs. In our experiments various growth parameters were changed in order to obtain a full knowledge of the doping behaviour; these variations included Si input concentration, growth temperature, overall growth rate and GaAs substrate orientation. Moreover, the dopant concentration as a function of substrate position in MOCVD reactor was investigated, using silane (SiH_4) as dopant source. It was found that the electron concentration n, defined as $N_D - N_A$, is superlinearly dependent on the input mole fraction (M.F.) of SiH_4. Above $n = 6 \cdot 10^{18}$ cm^{-3} the electron concentration begins to saturate. Careful study shows that the compensation ratio, defined as N_A/N_D, increases strongly above this point. The total carrier concentration $N_D + N_A$ can be obtained by n and the compensation ratio. $N_D + N_A$ also appears to be superlinearly dependent on the mole fraction of SiH_4, even above the saturation point at $n = 6 \cdot 10^{18}$ cm^{-3}. Apparently, in the neighbourhood of this point the incorporation of Si at As sites starts to increase, resulting in saturation. The superlinear behaviour of log $(N_D + N_A)$ versus log (M.F. $_{SiH_4}$) could be caused by a gas phase component which contains more than one Si atom. However, it is not known that such a component exists under the MOCVD growth conditions. Further work will be done to solve this problem. On the other hand it is quite clear that the decomposition of SiH_4 into SiH_2 plays an important rôle in the Si-doping. With this knowledge, experimental results of the Si incorporation as a function of growth temperature, substrate orientation and substrate position in the reactor can be explained.

2.2 Zn-doping of MOCVD-grown $Al_xGa_{1-x}As$.

In order to make a tandem solar cell, the first stage of its construction requires the combination of an $Al_xGa_{1-x}As$ top cell with a GaAs bottom cell. This requirement leads to the study of the doping behaviour of $Al_xGa_{1-x}As$. As the first step, Zn doping was studied which will result in p-type $Al_xGa_{1-x}As$. As in the case of Si-doping of GaAs, experiments were performed as a function of various parameters but in this case we also had to vary the parameter x, which stands for the AlAs fraction in the compound. As compared to Zn-doping of GaAs, the introduction of Al gives an overall decrease of the hole concentration. This is probably due to a weaker binding between Zn and Al as compared to that between Zn and Ga. From the photoluminescence (PL) experiments it appears that the Al concentration in $Al_xGa1_{-x}As$ is varying with the Zn input concentration at constant Al, Ga and As input concentrations. This phenomenon can be observed by the location of the PL peak position. However, the results may be influenced by the absolute Zn concentration in the samples. For highly Zn-doped GaAs samples, the PL peak positions are shifted, this is known as the Moss-Burstein shift. A similar behaviour is expected for

$Al_xGa_{1-x}As$. In $Al_xGa_{1-x}As$ this effect is further mixed with the fact that the activation energy of a Zn acceptor increases at increasing Al concentrations. Therefore a careful calibration of the PL-spectra is needed before a quantitative expression can be obtained for the Al concentration versus the Zn input concentration.

2.3 p-n GaAs solar cells by MOCVD.

With the MOCVD technique p-n junctions can be formed between GaAs epitaxial layers. Zn and Si dopants are used for p an n side respectively. A highly doped p-$Al_xGa_{1-x}As$ layer is grown on top of the structure in order to reduce surface recombination. Substrates used are n-type, Si-doped, GaAs with electron concentrations varying from $2 \cdot 10^{17}$ cm^{-3} to $2 \cdot 10^{18}$ cm^{-3}. No substantial differences are found for solar cells with different dopant concentration. Instead of the p-n junction in the epitaxial layer, a few solar cells were made in which the p-n junctions are formed between a p-GaAs epilayer and a n-GaAs substrate. These cells appear to give a poorer quality than solar cells based on p-n junctions formed in the epitaxial layers. The origin of this is believed to be a structural imperfection at the substrate surface as well as a contamination of the substrate surface with impurities, resulting in a large number of recombination centers, which decrease the photogenerated current. Although this type of solar cell has a simple structure, it was abandoned.
In addition to the cell structure, growth parameters too play an important rôle in determining the performance of solar cells. This is not only true in connection with dopant concentrations and various layer thicknesses, but it was also found that the actual configuration of the p-n junctions is changing to a great extend with the growth parameters. Evidence was obtained on the relation between solar cell performance and p-n junction configuration as measured by means of the C.V. profile technique. With non-optimal growth conditions, the solar cell can have a highly compensated region at the p-n junction where both electrons and holes will suffer from low mobilities. This results in a longer residence time in the junction region, leading to an increase of the recombination rate.
The best solar cell we made up to now has an uncorrected efficiency of 9.4 %. This cell contains a structure of p^+-GaAs/p^+-$Al_xGa_{1-x}As$/p-GaAs/n-GaAs/n-GaAs-substrate where the first p^+-GaAs has been introduced to lower the contact resistance. Metal contacts used for p- and n-side are Au/Zn/Au and AuNiGe respectively. Considering the application of an anti-reflection layer on top, together with the optimization of the top contact finger pattern, the efficiency of 9.4 % is expected to be above 14 %. Further increase of the efficiency requires careful optimization of the various parameters in every step during the processing of the solar cells. Key points may be substrate dislocation density, cell construction and actual p-n junction configuration, the application of contacts and the addition of the anti-reflection coating.
A study of the influence of structural defects, which are

present in the GaAs substrates, on the performance of solar cells has been started in a cooperation with CISE, Italy. As a first step a number of substrates containing characteristic and essentially different structural defects were selected and prepared for MOCVD growth of solar cells. This project is realized currently.

2.4 Solar cells by Ion Implantation - Rapid Thermal Anneal Techniques

In cooperation with IMEC, Belgium, a study has been made on the characterization of GaAs before and after Si implantation and rapid thermal anneal (RTA) using selective photoetching. It has been established by the DSL method that the defects which are present in the GaAs substrates, particularly dislocation cell walls, influence the structure of the implanted samples. Some arguments were found which point out that Ga clusters are formed during Si implantation. They cause formation of deep craters during DSL photoetching. On the base of the VLS (Vapour-Liquid-Solid) mechanism, the formation of peculiar surface growth features during RTA can be explained.
Also in cooperation with IMEC, Belgium, a begin has been made with the preparation of solar cells by the abovementioned technique. The aim of this project is to make a comparison between solar cells made by MOCVD and solar cells made by Ion Implantation followed by a Rapid Thermal Anneal treatment (II/RTA).
Up to now the solar cells we made, mostly are of the p-n configuration. This type of cell already has been described in the first part of the last section and consists meanly of a p^+ GaAs layer on top of a n-type GaAs substrate. These cells are grown in one run with MOCVD. As dopant for the p^+ GaAs layer Zn is used to a concentration of typically $5 \cdot 10^{18}$ cm^{-3}. The p^+ layer has a thickness of 0.6 μm.
The same type of cells can be made by II/RTA. In this method n-doped GaAs substrates are implanted with Zn to a dose of 10^{13} and 10^{14} atoms/cm^2 in order to obtain a concentration of presumably 10^{18} and 10^{19} cm^{-3} respectively. After implantation the samples are annealed by the method of RTA, viz., they are heated to 800 or 900°C for 3 seconds in a forming gas (H_2/N_2) atmosphere. This is needed to recrystallize the surface layer which was damaged by the implantation. A short heat treatment is preferred, especially in the case of Zn. Otherwise the implanted atoms will diffuse deeply into the substrate, giving rise to a very diffuse interface.
Another possibility for a solar cell is the np configuration i.e. a n^+-doped toplayer on to a p-doped substrate. An advantage of the np configuration in comparison to the pn configuration is that the thickness of the n^+-layer can be made much thinner than the 0.6 μm thick p^+-layer in the pn configuration. Consequently the efficiency will be higher as the diffusionlength of electrons in p-doped GaAs is about 5 μm, whereas the diffusionlength of holes in n-doped GaAs is only 1 μm. Therefore, if one makes a np configuration with a n^+ layer as thin as possible then nearly all carriers will be generated in the p-layer and relatively much more carriers

(electrons in the p-layer) are swept down the junction to contribute to the current then in the case of the pn configuration. Another aspect is that in the pn configuration carriers are generated in both layers and one must have an AlGaAs passivation layer on top of the p-layer to avoid or at least to diminish surface recombination. In the np configuration surface recombination is much less important and thus an AlGaAs passivation layer is not really needed.
The following experiments were started in this context: p-type GaAs substrates are implanted with Si up to concentrations of 10^{18} and 10^{19} cm^{-3}. The thickness of the implanted layer is varied between 0.05 and 0.3 μm. Then a RTA treatment follows and after that the samples are further processed to solar cells. These cells will be compared to MOCVD grown cells. Results from electrical characterization of MOCVD grown-up cells show that the open-circuit voltage (V_{oc}) and the filling factor (FF) hardly depend on the thickness of the n-type layer in the range of 0.05 to 0.6 μm. The short circuit current (I_{sc}) and efficiency (η) however show optima around 0.12 μm, a value which seems to be confirmed by theory. Below 0.05 μm V_{oc}, FF, I_{sc} and η drop sharply to zero. So far only a few solar cells have been examined and comparisons have not been made yet between MOCVD grown cells and cells made by II/RTA. An attempt will be made to correlate differences in cell parameters to the different processes by which the cells are made. Another future experiment will involve implantation in undoped layers that are grown by means of MOCVD. Up to now the solar cells were only studied by measuring the electrical characteristic (V_{oc}, I_{sc}, FF, η). In the future cells will also be examined by DSL in order to correlate bad cell parameters with structural defects such as dislocations or implantation damage, which has not been annealed away properly.

3. Characterization studies

3.1 High Resolution Photoluminescence on GaAs epitaxial MOCVD-grown layers on In-alloyed GaAs substrates.

Undoped GaAs epitaxial layers were grown by MOCVD on indium alloyed GaAs substrates. Misfit dislocations, created in the epitaxial layers due to lattice mismatch with the substrate, are visualised by the DSL photoetching method. A cross-hatched pattern of closely separated dislocations is observed. In order to investigate the interface between epilayer and substrate, the samples were beveled by chemical etching in a bromine/methanol solution. In the interface region the same pattern of misfit dislocations as in the epilayer can be discerned. However, some of them are clearly decorated, which indicates the gettering of impurities by these dislocations during the first stages of epitaxial growth. By spatially resolved photoluminescence (resolution 15 μm) 2-dimensional mappings of the distribution of shallow impurities were made. Due to the indium alloying of the substrate, which gives rise to a shift of all PL signals towards larger wavelengths, a distinction could be made between the signals arising from impurities in the epilayer and in the substrate. It is shown

that the epilayer exhibits a good spatial homogeneity with respect to the incorporation of shallow impurities. In the interface region however, strong PL contrasts have been observed. These contrast are much clearer than those which already exist in the bare substrate. From this it is concluded that an enhanced inhomogeneous distribution of shallow impurities exists in the interface region. It is not yet clear whether these impurities are built in from the MOCVD system, or are diffused out of the substrate towards the interface.
This study has been performed in cooperation with the Université Scientifique et Medicale de Grenoble, France. In order to do this type of high resolution PL measurements in our own laboratory (at 4.2 K and with a resolution of about 10 μm) an insert for a He bath cryostat, in which a focussing and translation mechanism will be mounted, is in course of construction presently.

3.2 PL-Determination of Al content in $Al_xGa_{1-x}As$ layers with $x > 0.4$.

In case of $Al_xGa_{1-x}As$ samples with an Al mole fraction x smaller than 0.4 it is possible to determine the Al concentration directly by means of PL measurements because of the direct nature of the PL transition in the material with the abovementioned composition. For $Al_xGa_{1-x}As$ samples with Al mole fraction x larger than 0.4 we are now able to measure photoluminescence transitions at 4.2 K in spite of the indirect character of the material. For these measurements a Ar-laser is focussed in such a way that excitation intensities as high as $3 \cdot 10^4$ W/cm^2 can be reached. It is assumed that the weak and broad (~ 50 meV) PL transition we measure can be attributed to an indirect X - Γ transition in which an optical phonon is active. For the compositional dependence of the energy gap in $Al_xGa_{1-x}As$ the formula suggested by Lee et al. can be used and the compositional dependence of the optical phonon frequencies may be taken from Kim and Spitzer. This procedure enables us to determine the Al mole fractions from the PL results. It appears that the results, obtained by this method for the determination of the Al mole fraction, are in good agreement with those arrived at by other methods such as elastic recoil detection (ERD) and ICP-AES as described in our report of March 1987. The discrepancy in literature on the compositional dependence of the indirect bandgap of course introduces an uncertainty but the PL-method has the advantage of non-destructivity in contradistinction to the abovementioned techniques.
An electroreflection equipment - optically modulated reflection measurements as a function of wavelength - has been constructed in this period and is tested presently. Probably this technique will be the best solution to this characterization problem.

3.3 EBIC measurements on substrate materials.

In order to investigate the occurrence of deep levels at dislocation cell walls, another characterization technique can be used. If one irradiates GaAs samples, on which a metal-semiconductor Schottky barrier is present, with an electron beam, the induced current gives detailed information on the local variation of the concentration of deep levels. Resolutions which can be reached are as good as 1 μm. A cooperation has been started with the University of Groningen, the Netherlands. The first results are very promising: defects revealed by PL and DSL-photoetching give a clear EBIC contrast as well. In cooperation with Maspec, Italy, growth inhomogeneities such as impurity striations and grown-in dislocations were also studied by EBIC coupled with DSL photoetching. Quantitative EBIC measurements of the concentration of n-type dopant across the growth striations- first revealed by DSL-photoetching - are in good agreement with the theory of DSL-photoetching.

3.4 Anodic oxide layers for cleaning GaAs-substrates.

To improve the quality of epitaxial layers, we studied the growth of anodic oxide layers on top of GaAs substrates. These layers can be etched away in situ in the reactor cell prior to epitaxial growth. From PL measurements on epitaxial layers grown after this procedure it can be concluded that there is less incorporation of C_{As} in the layer. This supports the idea that for normal growth conditions C_{As} is strongly incorporated at the interface between substrate and epilayer. This incorporation obviously can be reduced by means of the improvement of the cleaning of the substrate using the anodic layer method.
Unfortunately the epilayers which are grown after the abovementioned cleaning procedure show lower PL intensities. This phenomenon gives rise to the suspicion that the layer quality is deteriorated although the grown layers show a very good morphology. We will study this further by means of PL measurements on bevelled samples. By this method, special attention can be given to the epilayer - substrate interface. Further investigations on this type of samples, such as SIMS and Auger, are carried out presently at IMEC, Belgium.

3.5 Misfit dislocations in GaAs epilayers on In-doped GaAs.

In-alloying is known to be effective in the hardening of the GaAs matrix; thus it can be used in diminishing the density of dislocations in GaAs substrate materials. However, heavily In doping (2-3 %) of (001) oriented GaAs crystals appears to be deleterious for epitaxially grown layers as, due to the lattice mismatch, numerous misfit dislocations along [110] and [$1\bar{1}0$] directions are introduced in the epilayer parallel to the (001) oriented substrate-epilayer interface. It is to be expected that lowering the In-content of the substrate probably will diminish the tendency of formation of misfits in the epilayer while there is still hardening of the substrate matrix. Therefore a systematic study has been done in which GaAs layers (3 μm) were grown under optimal conditions on top of substrates with different In contents. The In concentrations in the substrate materials were determined by ICP measurements. By means of DSL etching - by which technique misfits are shown as extensive cross-hatched patterns - it was possible to estimate

the maximum In concentration at which no misfit dislocations were present in the epilayer. This concentration appears to be about 0.2 mole % of In.
Also TEM measurements were performed on two of the above-mentioned samples; one, sample A, had a substrate In content of 0.91 mole %; the other, sample B, of 0.18 mole % In. Sample A appeared to contain interfacial misfit dislocations, which were not observed in sample B, in agreement with the DSL experiments.
Plan view sections of sample A also showed that extensive strain remained between the substrate and the epilayer, implying that total relief of the mismatch had not been achieved by the generation of the misfit dislocations. However, the observed dislocation spacing of about 0.3 μm corresponds to 100 % accommodation of a mismatch of $7 \cdot 10^{-4}$, calculated for this sample. Thus the mismatch in these regions must be greater than calculated. Either, quite improbable, the epilayer was not stoichiometric, causing variations in the lattice parameter or, more probable the underlying In-doped substrate showed local variations in lattice parameter, which would not have been detected in lattice parameter measurements determined by the Debije-Scherrer X-ray powder technique.
From this study it may be inferred that - as misfit dislocations most probably will be deleterious for the efficiency of solar cells and also as low In concentrations appear to be quite effective in diminishing the dislocation density in GaAs material - the optimal concentration of In in GaAs substrate material for the growth of solar cells appears to be in the order of 0.2 mole %.

4. Modelling of MOCVD reactors

4.1 Flow visualization and numerical calculations on gas flows in horizontal MOCVD reactors.

For epitaxial growth of high quality, sharp semiconductor interfaces it is essential to be able to exactly control the gas flow patterns inside the reactor. One of the problems which hinders the growth of sharp interfaces is the occurrence of return flows, due to the heating of the cold gases in the entrance region of the reactor. Therefore memory cells will develope. As a solution to this problem we decreased the pressure in the reactor. Because of the larger molecular diffusionlength the return flows are less likely to occur at lower pressures. It is indeed shown by flow simulation experiments, - i.e. "smoke" experiments with TiO_2 particles- that we are able to avoid return flows if the pressure in the reactor is lower than a certain critical value, determined by the reactor size and the growth parameters. For typical flow rates, reactor sizes and temperature gradients in the reactor, this critical pressure is about 0.2 Bar.
Parallel to these flow simulation experiments numerical calculations on gas flows have been performed for us at the Technical University of Delft, the Netherlands. In comparing the results obtained by the flow simulation experiments and the

numerical calculations, an excellent one-to-one relationship between the two approaches is obtained. It has been shown that the occurrence of return flows is determined by only two dimensionless hydrodynamic numbers: the Grashoff number Gr and the Reynolds number Re. For every reactor it is possible to define a number α_{crit}, depending only on the temperature gradient and the reactor height, such that no return flow occurs for $Gr/Re < \alpha_{crit}$. This criterion is confirmed by both the TiO_2 experiments and the numerical calculations.

OPTIMIZATION OF HIGH EFFICIENCY MULTILAYER SOLAR CELLS BASED ON III/V COMPOUNDS

Authors : M. MEURIS, G. BORGHS, W. VANDERVORST

Contract number : EN3S-0079-B (GDF)

Duration : 30 months 1.7.1986 - 31.12.1988

Total budget : BF 15.457.000 CEC contribution : BF 7.520.000

Head of project : G. BORGHS

Contractor : IMEC

Address : Kapeldreef 75, 3030 Leuven Belgium

Reporting period : 1.4.1987 - 30.9.1987

Summary

SIMS measurements on solar cell structures grown by IMEC and by the KUN have been carried out. Measurements of layer thickness, doping concentration and contamination have been performed. The presence of oxygen in the epitaxially-grown layers has been extensively investigated. This is of importance mainly for characterisation of MBE grown AlGaAs layers. Due to the high reactivity of aluminium, oxygen is gettered from the vacuum of the MBE system and incorporated in the grown AlGaAs layers. The same mechanism occurs during SIMS-measurement, making the interpretation of the data complicated. A detailed study on oxygen sputtering and adsorption is therefore made. This systematic study is reported here.

Physical characterization of AlGaAs-GaAs solar cells by SIMS

1. Introduction

For the interpretation of solar cell behaviour, a good knowledge of the layer-thickness, alloy and doping concentration and possible contaminations is very important. SIMS analyses, sometimes combined with other techniques, is a very powerful technique for the qualification of epitaxial grown layers.
We have carried out a calibration of the sputter yield as a function of Al content, enabling us to determine precisely the layer thickness of various AlGaAs alloy compositions.
We have also made an extensive study of SIMS measurements of oxygen contamination in AlGaAs layers. The oxygen is not only detrimental for the quality of bulk material but also for the interface between the GaAs and AlGaAs layers. It is an important factor in the performance of solar cells as well as many other devices. The high reactivity of Al with oxygen-containing background species during growth of AlGaAs layers may induce oxygen-related defects, whose concentration should increase with aluminium content.

2. Determination of alloy-composition and dopant activation

The absolute concentration of Al in the $Al_xGa_{1-x}As$ layers has been measured and calibrated by use of optical techniques.
We extended the Al/Ga-signal ratio versus x/(1-x) up to x=1. Data up to x = .5 are published (1). The linear relationship is shown in Fig.1. The samples (2 µm thick) are calibrated using electron microprobe, electroreflectance and for x>0.4 photo-lumeniscence at 4K. Using a calibration standard, we can now obtain an accuracy of 1 % in Al-content by SIMS for a layer thickness as thin as 1 nm.
The electrical profile and the atomic concentration of dopants in GaAs are not always identical. This is shown in fig. 2. where the dopant activation of Be has been measured by chemical capacitance-voltage (polaron) and SIMS. The sample analysed here has been grown with different Be-concentrations (from $2x10^{17}$ up to $1x10^{19}$ cm^{-3}). The Be-doping up to $4x10^{18}$ cm^{-3} is completely incorporated and electrical active. The highest doping of $1x10^{19}$ is only 70 % active.

3. Quantitative analysis of oxygen-content in AlGaAs

The high affinity of aluminium for oxygen will not only cause an increasing O-incorporation with higher Al-content during growth, it will also influence the SIMS measurements due to the high gettering effect. To disentangle the bulk oxygen contribution from the background we made a detailed study of oxygen sputtering and adsorption. During sputtering through layers with a different matrix composition, SIMS-signals can vary because of changes in the sputterrate, ionization yield and also sticking coefficient if the measured species is also present in the vacuum (like H, C, O, N).

For an experimental determination of these parameters we have prepared a set of 8 samples with $Al_xGa_{(1-x)}As$-layers on a GaAs-substrate (x varying from 0.0 to 1.0 and with layer thickness of 1 to 2 µm). These samples were implanted with 10^{15} cm^{-2}, 100 keV of ^{16}O.

a. The sputter yield

The sputter yield, as determined from the crater depth measurements is found to decrease linearly with Al-concentration (Fig.3.)
In the high concentration part of the implantation-profile the O-signal is solely determined by the bulk-contribution so that the influence of background is neglible. As can be seen on Fig.3., the sputter yield decreases linearly with approximately a factor of 2 in going from x=0 to x=1. The sputter speed with the Ar^+ primary beam is also approximately a factor of 2 lower then under Cs^+ bombardement of the sample.

b. The ionization yield

In Fig.4. we show the ionization yield of O as a function of Al content, relative to the ionization-value in GaAs. This relationship is also linear and an increase of the ionization coefficient of almost a factor of 3 is seen when the Al-concentration ranges from x = 0.0 to x = 1.0. The ionization yield for a Cs^+ primary beam is increasing slightly more than for Ar^+ .

c. The sticking-coefficient

With the parameters known as described in parts a. and b., we could also determine the influence of O_2 adsorption on the AlGaAs-layers. The sticking of O_2 as a function of Al-content for Cs^+- and Ar^+-bombardment was determined using measurements under $^{18}O_2$-flooding conditions. The bleed-in allows one to discriminate against ^{16}O present in the sample and uncontrolled adsorption from the vacuum. The sticking coefficient as derived from these measurements is shown in Fig.5 for Cs^+ and Ar^+ bombardment. It is interesting to observe that for low Al contents the effect of oxygen adsorption is less during Ar^+ bombardment than during Cs^+ sputtering. This is a consequence of the different surface properties under different bombardments. The cesiated surface is hardly influenced by the aluminium concentration in the bulk. To minimize the adsorption of oxygen from the environement we can conclude that in cases of low Al-content ($x < .7$) it is better to use an Ar^+ primary beam and for $x > .7$ Cs^+ is better. The above conclusion is only valuable for the determination of O coming from the di-atomic molecule. When other background-molecules containing oxygen are involved, the measurements can differ considerably.

This is made clear in Fig.6, where three measurements are combined. The 3 profiles (a-b-c) are ^{16}O measurements on a sample, with an AlGaAs-layer of x = .65 on GaAs. The primary beam used is Cs^+. In case (a) the sample is measured 15 minutes after introducing the sample into the SIMS-chamber. Profile (b) gives the same measurement after a night pumping. The background of ^{16}O in (b) is much lower, showing that the main contribution to the profile (a) is due to adsorption of the residual vacuum. Profile (c) on the other hand shows the ^{16}O-signal under flooding $^{16}O_2$ conditions.

The decrease in O-signal of profile (a) going from AlGaAs into GaAs is much more pronounced than in case (c) (flooding condition). We conclude that in the case (a) the vacuumcomponent, with a large difference in sticking coefficient on AlGaAs as compared to GaAs is not due to the O_2 molecule in the residual vacuum.

4. Conclusion

We have demonstrated that for the quantitative determination of oxygen in AlGaAs epitaxial layers a careful analysis has to be made of perturbing influences during SIMS measurements. A low abundant oxygen-contamination is unmeasureable due to the high sticking-coefficient of oxygen present in the vacuum system.
Recently Achtnich et al. showed that for the MBE layers grown in their system oxygen contamination could be attributed to the layers at the interface AlGaAs/GaAs with the $Al_xGa_{1-x}As$ layer below a GaAs layer (substrate is lowest) for x>0.35 (2). This has important consequences for solar cells and device applications.

5. References

1. Bouwdewijn P.R., Leys M.R., Roozeboom F., Surface and Interface analysis 9, 303 (1986)

2. Achtnich T., Burri G., Py M.A., Ilegems M., Appl. Phys. Lett. 50, 1730 (1987)

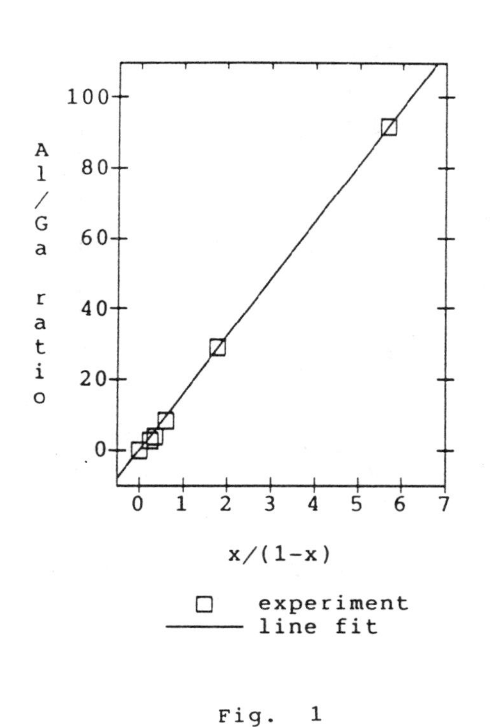

Fig. 1

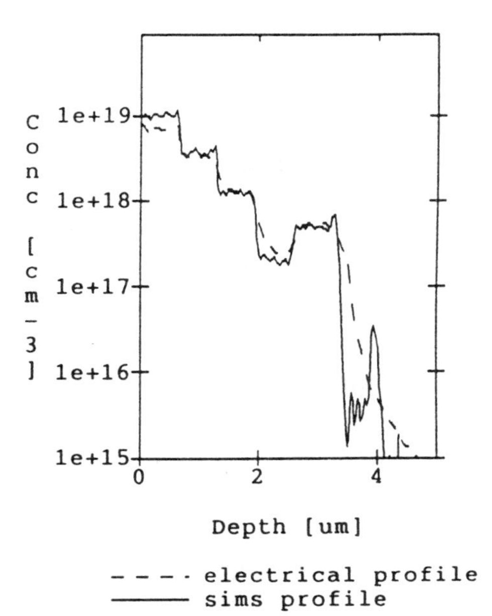

Fig. 2

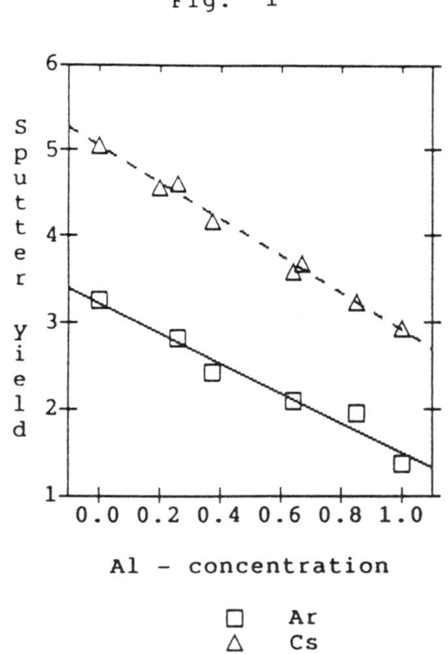

Fig. 3

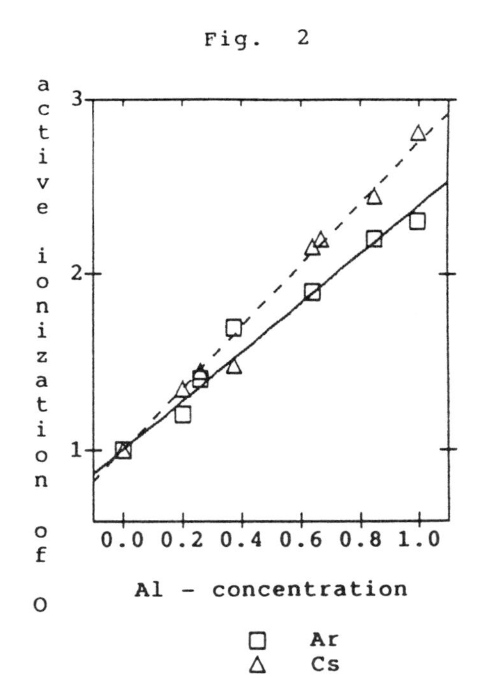

Fig. 4

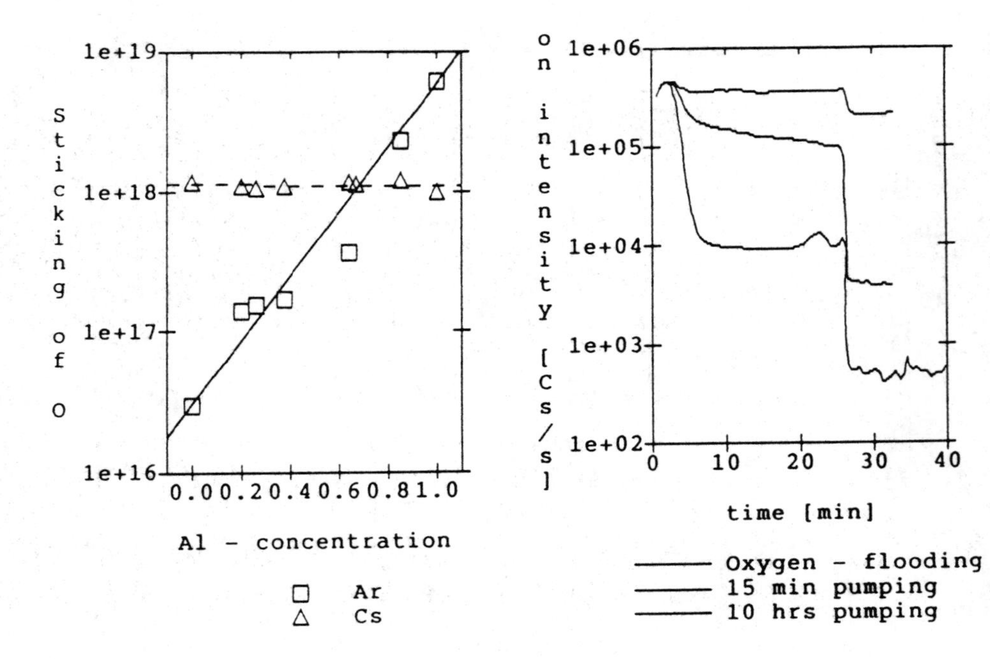

Sticking of O
1e+19
1e+18
1e+17
1e+16
0.0 0.2 0.4 0.6 0.8 1.0
Al - concentration
Ar
Cs
on intensity [Cs/s]
1e+06
1e+05
1e+04
1e+03
1e+02
0
10
20
30
40
time [min]
Oxygen - flooding
15 min pumping
10 hrs pumping

Fig. 5

Fig. 6

INHOMOGENEITIES IN LEC GaAs SUBSTRATES FOR SOLAR CELL APPLICATIONS

Authors : C. Frigeri, G. Frigerio and L. Zanotti

Contract Number : EN 3S / 0080

Duration : 36 Months 1 July 1986 - 30 June 1989

Total Budget : 180000 ECU

Head of Project : Dr. L. Zanotti, Istituto Maspec CNR, PARMA

Contractor : Consiglio Nazionale delle Ricerche, ROMA

Address : Consiglio Nazionale delle Ricerche,
P.le Aldo Moro 7
00100 ROMA
ITALY

Summary

Heavily doped, n-type GaAs crystals have been grown in order to obtain low-dislocated substrates for solar cell applications. In this paper we briefly summarize the recent findings of investigations (by means of Capacitance-Voltage measurements, chemical etchings, and Electron Beam Induced Current measurements in an SEM) concerning the presence of structural defects, compositional inhomogeneities and electrical properties. In particular we report on the inhomogeneous distribution of dopant concentration in correspondence to growth striations, on the decoration of dislocations by impurities and on the inhomogeneity of donor concentration on millimetric scale.

1. INTRODUCTION

During the past few years a great effort has been made to grow gallium arsenide single crystals with very low density of extended defects, in particular dislocations. The hardening effect obtained by addition of n-type dopants (Si, S, Se, Te) in high concentration has been proved to be effective in reducing the dislocation density. However, up to now only little is known about possible side effects produced by this heavy doping, as the formation of precipitates, complexes, inhomogeneities and so on, that can be detrimental in view of applications of the crystals as substrates for optoelectronic and solar cell technologies. Aim of the present work is to study the defect content and the electrical homogeneity of these substrates.

2. EXPERIMENTAL

2.1. MATERIAL PREPARATION - The GaAs crystals were grown by LEC technique, starting from polycrystals separately prepared by direct synthesis in a high-pressure vessel. Seeds were <100> oriented with average dislocation density of $1\text{-}5 \times 10^4$ cm^{-2}. Growth runs were performed in a quartz crucible, by using dry B_2O_3 (J. Matthey, water content less than 250 ppm), 10-12 mm thick, as an encapsulant. Crystals were pulled at a rate of 10 mm/h, with the crucible and the seed kept in counter-rotation (15 rpm and 5 rpm, respectively). The axial thermal gradient of the puller was about 115 °C/cm. After the growth, the crystals were slowly cooled down (50 °C/h from 1240 to 800 °C, 100 °C/h from 800 °C to room temperature). Different doping procedures were carried out in order to find the most appropriate concentration and type of dopant.

2.2. ELECTRICAL MEASUREMENTS - Schottky diodes for C-V measurements were prepared by evaporating gold dots. Immediately before the metal deposition the chemically-polished samples were etched by H_2SO_4:H_2O_2:H_2O (5:1:1), washed in distilled water, dipped for 20 sec in HCl and finally washed in isopropyl alcohol.

2.3. CHEMICAL ETCHINGS - Standard KOH etching was employed for dislocation density measurements. Polished samples were etched for 10′ into molten KOH (400 °C), then rinsed in ethylene glycol and water. Diluted Sirtl-like photoetching (DSL) (1) was used for further defect and inhomogeneity investigations. Chemically-polished samples were dipped into CrO_3:HF:H_2O (2.3M HF and 1.8M CrO_3) solution and illuminated by a halogen lamp for about 2′. Etch pit counting and defect investigations were performed by using a standard metallographic microscope with a Nomarski interference contrast attachment.

2.4. EBIC MEASUREMENTS - Schottky diodes for Electron Beam Induced Current (EBIC) investigations in an SEM were prepared in the same way as for C-V measurements. Quantitative evaluation of dopant concentration with a spatial resolution of about 10 μm was achieved by using a method (2) that is based on the measurement of the EBIC efficiency as a function of beam energy. Electron beam currents smaller than 1 nA were used in order to work in the low injection regime.

3. RESULTS AND DISCUSSION

3.1. DISLOCATION DENSITY - From the study of different dopants we obtained that in GaAs the dislocation reduction via hardening effect is achieved only when n-type heavy doping is used (Fig. 1). In particular S and Si doping processes were investigated. Si has been found to be more advantageous because it gives a reduction in dislocation density while avoiding the formation of undesidered precipitates which are common with heavy S-doping (3,4). All results reported in the following refer to Si-doped samples.

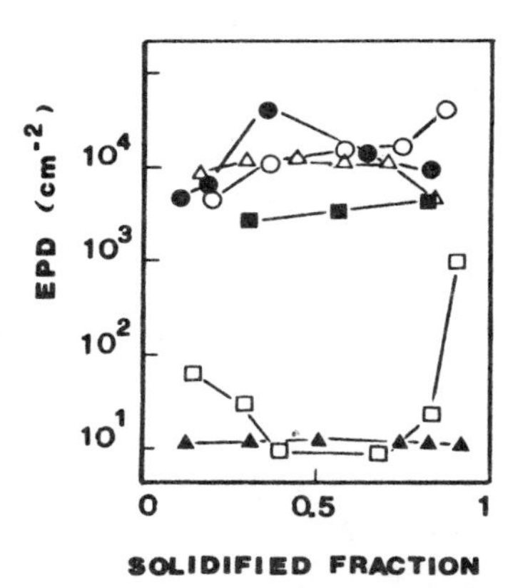

△ Si 1-5 x 10^{17} cm^{-3}

● Se 1-3 x 10^{17} cm^{-3}

■ Cr 5-10 x 10^{17} cm^{-3}

○ Zn 1-2 x 10^{18} cm^{-3}

□ Si 1-5 x 10^{18} cm^{-3}

▲ S 1-3 x 10^{18} cm^{-3}

Fig. 1 - Axial dislocation density estimated from etch pit counting after KOH etching in differently doped GaAs crystals.

3.2. C-V RESULTS - Here we report the results on net ionized donor distribution measured on samples doped in the range $1x10^{16}$ - $1x10^{18}$ cm^{-3} by means of C-V on linear arrays of diodes deposited onto stripes (about 5x40 mm^2) cut along wafer diameter. Fig. 2 shows the percent deviation of the value measured on each diode from the mean value on the whole stripe for three different samples. As one can see, at low doping levels the scattering of data is not higher than a few percent, while it increases as the carrier concentration increases.

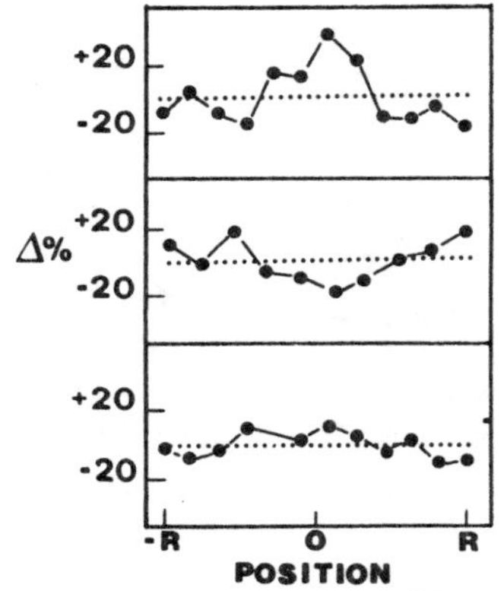

Sample 47
$<N_D-N_A>=7.9x10^{17}cm^{-3}$

Sample 24
$<N_D-N_A>=1.8x10^{17}cm^{-3}$

Sample 48
$<N_D-N_A>=1.4x10^{16}cm^{-3}$

Fig. 2 - Per cent deviation from net ionized donor mean value, obtained from C-V measurements, for differently Si-doped samples.

3.3. EBIC AND DSL RESULTS - A microscopic assessment of the electrical inhomogeneities induced by structural defects and compositional inhomogeneities was performed by DSL and EBIC. Figs. 3 and 4 show typical defects which are present in GaAs substrates, namely growth striations and grown-in dislocations. The one-to-one correspondence between DSL and EBIC pictures is quite good, and one should note that in the surroundings of the dislocations a zone with higher recombination efficiency is present that may be related to the so-called Cottrell atmosphere formed by impurities gettered by the dislocations. In order to quantitatively evaluate DSL etch rate dependence on dopant concentration, the profile of the etched surface of samples having growth striations has been measured with a step profiler and compared with the dopant density as measured by EBIC. An example of such measurements is shown in Fig. 5 for a sample with an average doping level of 1.8×10^{17} cm^{-3}. The etching has actually removed the material in an inhomogeneous way, grooves corresponding to regions of lowest dopant density. These data support the following:

- The etch rate dependence on dopant concentration is confirmed;
- Very large dopant density differences can occur between striations and 'matrix' and among striations themselves;
- Grown-in dislocations are surrounded by excess of impurities whose nature (dopant and/or residual impurities) is not yet clear.

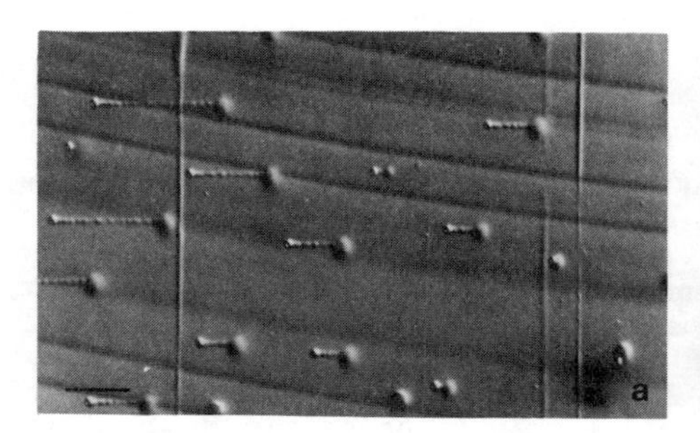

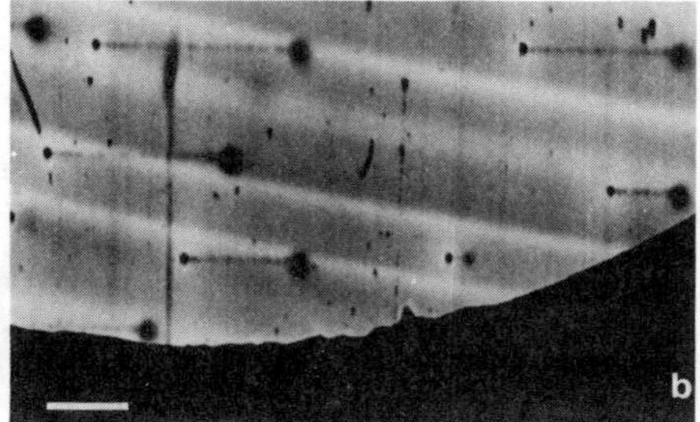

Fig. 3 - Comparison between DSL (a) and EBIC (b) images of the same zone of a Si-doped sample with striations and dislocations. Bar is 100 μm.

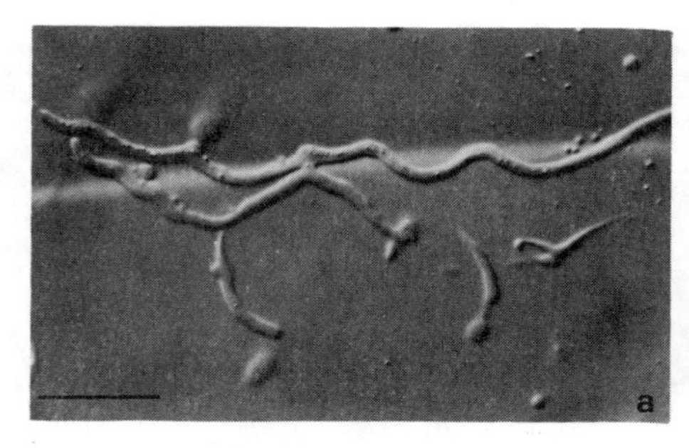

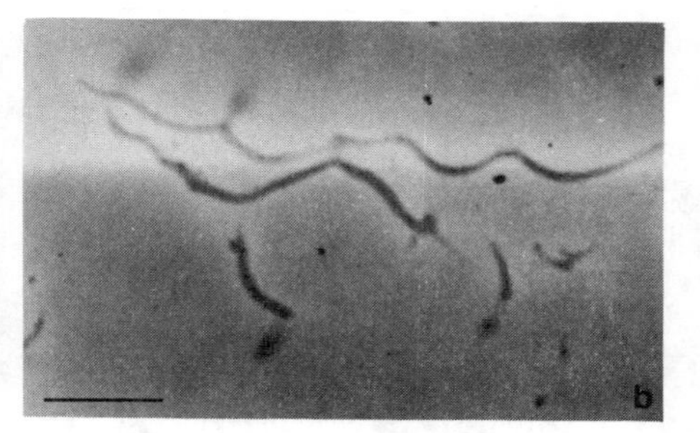

Fig. 4 - Dislocations lying parallel to the (100) plane of the sample as revealed by DSL (a) and EBIC (b). Bar is 100 μm.

4. CONCLUSIONS

Results reported here are evidence that the usual LEC process for GaAs single-crystal growth induces strong inhomogeneities in the crystalline boule. They can be summarized like this:

- Variations of electrical properties on millimetric scale, due to inhomogeneous dopant incorporation;
- Growth striations induced by free convection in the melt. They cause variations of the electrical properties on a micron scale;
- Impurity and/or dopant atmospheres around dislocations.

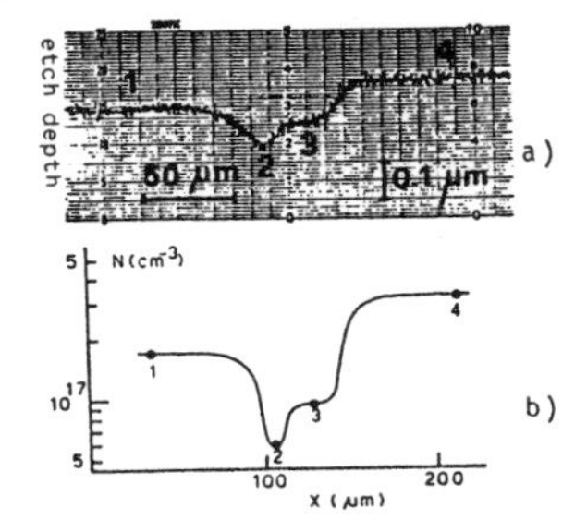

Fig. 5 - a) Etch depth profile after shallow DSL etching across two growth striations; b) plot of dopant density N in the same region as measured by EBIC. Differences in N even up to a factor 5.

These inhomogeneities are of course detrimental in view of solar cell applications, because they can limit the efficiency of the final device. Local decrease of dopant density, for instance, causes an increase both of the substrate resistivity and of the depletion region width thus increasing the series resistance and the recombination current, respectively (5). Furthermore, it is well known (6) that dislocations can propagate into the top layer and reduce the carrier lifetime therein because of their high electrical activity. It seems, however, that not always the Cottrell atmospheres are able to keep up with the threading dislocations to the layer (1). It is thus necessary to develop new processes that avoid the evidenced defects.

For instance, whole-ingot post-growth annealing may be a powerful way to improve the uniformity of dopant distribution, so reducing the macroscopic inhomogeneity evidenced by C-V measurements.

Further, growth striations may be removed by two means:
- elimination of free convection in the melt during growth by application of magnetic fields. By this way the thickness of the boundary layer is as large as the liquid mass so that the growth process is controlled only by diffusion;
- generation of forced convection by appropriate rotation (constant or accelerated) of the crucible. In this case the boundary layer is extremely thin and the temperature at the growing interface (a parameter strongly affecting dopant incorporation) is much more constant.

On the other hand, grown-in dislocation density can be effectively reduced by an accurate in situ control of melt composition. This can be achieved by continuous injection of arsenic into the liquid during the pulling from a separate container mantained at appropriate temperature. Also very promising is the procedure of pulling in hot walls systems in which the crystal is grown in a closed quartz ampoule from an encapsulant-free melt in accurately controlled arsenic atmosphere.

REFERENCES

(1) J.L. Weyher and J.van de Ven, J. Cryst. Growth 78, 191 (1986).
(2) C. Frigeri, Proc. V Oxford Conf. on Microscopy of Semiconducting Materials, Oxford 1987, Adam Hilger, in press.
(3) L.J. Giling, J.L. Weyher, A. Montree, R. Fornari and L. Zanotti, J. Crystal Growth 79, 271 (1986).
(4) R. Fornari, P. Franzosi, G. Salviati, C. Ferrari and C. Ghezzi, J. Crystal Growth 72, 717 (1985).
(5) S.M. Sze, Physics of Semiconductor Devices, J. Wiley, N.Y., (1969).
(6) D.B. Darby and G.R. Booker, Inst. Phys. Conf. Ser. 56, 595 (1981).

HIGH EFFICIENCY MULTISPECTRAL CELLS BASED ON III-V COMPOUND SEMICONDUCTORS

Contract Number: EN35/C2/175/F

Duration : 32 months 10 October 1986 - 30 juin 1989

Total Budget : FF 16,805,000 CEC contribution: FF 2,422,000

Head of Project: Dr. C. VERIE, Laboratoire de physique du solide et énergie solaire

Contractor : Centre National de la Recherche Scientifique

Address : 15 quai Anatole France
75700 PARIS
FRANCE

Summary
The main objective of this work is the achievement of high efficiency multigap solar cells. The implementation of two fundamental concepts-multispectral conversion and monolithic integration-is expected to give a conversion efficiency equal to or greater than 30% under a concentration (C) in the 100-500 range. In this project two tandem cells, made of a two gap monolithic structure, are optically coupled by a dichroic filter. Constituting materials of these so called tandems are $Ga_xAl_{1-x}As$-GaAs and $Ga_xIn_{1-x}As_yP_{1-y}$-$Ga_xIn_{1-x}As$ epitaxied on GaAs and InP substrates respectively. Up to now material science problems connected to doping and growth conditions of the multilayered structures have been solved. Following this study good conversion efficiencies (compared with the maximum theoretical values) have been obtained for GaAs and $Ga_{0.53}In_{0.47}As$ single cells (17% and 6% respectively under AM0 and C=1). The study of the influence of deep centers on the photovoltaic qualification of $Ga_xAl_{1-x}As$ has been carried on and a conclusion can be given now. Concerning the low energy tandem, preliminary results concerning $Ga_{0.53}In_{0.47}As$ cell concentration behavior have been obtained; the first InGaAsP structures have been elaborated and characterized. Concerning new photovoltaic concepts, a theoretical modelling of a GaAs point contact cell has been recently undertaken. The other aim of this contract is to demonstrate the possibility of replacing GaAs substrates by less expensive Si substrates using an epitaxial matching structure. Direct growth of GaAs on Si substrates has been performed by MOVPE. Structural studies of 2 µm thick layers by Raman spectroscopy and X-ray double crystal diffractometry have shown crystalline quality sufficient for use in a fair efficiency solar cell.

1.1 Introduction

Terrestrial applications of solar energy photovoltaic conversion require high conversion efficiencies which can only be obtained by using multispectral conversion associated with concentrated light.The "rainbow project" proposes the study of a quadrispectral conversion system combining monolithically stacked two bandgap tandems made of appropriate semi-conductors. Both junctions of each tandem are connected by a tunnel junction.
The aim is to demonstrate an overall efficiency potentiality of at least 30% at 100-500 suns. A theoretical modelization was performed which analyses all the possible configurations (bi,tri and quadrispectral conversion) and gives a selection of III-V compound alloys currently used in optoelectronic devices: GaAs and GaAlAs, GaInAs, GaInAsP derived alloys [1]. Intermediate experimental goals have been defined: material and structure growth control, state of the art single cell elaboration,material photovoltaic qualification, demonstration of a bicolor conversion system (GaAs/GaInAs) which represents an important step towards more efficient but more complicated systems.

Besides, a preliminary study of different matching structures for GaAs epitaxial growth on silicon substrate has allowed to orientate this activity towards the direct growth of GaAs on silicon substrate by MOVPE.

1.2.1 High Energy tandem: GaAs-GaAlAs/GaAs.

MOVPE and MBE epitaxial techniques are used for this tandem elaboration. GaAS solar cells have been obtained by both growth methods. Following this an important work has been developped to appreciate GaAlAs photovoltaic qualification.

1.2.1.1 GaAs solar cells.

Standard GaAs solar cells are composed of the following sequence: A.R. coating/$Ga_{.2}Al_{.8}As$ window (p+, 500Å)/GaAs(p+,.4µm)/GaAs(n+, 4µm)/GaAs substrate(n+). Main electrical characteristics are reported in table 1.

	Voc (Volt)	Isc (mA)	FF	η (%)
MOVPE	.99	.91	81.8	18.1
MBE	.94	.90	76.5	15.8

Table 1---AM1.5 GaAs solar cell characteristics.Active area=.041cm^2 .MOVPE cell A.R.coating is a TiO_2 layer with a 10% average residual reflectivity MBE cell has no A.R. coating.

1.2.1.2 (GaAl)As photovoltaic qualification.

It is of great technological importance to be able to grow high quality $Ga_{1-x}Al_xAs$. However two main problems arise when growing $Ga_{1-x}Al_xAs$ layers.
-DX centers do exist whatever the epitaxial method and the nature of n dopants for $x \geq .2$.
-In MOVPE, due to the high reactivity of Al, the residual amount of O_2 and H_2O in the vapour phase is of critical importance and the amount of O and C

in $Ga_{1-x}Al_x$ As is usually high. To get a better insight on the nature and properties of DX centers some basic experiments were done.

1.2.1.2.1 High field high pressure transport data (SNCI Grenoble)[2].

Magnetoresistivity was measured in $Ga_{1-x}Al_xAs$ epilayers at 4.2K under high magnetic fields in the pressure range 0-1GPa. In addition the permanent photoconductivity was produced on $Ga_{.7}Al_{.3}As$ samples. At 4.2K Shubnikov de Haas (S.d.H) oscillations appear which give the Γ carrier density. As the hydrostatic pressure increases the band structure is modified (and this simulates more or less an increase of the x value, about .1 per GPa). Under pressure the S.d-H oscillations gradually disappear until .8GPa. In these particular experimental conditions (.8 GPa,4.2 K), under stepwise light illumination, the $n_Γ$ value deduced from S.d-H oscillations increases up to the normal pressure value (fig 1). These features are understood asssuming that electrons are transferred from the DX centers to the Γ valley. More generally, these kind of experiments show that DX centers are resonant states located 50-100meV (depending of x and doping level) below the X minimum.

1.2.1.2.2 Photoluminescence decay measurements (PLD)[3].

As the band structure of GaAs under hydrostatic pressure is quite similar-although non completely equivalent to that of (GaAl)As-PLD experiments were performed under pressure to check whether or not the reduction of minority carrier lifetime results from a band structure. A strong decrease (from 9 ns to below 1 ns) of the decay time was observed on both MBE and MOVPE n-type samples with hydrostatic pressure up to 4 GPa. This feature was not observed in p-type GaAs.

1.2.1.2.3 Mössbauer spectroscopy [4]

Mössbauer spectroscopy, although scarcely used in semiconductor physics, is a spectroscopic tool which can give information on the nature of the chemical bond on a particular atom in a solid together with the local symmetry. This is henceforth a possibility to investigate the nature of shallow and deep donors provided a donor atom with Mössbauer isotope can be found. Thick layers of $^{119}Sn-Ga_{1-x}Al_xAs$ were grown by MOVPE and investigated by Mossbauer spectroscopy on ^{119}Sn. The following features were shown:

-for x $\leq$.2, Sn appears in Ga(Al) site as a shallow donor only. No Sn_{As} appears to a concentration high enough to be seen by Mossbauer spectroscopy.

-for x $\geq$.2, two kinds of Sn atom both substitutional simultaneously do exist. In addition to the shallow donor, a second type of Sn atom with a different electronic structure and non cubic local symmetry appears. This particular Sn atom corresponds to DX centers. In other words for x$\geq$ .2 any donor atom appears under two electronic states: a shallow hydrogen like donor and a deep donor.

1.2.2 Low energy tandem: GaInAsP-GaInAs/InP.

LPE technique is used for this tandem cell. GaInAs and GaInAsP single cells have been obtained and following structures have been tested.

$Ga_{.27}InAs_{.63}P$/ GaInAs (p)/ GaInAs (n) /InP

$Ga_{.27}InAs_{.63}P$ (p)/ $Ga_{.27}InAs_{.63}P$ (n)/ InP

Their main electrical characteristics are reported in table 2

	Voc (Volt)	Isc (mA)	FF	η (%)
GaInAs	.29	.31	58	5.6
GaInAsP	.40	.32	63	6

Table 2- AM0 C=1 GaInAs and GaInAsP solar cell characteristics. Respective active areas: 0.04 and 0.01 cm^2. No A.R. coating.

GaInAs cell I-V characteristics measured under full AM1 (93 mw/cm2), for concentration ratio ranging from 13 to 80 are shown in figure 2. The expected contribution of the GaInAs cell to the GaAs/GaInAs tandem efficiency, derived from these measurements are given in table 3:

C AM1 93mW/cm²	η GaInAs	
	no A.R.	A.R.
165	5.9	8
113	5.6	7.6
85	5.5	7.4
65.5	5.2	7
48.4	5.1	6.9
27	4.8	6.5

Table 3-Efficiencies measured without A.R. coating are corrected to take into account the possible increase of the photocurrent densities given by a two layers A.R. coating. Using experimental results on GaAs (η_{GaAs}=24% under C=100, ref 2) a combined efficiency exceeding 31% could be obtained.

High level doping of $Ga_{.27}InAs_{.63}P$ alloy have been tried using (Sn, Te) and (Zn,Mg) as dopants for n and p type respectively in view of the realization of tunnel junctions. Incorporation of donors and acceptors have shown the following carrier densities limits:

n type Sn= 5 10^{18} cm^{-3} p type Mg= 5 10^{18} cm^{-3}
Te= 3 10^{19} cm^{-3} Zn= 1 10^{19} cm^{-3}

1 2.2 Use of Silicon substrate for GaAs epilayer elaboration[5](Fig.3).

The direct growth of GaAs on misoriented (3° off) (100) Si substrates has been performed in a MOVPE reactor. The two steps method has been adopted and optimized. The optimization concerns the choice of the temperatures at the different steps of the growth and a careful correlation between moisture content in the reactor and the crystalline quality of the GaAs layers. A detailed study by double X ray diffraction, confirmed by Raman spectroscopy showed that the crystalline quality is the higher the lower the moisture content and that this parameter is a much more critical one than for usual GaAs on GaAs growth. The depth dependence of the crystalline quality is far from beeing uniform but this quality increases rapidly and beyond ≈ 1 μm from the GaAs interface the density of dislocations ranges between 10^6-10^7 cm^{-2}. Photoluminescence spectra have been obtained which confirm the parallel improvement in crystalline quality and in electronic properties.

1.3 Analysis of results and comments

State of the art GaAs and GaInAs single cells have been elaborated*. Tests under concentration are in progress and results obtained on GaInAs cells reveal clearly the effect of a serial resistance induced by the process technology.

Concerning GaAlAs photovoltaic qualification, different experiments were carried out and give clear evidence that the occurrence of DX centers results from the particular band structure of GaAlAs alloys. However, in addition to better purity, solar cells using GaAlAs could probably be designed provided the emittor is made of p−GaAlAs.

Photovoltaic qualification of GaInAsP as a constituting material of the low energy tandem tunnel junction is not obvious because of p doping limitation.

Preliminary results concerning GaAs point contact cell show an asymptotic limitation of the efficiency by radiative recombination and gives calculated conversion efficiency greater than 30% AM1.5, C=500.

Crystalline quality and electronic properties of GaAs layers grown on Si substrate are at the very top level of the international scene. Such a material is certainly suitable for growing solar cells in the 15 % efficiency range at AM1.

1.4 Conclusions

The final goal of this project has been approached by the demonstration of a bicolor conversion system associating a GaAs cell with a GaInAs cell. Some progress are still to be made to give an entirely experimental demonstration.

GaAlAs and GaInAsP photovoltaic qualification has been precised and limitations have appeared.

The direct growth of GaAs on Si substrate permits to obtain a matching structure well adapted to photovoltaic applications.

At last ,a realistic modelization of a GaAs point contact cell is in progress.

References.

1−B. Beaumont, G. Nataf, F. Raymond and C. Vèrié, Proc. 16th IEEE Photov. Spec. Conf. San Diego 1982, 595

2−P. Basmaji, J.C. Portal, R.L. Aulombard and P. Gibart, Solid State Comm, 63, 73 (1987)

3−A. Saletes, A. Rudra, P. Basmaji, M. Leroux, J.P. Contour, P. Gibart and C. Vèrié, 19th IEEE Phot. Spec. Conf. 1987

4−P. Gibart, D. L. Williamson, B. El Jani, P. Basmaji (to be published)

5−A. Freundlich, A. Leycuras, and C. Vèrié, Ann. Chim.Fr., 1986,11,625,632 and references therein.

* One MBE GaAs cell and one LPE GaInAs cell have been set up aboard spacecraft LIPS III for a study in spatial environment. LIPS III was launched into its expected orbit some weeks ago and data analysis is in progress now.

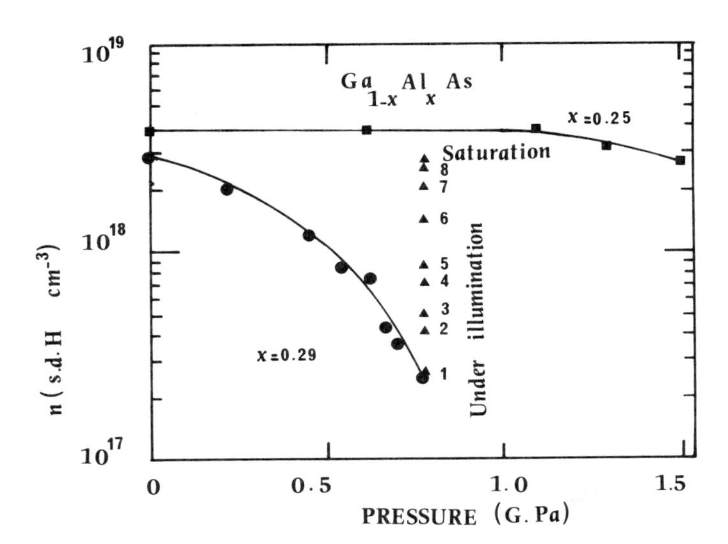

FIGURE 1 : PRESSURE DEPENDENCE OF n_Γ AS DEDUCED FROM SHUBNIKOV-DE-HAAS OSCILLATIONS. THE SOLID LINE CORRESPONDS TO THE THEORETICAL FIT ASSUMING TRANSFER OF Γ ELECTRONS TO DEEP CENTERS. THE DATA POINTS GIVE n_Γ (S.-D.-H.) UNDER ILLUMINATION.

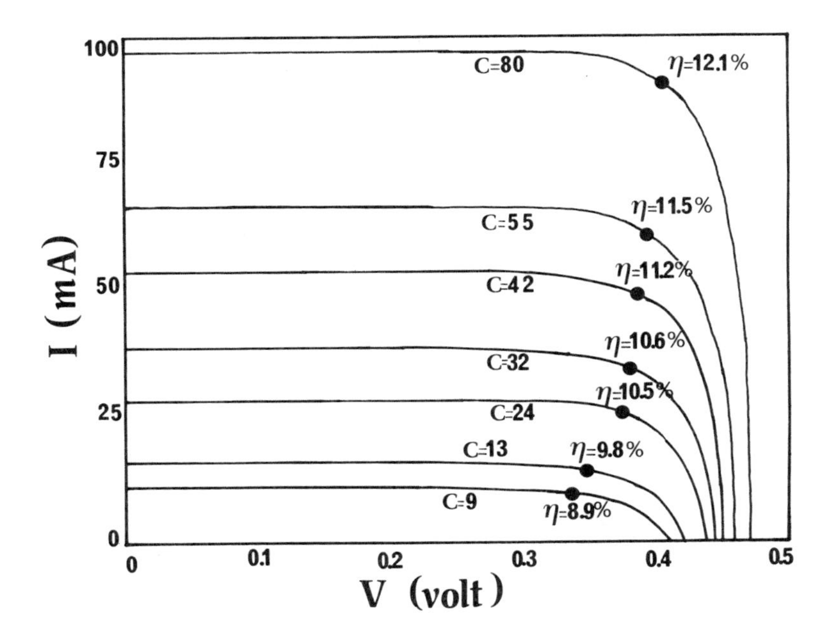

FIGURE 2 : GaInAs CELL I-V CHARACTERISTICS AND CORRESPONDING EFFICIENCIES η UNDER AM1 (93 $mwcm^{-2}$) FOR DIFFERENT CONCENTRATION RATIOS C. CELL ACTIVE AREA = 0.04 cm^2.

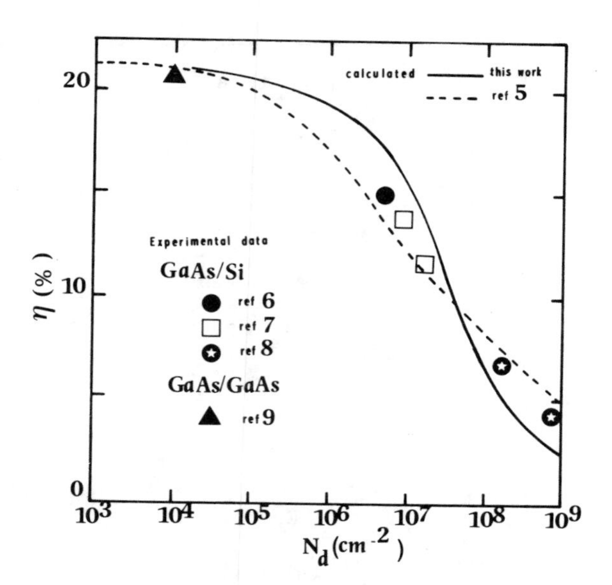

FIGURE 3 : CALCULATED EFFICIENCY η AT 300 K FOR np^+ GaAs THIN FILM SOLAR CELL WITH GaAlAs WINDOW VERSUS THE DISLOCATION DENSITY N_d.

"GaAs THIN FILM SOLAR CELLS OBTAINED BY VAPOUR PHASE HOMOEPITAXY (VPE) ON REUSABLE SUBSTRATES"

Author : M. GAROZZO

Contract number : EN3S/0072 and 0073 (I)

Duration : 36 months 1 January 1986-31 December 1988

Total Budget : ECU 1260000,

CEC Contribution: ECU 500000

Head of Project : Dr. M. Garozzo, ENEA - FARE

Contractors : ENEA

: CISE S.p.A.

Addresses : ENEA Casaccia, Via Anguillarese 301
00060 ROME (ITALY)
CISE S.p.A., Via Reggio Emilia 39
20090 SEGRATE (MI) (ITALY)

SUMMARY

This paper describes the activities carried out at ENEA and CISE laboratories to fabricate thin film single crystal GaAs solar cells reutilizing the same GaAs substrate after each growth run. An $As-Cl_3$ VPE reactor was set up and optimized to grow p-and n-doped GaAs films. Diethylzinc (DEZ) and silane were used in the gas phase for respectively p and n doping. Carrier concentration and mobility of the films are reported as a function of the doping gas molar fraction. Both p-type and n-type films resulted as suitable to be employed respectively as base and emitter in homojunction cells, even if some limitation in silane incorporation was observed at very high doping level ($>4\ 10^{18}$ cm^{-3}). The fabricated homojunction devices showed, after a thinning procedure, a short circuit current density as high as 22 mA/cm^2 without ARC, while open circuit voltage values (700-850 mV) and fill factors (0.65-0.68) seem to be limited by some not well understood shunt effect. Some preliminary lateral growth process has been carried out on properly masked <110> oriented substrates. A lateral to vertical growth rate ratio of 2-3 was observed, showing the feasibility of this approach, even if the optimum growth conditions have to be still recognized.

1. INTRODUCTION

Material cost is the main drawback of GaAs solar cells; for this reason mainly concentrator cells have been previously studied and experimented.
However another possibility to reduce the cost contribution is the fabrication of monocrystalline GaAs cells with a thickness ranging from 5 to 10 μm and suitable for flat panel solar arrays. This kind of thin devices can be made only with direct-bandgap semiconductors like GaAs, in which the light absorption occurs within few micrometers below the surface and the substrate only acts as a mechanical support. Aim of this work is the fabrication of single crystal GaAs homojunction solar cells, few micrometers thin with a conversion efficiency similar to that of thick devices. Moreover the fabrication process will allow to reutilize the same GaAs substrate for several times permitting a drastic reduction of the cell cost.
This paper describes the activities carried out by ENEA and CISE to fabricate low cost, high efficiency single crystal GaAs solar cells using the technique developed by MIT (1,2) and named Cleavage of Lateral Epitaxial Film for Transfer (CLEFT). This work is partially financed by the Commission of the European Communities on the basis of a 3 years programme. The CLEFT process allows to fabricate several thin film single crystal GaAs solar cells reutilizing the same GaAs substrate after each run. The process is based on the epitaxial lateral overgrowth of GaAs film over a photolithographically masked monocrystalline substrate (see Fig.1). The epitaxial growth starts at first vertically on the GaAs surface exposed through the openings of the mask and then it follows laterally over the mask. In order to obtain a continuous single crystal film the lateral growth rate must be much higher than the vertical one. That can be achieved by taking advantage from the dependence of the growth rate from the crystallographic orientation. Since both Liquid Phase Epitaxy and Molecular Beam Epitaxy do not show any preferential growth direction, Vapour Phase Epitaxy (VPE) must be used. Particularly the proper growth technique is the $AsCl_3$-Ga VPE, where the growth rate is much more dependent on the crystallographic orientation (3) than in Metallorganic VPE (4).
The <110> oriented single crystals are the more suitable substrates for the CLEFT process because the <110> direction shows a minimum on the GaAs vertical growth

rate by $AsCl_3$-Ga VPE, while it allows a lateral growth rate 10-20 times larger than the previous one.
Another very important regard is that the <110> plane is a cleavage plane for GaAs so it can be used for cleaving the grown film from the substrate.
After the growth of the p-n thin film homojunction, the grid fabrication and the Anti-Reflection Coating (ARC) deposition take place before the cleavage.
The upper surface of the cell is then bonded to a secondary substrate, typically glass, and cleaved from the GaAs substrate that can be so reutilized for a new growth process.
The ohmic back contact is deposited after sticking the film to the coverglass.
The activities described in this paper, carried out by ENEA and CISE laboratories in close connection, concern the set up of the VPE equipment, the VPE growth and the characterization of both n-and p-doped epilayers and the fabrication of several n^+ on p homojunctions (5).
Some preliminary results on the lateral growth process are furthemore presented.

1. VPE GROWTH OF GaAs EPILAYERS

The $AsCl_3$-Ga-H_2 chemical vapour epitaxy system (6,7) is at present a widely used technique to grow high quality GaAs epilayers. In particular VPE is extensively utilized to grow n-type layers for microwave devices. Few experiences are however reported concerning the growth of p-type materials as well as the grow of p-n junctions. An $AsCl_3$-Ga-H_2 VPE system has been designed and fabricated with the following specifications:

- possibility to grow n-type layers with a carrier concentration higher than 10^{18} cm^{-3};
- possibility to grow p-type layers with a carrier concentration lower than 10^{17} cm^{-3};
- high uniformity of the grown layers also for a thickness lower than 0.1 μm

The VPE apparatus is described in details in the paper presented by C. Flores in this Contractors Meeting.
Several GaAs films, unintentionally as well as p and n doped, have been grown with thickness ranging from 3.5 to 20 μm. All films were obtained with the following growth parameters:

Source temperature	815°C
Substrate temperature	755°C
H_2 flux	1 l/min
$AsCl_3$ molar fraction	$3\ 10^{-3}$
Growth rate (on ⟨100⟩)	0.3 μm/min

The growth conditions have been selected taking into account the requirements for a large lateral to vertical growth rate ratio.

Doping level, mobility and consequently resistivity have been measured for each film by the Van der Pauw method. As it is well known this method allows the determination of both the Hall mobility and the doping level without any particular requirement on the sample geometry and size.

The measure accuracy is affected by the ratio between the sample area and that of the four contact dots. It also depends on their distance from the sample edges, while it is independent on their reciprocal distance. Furthermore very low contact resistance is needed, to minimize the noise in the electrical measurements.

The area of the grown films was typically 2-3 cm^2; from each film two or three samples of about 1 cm^2 area were obtained and characterized to estimate the electrical homogeneity in a single growth run.

The area of each metallic dot was photolithographically defined at a value of $3.8\ 10^{-3}\ cm^2$.

Au-Zn-Au on p-type material and Au-Sn-Au on n-type and unintentionally doped films was vacuum evaporated.

Hall measurements were performed with a current not exceeding 1 mA, in order to avoid thermal drifts.

Unintentionally doped films resulted always as n-type with typical values $N_D - N_A = 2.5\ 10^{15}\ cm^{-3}$ for the doping level. The room temperature mobility was 5000 $cm^2 V^{-1} s^{-1}$, while at liquid nitrogen temperature its value ranged from 30000 to 45000 $cm^2 V^{-1} s^{-1}$, corresponding to a compensation ratio $(N_D+N_A)/(N_D-N_A)$ in the range 2-3.

The residual doping level in VPE growth processes is usually due to the incorporation of Si from the reactor walls; in our case this phenomenon seems to be limited, as results from the low found value of N_D-N_A. The high mobility values demonstrate that the oxigen incorporation during the growth is negligible.

These characteristics show that our growth conditions allow to obtain very high values for the minority carrier diffusion lenghts in the cell base.

Diethylzinc (DEZ) was used as doping source for p-type material. The liquid source was kept at the well

controlled temperature of -10°C. The zinc molar fraction in the gas phase was changed from 1 10^{-7} to 1 10^{-5}. The acceptor concentration as a function of zinc partial pressure in the reactor is shown in Fig.2. Its behaviour seems to be linear, but divided in two regions with different slopes. That is likely related to the well known different doping mechanism of zinc in GaAs.
The room temperature mobility is reported in Fig.3. As expected mobility lightly increases decreasing the acceptor concentration and its values are quite high and very close to those usually reported in the literature.
As foreseen from the measurements on unintentionally doped material, the p-type material, to be employed as cell base, shows very good transport properties and the covered range of the doping level is large enough to satisfy any requirement.
The n-doped films were obtained adding to the gas flux a mixture of SiH_4 in H_2 (250 ppm). The SiH_4 partial pressure was changed from 2.5 10^{-7} to 2 10^{-6}.
Donor concentration (Fig.4) and mobility (Fig.5) in n-doped films are reported as funtion of Si molar fraction. Also in this case the mobility values are quite high.
Increasing the Si molar fraction over 2 10^{-6} the donor concentration shows a sublinear behaviour and then saturates. A further increase affects very strongly the sample morphology and it can in some case prevent the growth. The cause of such a behaviour is not yet clear; it can be attributed or to a too high impurity concentration in the doping source or to the not well understood mechanism of Si incorporation in GaAs during the VPE process. Some experimental work is in progress to clarify this question.
This limitation of silane incorporation in VPE is however acceptable for the solar cell fabbrication in which a n^+-type layer with a donor concentration of 10^{18}-4.10^{18} cm^{-3} is sufficiently high.
An extension of this analysis will be carried out by studying the effect of the $AsCl_3$ molar fraction on the silane incorporation.
Some preliminary growth processes have been carried out on <110> oriented GaAs substrates.
These substrates were SiO_2 coated through a proper photolitographically defined mask. The mask was designed and optimized in order to verify the dependence of the lateral growth rate on the crystallographic orientations on the substrate surface. A lateral to vertical growth

rate ratio of 2 or 3 was observed, showing the feasibility of the process even if the optimum conditions to enhance this phenomenon have to be still studied.

2. n^+ ON p^+ HOMOJUNCTION DEVICES : GROWTH AND CHARACTERIZATION

In the solar cells obtained by CLEFT process the use of GaAlAs as top layer is not allowed because the aluminum compounds strongly react with the hot walls of the VPE reactor and moreover the mechanical stress between GaAs and GaAlAs makes less successful the cleaving process.
Therefore p-n shallow homojunction devices have to be fabricated without passivating top layer. In our activity we preferred the n^+/p structure to have larger values of the minority carrier diffusion length in the base region and to achieve ohmic contact without annealing.
Shallow homojunctions require very thin and heavily doped emitter layers. Very good transport properties in the base region are also necessary for obtaining high values of both open circuit voltage and short circuit current, taking also advantage from an additional Back Surface Field, easily achievable by means of a proper p^+ buffer inter-layer in between the substrate and the p layer (base), see Fig.6. This p^+ layer can also strongly reduce the contaminations diffusion from the substrate into the active layer during the growth process.
As a first step, our work was concerning up today the more simple structure without buffer layer.
Several n^+ on p junction have been epitaxially grown on zinc doped commercial GaAs substrates. The substrates had a doping level of 10^{19} cm^{-3} and were <100> oriented, 2° off toward the <110> direction.
The most delicate aspect of the growth cycle is related to the growth of the n^+ layer. n^+ layer must be indeed very thin ($\lesssim 0.2\ \mu m$) and uniform in thickness, in order to obtain high photocurrent values without any appreciable shunt effect. Furthemore, in order to avoid a too high sheet resistance, high doping levels are also required.
Fig.7 shows the cross section of a homojunction structure at 1000 of magnification. The p^+-p^- interface is very clear owing to the effect of a chemical staining (A-B etch); the n^+-p^- interface is less evident owing to the low thickness of the n^+ layer. The CLEFT devices show some meaningful differences respect to the usual GaAs solar cells also from the point of view of the ohmic

contacts.
A shallow homojunction can not indeed tolerate the grid contact thermal annealing, owing to the occurence of shunts through the very thin n^+ layer.
The back contact is deposited after sticking the film to a secondary substrate: also in this case too high temperatures are not allowed. The device fabrication has therefore required a non-alloyed process consisting in the electrochemical deposition of gold on both sides of the cell.
The front grid pattern was photolitographically defined.
In order to understand the proper growth conditions for n^+ layer and the ultimate potentiality of the n^+/p structure in terms of photocurrent density, we investigated on the behaviour of this parameter as a function of the junction depth.
Before the Anti-Reflection Coating (ARC) deposition the thickness of the n^+ layer was reduced in a well controlled way by means of a chemical etching with a very slow etch rate : H_2O : H_2O_2 : H_3PO_4 (50:1:3).
The etch rate was adjusted and calibrated at 600 A°/min at 25°C. After each etching run the short circuit current was measured by means of a solar simulator at AMO conditions. The effective absolute total value of the n^+ layer thickness was obtained by means of a "talistep" after the complete remotion of the n^+ layer. The occurance of the complete remotion is indeed releaved from the fact that the I-V curves show the typical behavior of the two series connected back polarized diodes. In Fig.8 is reported the increase of the short circuit current obtained by decrasing the n^+ layer thickness. As shown, the values obtained for n^+ layers thinner than 0.3-0.2 μm are remarkably high, taking also into account that no ARC treatment was carried out.
Several n^+ on p shallow homojunctions 1 cm^2 area GaAs solar cells have been fabricated.
The short circuit current density at AMO before the thinning procedure and without ARC ranged from 4.4 mA/cm^2 to 10.3 mA/cm^2 for n^+ layer thickness in the range from 0.80 um to 0.25 um. This behaviour seems to demonstrate that proper photocurrent values can be easily obtained if the n^+ layer thickness is well carefully controlled.
Open circuit voltage was typically 700-850 mV. The limitation to this parameter can arise or from a non-optimized surface morphology or from some microshunts occuring during the ohmic contacts fabrication or from the shunt at the cell boundaries during the cleaving

procedure. Some investigations are in progress to increase the value of this parameter up to the usual value in GaAs solar cells (about 1 V). In Fig.9 are shown the typical light (AMO) I-V curve.
The fill factor has a value of 0.67, typical in our devices.

4. CONCLUSIONS

An $AsCl_3$-VPE apparatus was set up and optimized to grow p- and n-doped GaAs epi-layers with the proper transport properties to be employed respectively as base and emitter in homojunction cells.
Several homojunction devices have been fabricated with good values of short circuit current. Open circuit voltage and fill factor are still limited by some not yet clear shunt phenomenon.
Some preliminary lateral growth process on properly masked ⟨110⟩ oriented substrates showed a vertical to lateral growth ratio of 2-3, demonstrating the process feasibility.

5. REFERENCES

1. McClelland and B.D. King, 17th IEEE Photovoltaic Specialists Conference, Orlando 1984.
2. J.C.C. Fan, C.O. Bozler and R.L. Capman, Appl. Phys. Lett., 32, 1978, 390.
3. L. Hollan and C. Schiller, J. Crystal Growth, 22, 1984, 175.
4. D.H. Reep and S.K. Ghandhi, J. Electrochem. Soc.,130, 1983, 175.
5. M. Garozzo, A. Parretta, G. Maletta, V. Adoncecchi and M. Gentili, Solar Energy Materials, 14, 1986, 29.
6. D. Effer, J. ELectrochem. Soc. 112 (1965) 1020.
7. J.U. Dilorenzo, J. Crystal Growth, 17, 1972, 189.

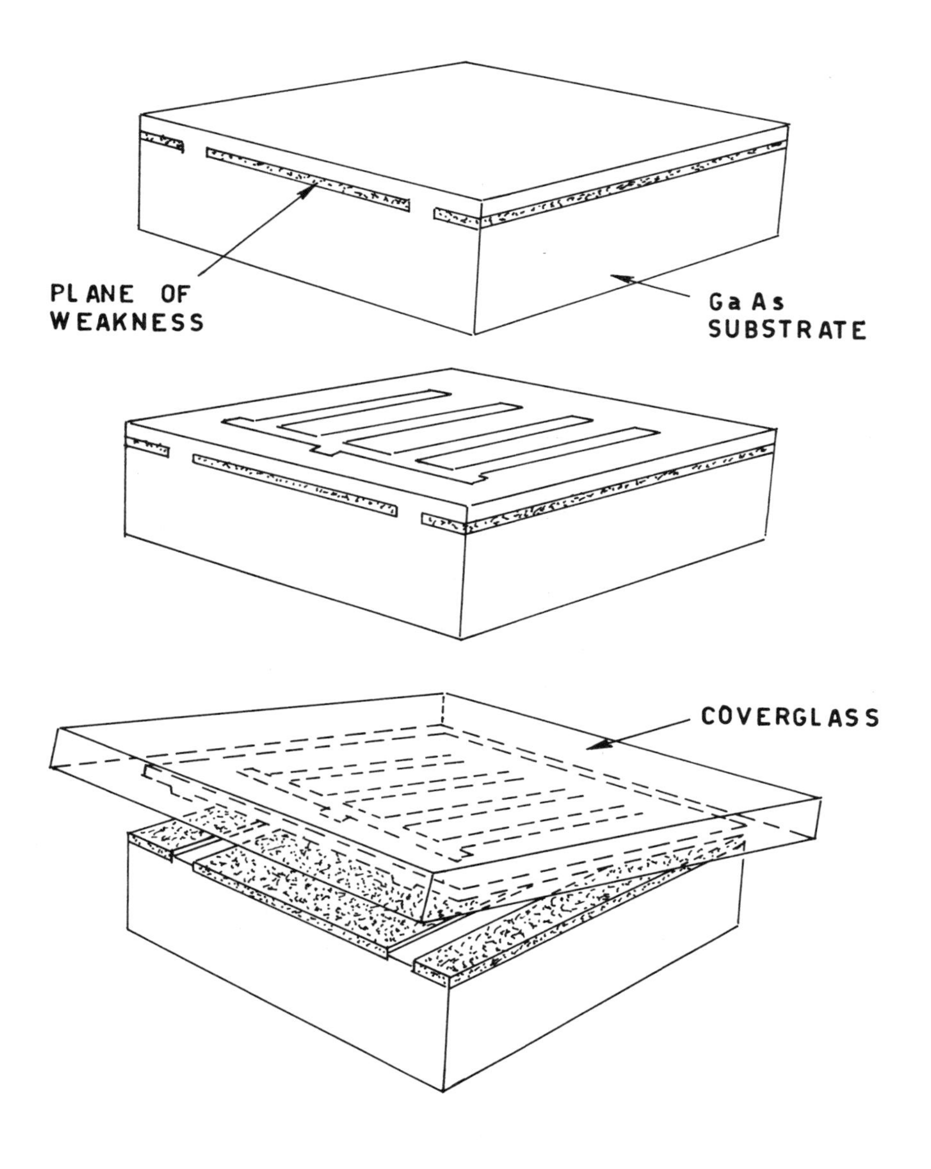

Fig.1 Schematic diagrams of the CLEFT process illustrating the separation of a cell from its substrate.

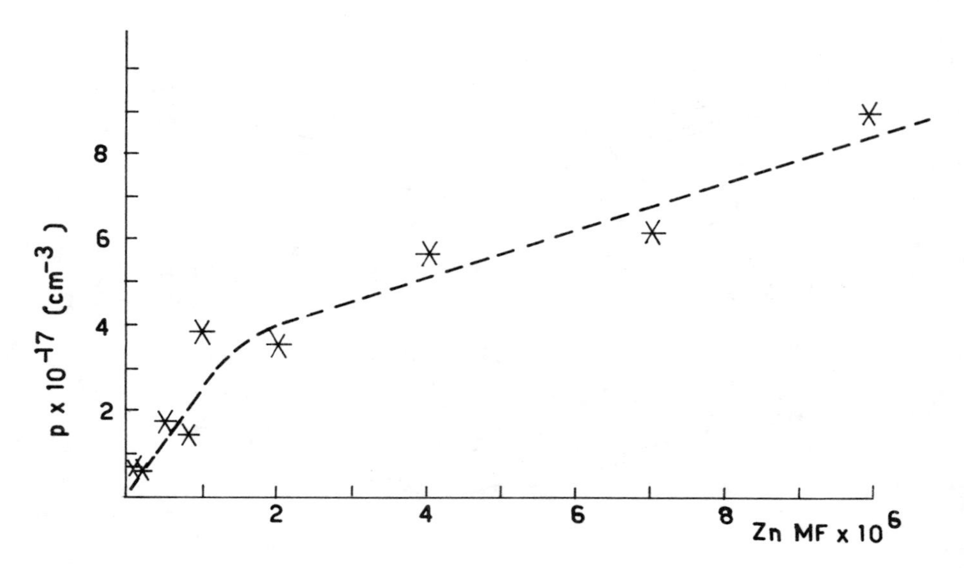

Fig.2 p doping vs. Zinc Molar Fraction.

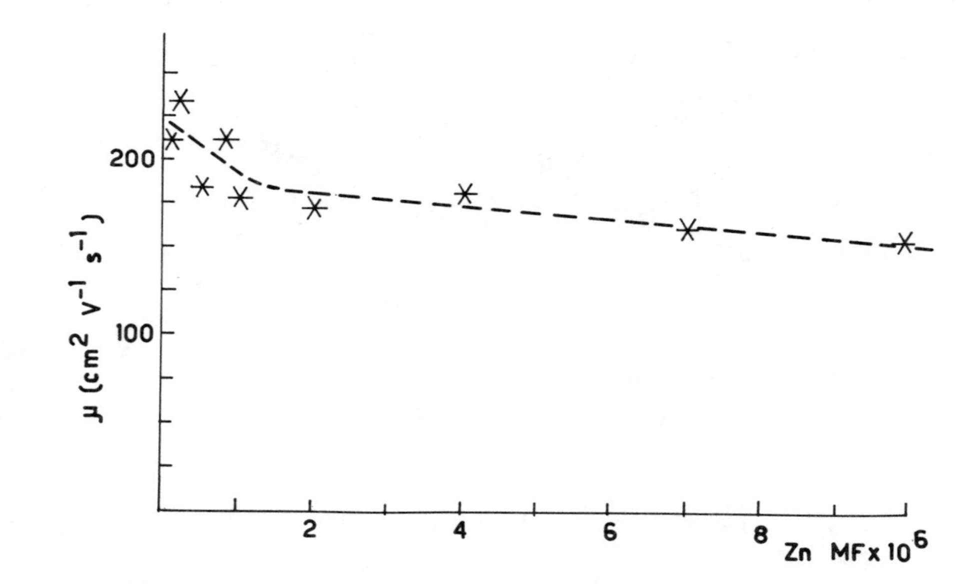

Fig.3 Mobility vs. Zinc Molar Fraction.

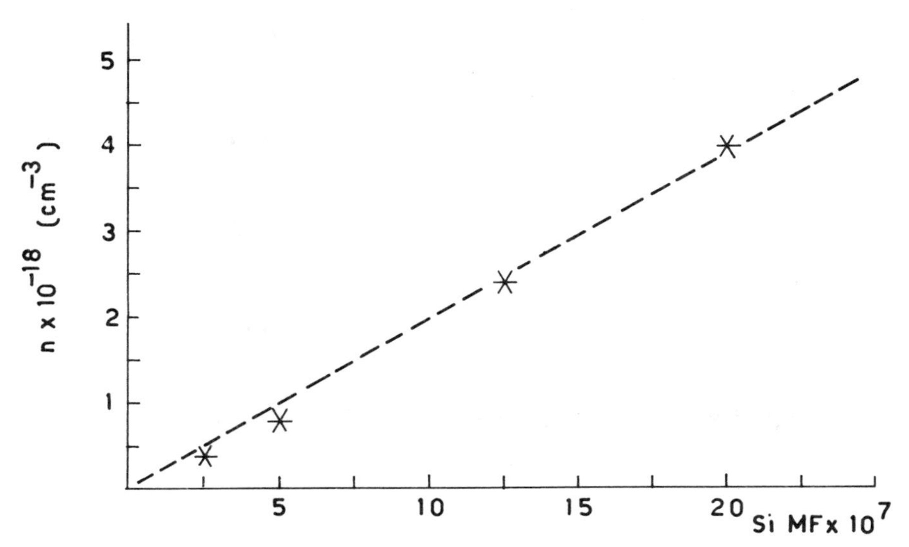

Fig.4 n doping vs. SiH_4 Molar Fraction.

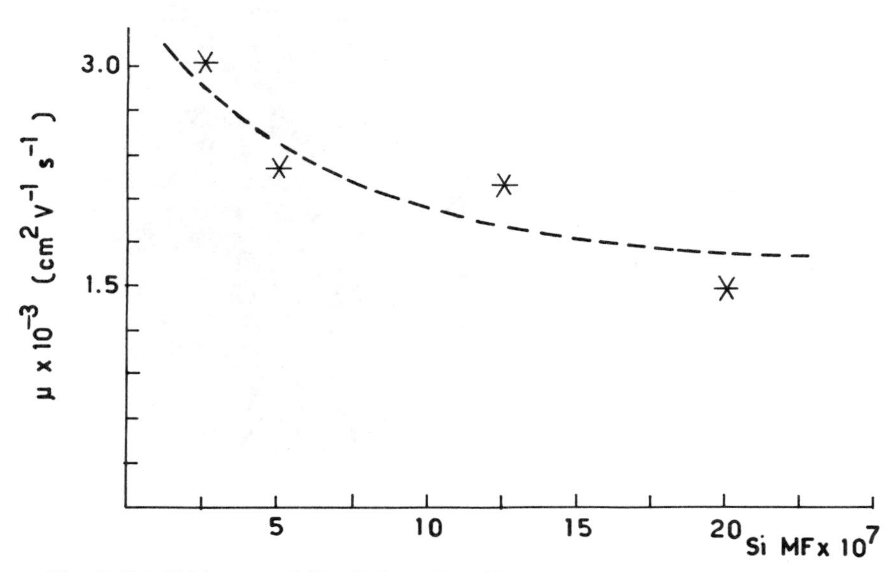

Fig.5 Mobility vs. SiH_4 Molar Fraction.

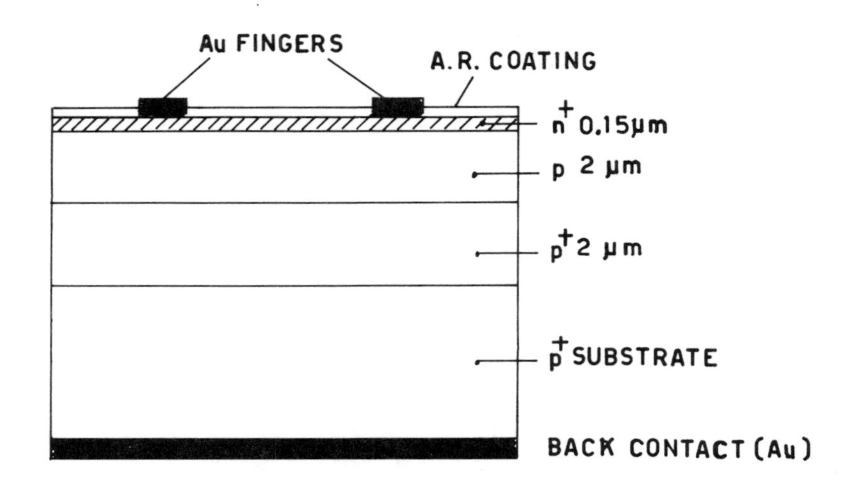

Fig.6 n^+-p-p^+ shallow GaAs homojunction solar cell.

Fig.7 Cross section of a n^+-p homojunction structure (1000x).

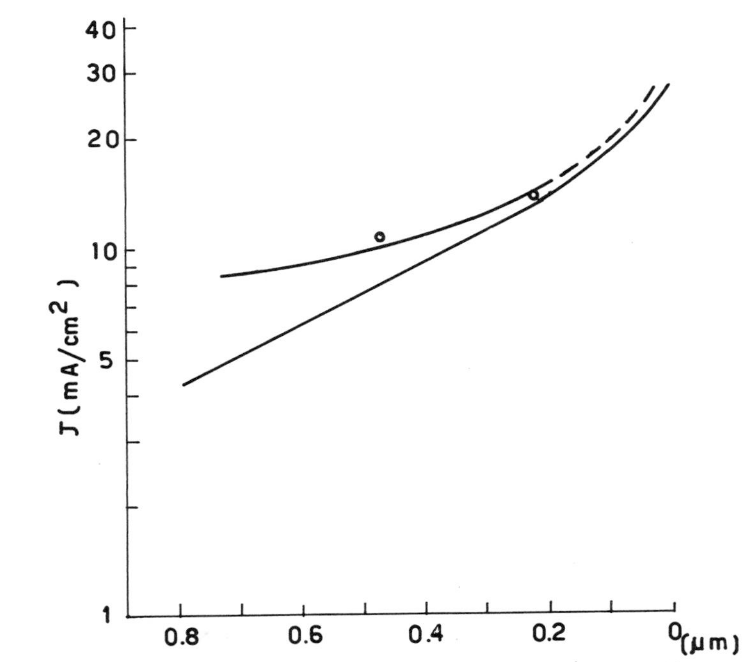

Fig.8 Short circuit current density in a n^+-p shallow homojunction vs. n^+ layer thickness.

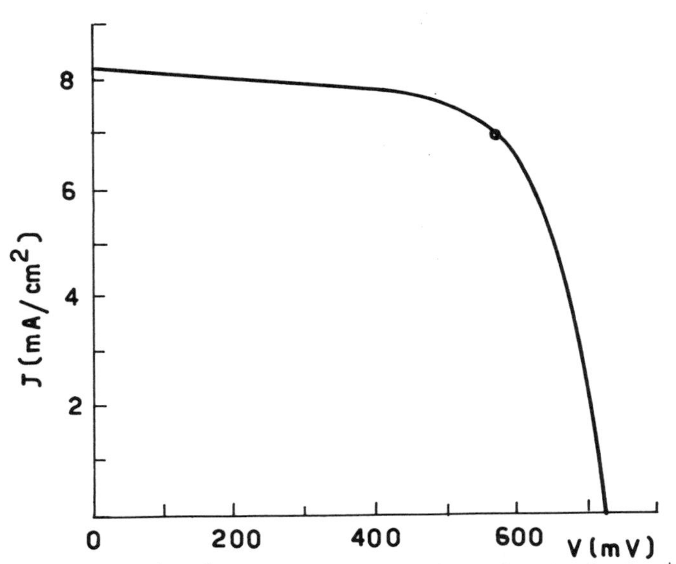

Fig.9 Light (AMO) J-V characteristics of a typical n^+-p shallow GaAs solar cell (F.F.= 0.69, No ARC).

"GAAS THIN FILM SOLAR CELLS OBTAINED BY VAPOUR PHASE HOMOEPITAXY (VPE) ON REUSABLE SUBSTRATES"

Authors: C. Flores, B. Bollani, D. Passoni

Contract number : EN3S-0072-I(S)

Duration : 36 months 1 January 1986 - 31 December 1988

Total Budget : 684.000.000 L. CEC contribution: 342.000.000 L.

Head of Project : C. Flores

Contractor : Centro Informazioni Studi Esperienze (CISE)

Address : Via Reggio Emilia 39, 20090 Segrate (MI), Italy

Summary

This paper describes the activities carried out by CISE in the frame of a collaboration with ENEA and CEC to develope low cost high efficiency GaAs solar cells fabricated using the CLEFT technique. In particular this paper reports the set-up of the vapour phase epitaxial equipment (VPE), the growth of epilayers and finally the fabrication of N^+-P homojunction GaAs solar cells.
The characterization of epitaxial layer and solar cells is described in the paper presented by M. Garozzo (ENEA) in this Contractors' Meeting.

1. INTRODUCTION

The goal of the activity, described in this progress report, is to obtain a drastic reduction of costs for GaAs solar cells.
There are several approaches to reach this goal: one of these is the fabrication of monocrystalline GaAs cells with a thickness ranging from 5 to 10 μm and conversion efficiency similar to that of thick devices. This technique named CLEFT, from the acrostic of Cleavage of Lateral Epitaxial Film for Transfer, was first developed at Lincoln Labs. of MIT and allows to fabricate several thin film single crystal GaAs solar cells reutilizing the same GaAs substrate. This technology is based on the epitaxial growth of GaAs using the vapour phase epitaxial technique (VPE). A VPE equipment has been fabricated and optimized to grow both N and P-type GaAs layers. Some N^+-P structures have been deposited and utilized to fabricate homojunction solar cells.

2. SET UP OF THE VPE EQUIPMENT

Fig. 1 shows the epitaxial vapour phase apparatus especially made to grow homojunction solar cells. The N-type doping is made using silicon that comes from a dilute mixture of Silane (250 ppm) and Hydrogen. SiH_4 is better than other type of dopants, like H_2S or H_2Se, owing to its high solubility in GaAs. Using Silane a carrier concentration up to 10^{19} cm^{-3} is in principle obtainable. Moreover SiH_4 is more stable during time and has a small interaction with the gas cylinder. The main disadvantage of Si as a dopant is that its incorporation is slight dependent by $AsCl_3$ molar fraction. As a P-type dopant a metallorganic compound of Magnesium $(C_5H_5)_2Mg$ has been considered. Mg offer several advantages in comparison with other elements, however its main drawback is that it reacts with the quarz tube. For this reason diethilzinc (DEZ) has been also utilized in spite of its high vapour pressure and doping efficiency.
Zinc can be obtained both from a metallorganic source and from a dilute mixture of H_2 and DEZ (100-1000 ppm). The main drawback, using a dilute gas mixture is the low reproducibility of the commercial supplies.
For this reason a solid source of metallorganic has been prefered.
To get low zinc concentrations from the metallorganic source, a dilution system has been fabricated. The system is made with three mass flow controllers (N. 5,6,8) and a mixing camera. Varing the fluxes is possible to obtain a dilution D given by

$$D = \frac{F_8}{F_5+F_4}$$

The dilution is essential to obtain low zinc concentration in the grown layer. In fact cooling down of DEZ does not guarantee a sufficient degree of freedom in doping variation. The dependence of DEZ vapour pressure versus temperature is given by

$$P_{DEZ} = 8.28 - 2190/T$$

while the melting point of DEZ is $-28^{\circ}C$. The zinc molar fraction (M.F.) is therefore:

$$M.F. = \frac{P_{DEZ} \cdot F_4}{F_{tot}} \cdot D$$

Both $AsCl_3$ and DEZ are dipped in thermostatic baths that can be varied

from -30° up to 20°C. Silane is maintained at ambient temperature. The Gallium comes from a solid source made with a high purity undoped polycristalline. Gas hydrogen is purified by means of palladium diffuser. The growth reactor and the source boat are made with high purity Pursil quartz. The furnace has two independent zones that can be supplied independently in order to obtain two flat thermal planes at different temperatures and separated by a zone with a constant thermal gradient.
The accuracy of the thermal profile is a key point of this technique.
A growth rate of about 0.3 μm/' has been achieved using the thermal profile in which the GaAs source is held at 815 °C and the substrate at 755 °C. These values however are not the optimum to grow very thin layers and at the same time are not optimized to achieve the best lateral growth. In fact it is reported in literature that the lateral growth rate tends to increase as the substrate temperature increases from 700 °C to 750 °C, but at this high temperatures the lateral growth becomes unstable.

We have therefore reduced the growth temperature and, to reduce the growth rate, the temperature difference between the two zones. With these new thermodinamic conditions a growth rate of 0.15 μm/' has been obtained. The growth morphology is not affected by the new temperature conditions. The morphology of the layers grown with the VPE technique is strongly affected by the environment cleanliness present in the growth reactor and by the substrate preparation before the epitaxial growth. This last step is in particular very crytical. Table 1 shows the cleaning procedure that has been optimized for this VPE process. In particular it is very important to utilize deionized water with a resistivity higher than 18 M Ωcm. Another crytical point is the drying procedure that must be done avoiding oxidation of the surface induced by the humidity condense.

3. GROWTH OF HOMOJUNCTION STRUCTURES AND SOLAR CELL FABRICATION

Some P-N junctions have been grown on GaAs substrates, zinc doped with a carrier concentration of 10^{19} cm^{-3} and an orientation (100) 2° off, toward (110). The typical structure is obtained by growing a 10^{17} cm^{-3} P-type layer about 5 μm thick and a $3 \cdot 10^{18}$ cm^{-3} N^+-type layer 0.2 μm thick. The crytical point of this epitaxial process is the growth of the N^+ layer that has to be very thin (<0.2 μm) and highly doped. We have calculated the silane deadtime that is the difference between the time in which the dopant enters the reactor and the switch-on time. This parameter is related to the lenght and to the diameter of the lines and to the flux of dopant. For a flux of 50 cc/' the dead time is about 40". Owing to this effect the zinc flux is generally switched off about 40"-50" after the silane switching on. The exact thickness of the N^+ layer cna be determined by measuring the I-V characteristic between two metallizations and by removing with a slow chemical etch, the N^+ material in between. As the N type material is completely removed, the I-V curve changes from a resistance to a back-to-back diode shape. After this change, it is possible using a "talistep", to measure the thickness of the removed material. This method has been utilized to set up the right growth conditions for the homojunction solar cells.

The homojunction device fabrication is quite different respect to the typical process utilized for the heterostructure solar cells. The main difference is that the homojunction can not tolerate the ohmic contact thermal alloying owing to the very thin N^+ layer. The device fabrication is therefore based on a non-alloyed process that consists in the electrochemical deposition of gold both on the front and on the back side of the cell. The front side is preventively masked with a photoresist layer in

order to define the grid pattern. Before the antireflection coating deposition, the thickness of the N^+ layer is adjusted by means of a very controlled chemical etching and by measuring the increase of the short circuit current as the N^+ layer thickness decreases.
Table 2 shows the device fabrication procedure.
A chemical etching with a very slow etching rate has been optimized; the composition is: $H_2O : H_2O_2 : H_3PO_4$ / 50 : 1 : 3 that shows an etching rate at 25°C of about 0.006 μm/min.
Fig. 2 shows the picture of some homojunction solar cells.

4. CONCLUSIONS

The VPE equipment has been optimized to grow homojunction solar cells. Some photovoltaic devices have been fabricated. The thermodinamic conditions for the best lateral growth have been found.

TABLE 1 - Substrate preparation

Substrate inspection
Whashing in hot trichloroethylene
Whashing in hot acetone
Whashing in hot iso-propyl alcohol
Rince in deionized water
Etch in H_2SO_4-H_2O_2-H_2O 5:1:1
Rince in deionized water
Drying with nitrogen

TABLE 2 - Homojunction solar cell fabrication process

Epitaxial growth
Lapping of the back side
Photoresist deposition
UV esposure
Developing
3 μm Au deposition
Cell cutting
N^+ layer adjustment
A.R. coating

FIG. 2 - Homojunction solar cells

STUDY ON THE APPLICABILITY OF THE IONIZED CLUSTER BEAM DEPOSITION TECHNOLOGY FOR GaAs THIN FILM SOLAR CELLS

Contract Number : EN3S/0095

Duration: 1 August 1987- 30 September 1989

Total Budget: 1.600.000 ECU, CEC contribution: 100.000 ECU

Head of Project: Dr. Dieter Bonnet

Contractor: Battelle Institut e.V.

Address: Battelle Institut e.V.
Am Roemerhof 35
D-6000 Frankfurt 90

Summary

GaAs deposition technologies recently have found increased interest under the aim of monolithically integrating GaAs and Si technologies for electronic and optoelectronic applications, including solar cells.

The potential for achieving above 10% efficiency for poly-crystalline films, close to 20% for films on Si and Ge and well above 20% for films on bulk GaAs has recently been demonstrated.

The process to be used in this project, namely the Ionized Cluster Beam (ICB) deposition gives an additional vector ("virtual temperature") to the proven processes of evaporation (MBE) and vapour deposition (MOCVD).

The project has started in August 1987. A vacuum system is being purchased, ICB sources have arrived and have been successfully tested. After set-up of the equipment, including safety precautions, deposition parameters will be elaborated, films will be deposited onto GaAs, Si, and neutral substrates in succession, evaluated and studied regarding photovoltaic properties.

Introduction

Currently the deposition of GaAs layers and epitaxial films is finding increased interest for achieving monolithic integration of GaAs and Si electronic and optoelectronic devices, and for the fabrication of high efficiency LEDs and solar cells. Worldwide work in this field has led to impressive success, regarding quality of GaAs films on foreign substrates, such as silicon.

Therefore it has to be considered worthwhile to study GaAs film deposition with the aim of high-efficiency thin film solar cells for use as stand-alone or tandem cells in conjunction with, e.g. silicon cells.

The present work is concerned with a new technology for depositing GaAs layers, which combines advantages of present technologies with a new parameter "virtual temperature" offered by the ionized cluster beam deposition process.

GaAs Layer Deposition Technology - Sate of the Art

Present deposition techniques for GaAs are based on two different process lines, both of which lead to the synthesis of GaAs from the elements on the substrate to be covered:

Molecular Beam Epitaxy (MBE) /1/ uses high or ultrahigh vacuum as the ambient for evaporating the elements Ga and As and reactively condensing them on the heated substrate. Impressive results have been achieved by this technique, including highly sensitive field effect transistors on Si and double heterostructure laser diodes on GaAs.

Metal-Organic Chemical Vapour Deposition (MOCVD) /2/ uses Ga- and As compounds, which are transported to the substrate surface by an inert gas stream at normal or reduced pressure, where the compounds are thermally decomposed and the liberated Ga and As species react forming GaAs layers.

Both technologies have been intensely studied by numerous groups worldwide and various improvements have been added, such as using gas sources instead of Ga and As sources for MBE /3,4/, introducing, e.g., AsH_3 through a suitable gas inlet, cracking the gas into H_2 and As_2 and directing the arsenic towards the substrate in high vacuum /5,6/. In the case of MOCVD, pre-cracking of the As-compound also has found interest /7/, as well as plasma activation /8/. This hes led to an interpenetration of specific aspects of both technologies. Today the main difference is the mode of arrival of the reacting species: Free-flight in high vacuum (MBE) vs. diffusion out of an inert atmosphere in front of the substrate (CVD, MOCVD).

Recently, significant progress has been made in depositing GaAs epitaxially on Si /9 - 11/, allowing the deposition of quantum well structures and the observation of room-temperature laser emission in such layers.

Ionized Cluster Beam Deposition Technology

This technology improves the MBE vacuum deposition process by improving the film synthesis process not via increased substrate temperature, but by using accelerated, clusters of 100 to 1000 atoms of Ga and/or As /12, 13/. This activation by kinetic energy - a "virtual" substrate temperature increase - has lead to improved deposition conditions for various materials, including GaAs epitaxial films /14/.

Fig. 1 shows the ionized cluster beam source fabricated in Japan. It is expected, that polycristalline thin film cells can be produced on neutral substrates at significantly decreased deposition temperatures. Fig. 2 puts ICB into the context of the other deposition processes and their different modifications.

GaAs Thin Film Solar Cells - State of the Art

A modified MOCVD process has led to GaAs solar cells made from epitaxial films on single crystalline GaAs having efficiencies of around 25% under concentrated sunlight /15 - 17/. GaAs on polycrystalline Ge has led to an efficiency of above 10% /18/ and efficiencies of around 15% have been achieved in GaAs deposited epitaxially on silicon wafers /19/.

Thin film polycrystalline solar cells deposited on neutral subtrates have achieved an efficiency of close to 10% /20/.

These - recent - results give hope for achieving improved results by the use of the ICB technology.

State of the Project

The project has only been started in August 1987. No experimental results are yet available for GaAs layers.

A high vacuum system is being built and the ICB sources have been acquired and are being tested presently. It is planned to first only deposit Ga by ICB and introduce the As by thermal evaporation, using a cracker stage for decomposing the AS_4 molecules into the more reactive As_2 molecules. This latter process has proven very advantageous in MBE epitaxy recently. Substrates will be: GaAs wafers for reference, silicon and neutral substrates (metals, glass, quartz). Solar cell structures will be made from qualified films primarily in the form of Schottky barriers and homojunctions. Comparison of their properties with state of the art results at other groups will be a major quality-criterion for the layers.

Furthermore, standard analysis, such as photoluminescence, photocapacitance, Hall-effect, deep level transient spectroscopy (DLTS) will add qualification criteria.

Literature

/1/ W.T.Tsang: Molecular Beam Epitaxy for III-V Compounds Semiconductors. Semiconductors and Semimetals, Vol 22, Pt. A, 94 - 207, 1985 (Eds. R.K.Willardson, A.C.Beer)

/2/ M.J.Ludowise: Metalorganic Chemical Vapour Deposition of III-V Semiconductors. J. Appl. Phys. 58 (1985) R31 - R55

/3/ W.T.Tsang: Chemical Beam Epitaxy of $Ga_{0.47}In_{0.53}As$/InP Quantum Wells and Heterostructure Lasers. J. Cryst. Growth 81 (1987) 261 - 269

/4/ M.B.Panish, H.Temkin, S.Sumski: Gas Source MBE of InP and $Ga_xIn_{1-x}P_yAs_{1-y}$: Materials Properties and Heterostructure Lasers. J. Vac. Sci. Technol. B3 (1985) 657 - 665

/5/ A.R.Calawa: On the Use of AsH_3 in the Molecular Beam Epitaxial Growth of GaAs. Appl. Phys. Lett. 38 (1981) 701 - 703

/6/ M.B.Panish: Gas Source MBE of GaInAs(P): Gas Sources, Single Quantum Wells, Superlattice p-i-n's and Bipolar Transistors. J. Cryst. Growth 81 (1987) 249 - 260

/7/L.M.Fraas, P.S.McLeod, R.E.Weiss, L.D.Partain, J.A.Cape: GaAs Films Grown by Chemical Epitaxy Using Thermally Precracked Trimethy-Arsenic. J. Appl. Phys. 62 (1987) 299 - 302

/8/ A.D.Huelsman, R.Reif, C.G.Fonstadt: Plasma-Enhanced Metalorganic Chemical Vapor Deposition of GaAs. Appl. Phys. Let. 50 (1987) 206 - 208

/9/ S.K.Shastry, S.Zemon, M.Oren: Epitaxial GaAs Growth Directly on (100) Si by Low Pressure MOVPE Using Low Temperature Processing. J. Cryst. Growth 77 (1986) 503 - 508

/10/ S.M.Koch, S.J.Rosner, R.Hull, G.W.Yoffe, J.S.Harris: The Growth of GaAs on Si by MBE. J.Cryst. Growth 81 (1987) 205 - 213

/11/ R.M.Lum, J.K.Klingert, B.A.Davidson, M.G.Lamont: Improvements in the Heteroepitaxy of GaAs on Si. Appl. Phys. Lett. 51 (1987) 36 - 38

/12/ I.Yamada, T.Tagaki, P.R.Younger, J.Blake: Ionized Clusters: A Technique for Low Energy Ion Baem Deposition. Proc. SPIE, 530 (1985) 75 - 83

/13/ T.Tagaki: Ionized Cluster Beam Technique. Int. Conf. Ion and Plasma Assisted Techniques - IPATT (1985)

/14/ T.Ueda, H.Takaoka, J.Ishikawa, T.Tagaki: Epitaxial Growth of GaAs Films by the ICB Technique. Proc. 11th Symp. on ISIAT, 1987, 325 - 328

/15/ L.M.Fraas, P.S.McLeod, L.D.Partain, J.A.Cape: Epitaxial Growth from Organometallic Sources in High Vacuum. J.Vac.Sci.Technol. B4 (1986) 22 - 29

/16/ S.P.Tobin et al.: A 23.7 Percent Efficient One-Sun GaAs Solar Cell. 19th IEEE Photovoltaic Specialists Conf. 1987

/17/ S.M.Vernon, V.E.Haven, S.P.Tobin, R.G.Wolfson: Metalorganic Vapor Deposition of GaAs on Si for Solar Cell Application. J. Cryst. Growth 77 (1986) 530 - 538

/18/ Sh.S.Chu, T.L.Chu, Y.X.Han: Thin-Film Gallium Arsenide Homojunction Solar Cells on Recrystallized Germanium and Large-Grain Germanium Substrates. J.Appl. Phys. 60 (1986) 811 - 814

/19/ Y.Itoh, T.Nishioka, A.Yamamoto, M.Yamaguchi: 14.5% Conversion Efficiency GaAs Solar Cell Fabricated on Si Substrates. Appl. Phys. Lett. 49 (1986) 1614 - 1616

/20/ Sh.S.Chu, T.L.Chu, W.J.Chen, H.Firouzi, Y.X.Han, Q.H.Wang: Large Grain GaAs Thin Films. Proc. 17th Photovolt. Spec. Conf., 1984, 896 - 899

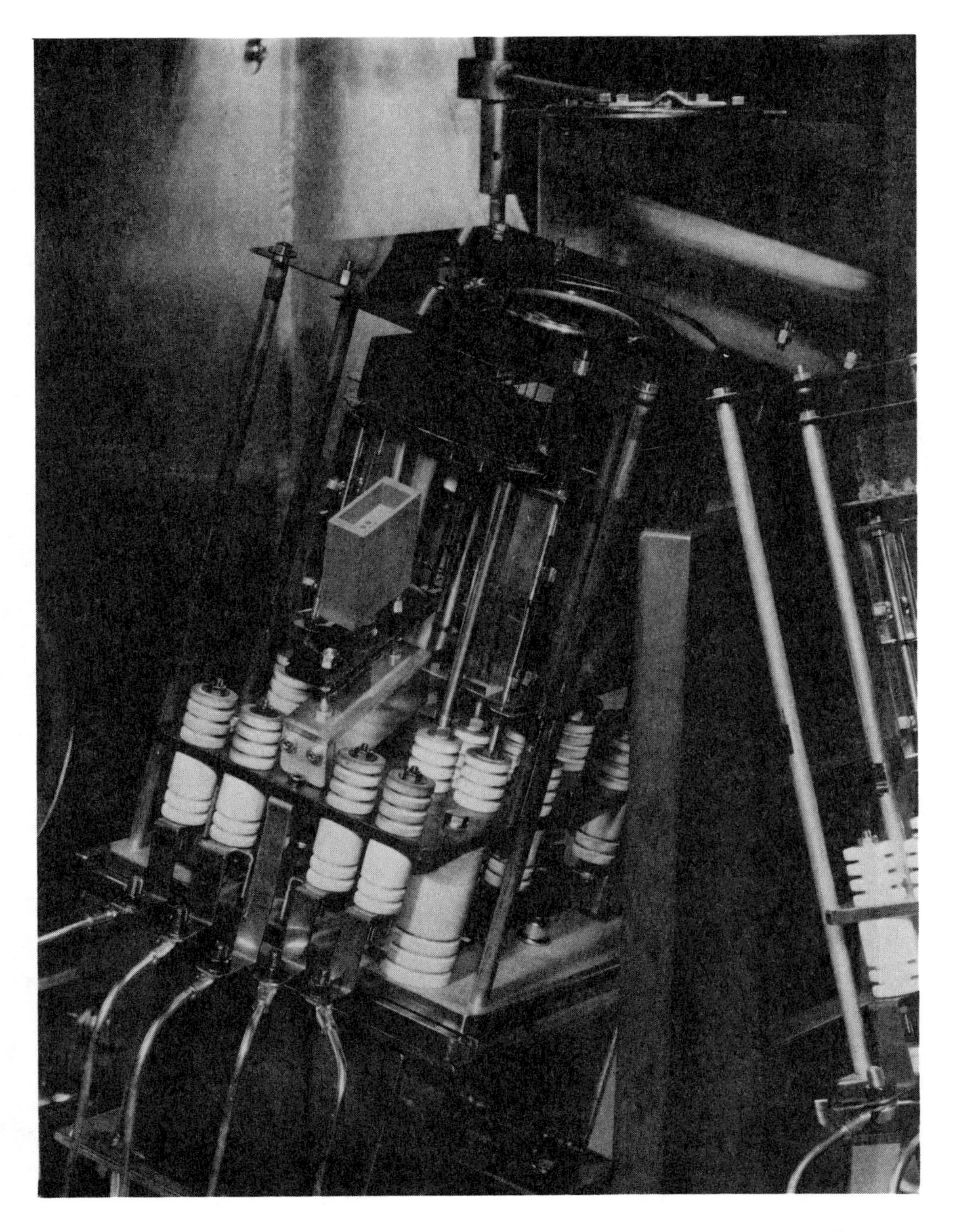

Fig. 1 Ionized Cluster Beam Source (Crucible tilted-out for better visualisation)

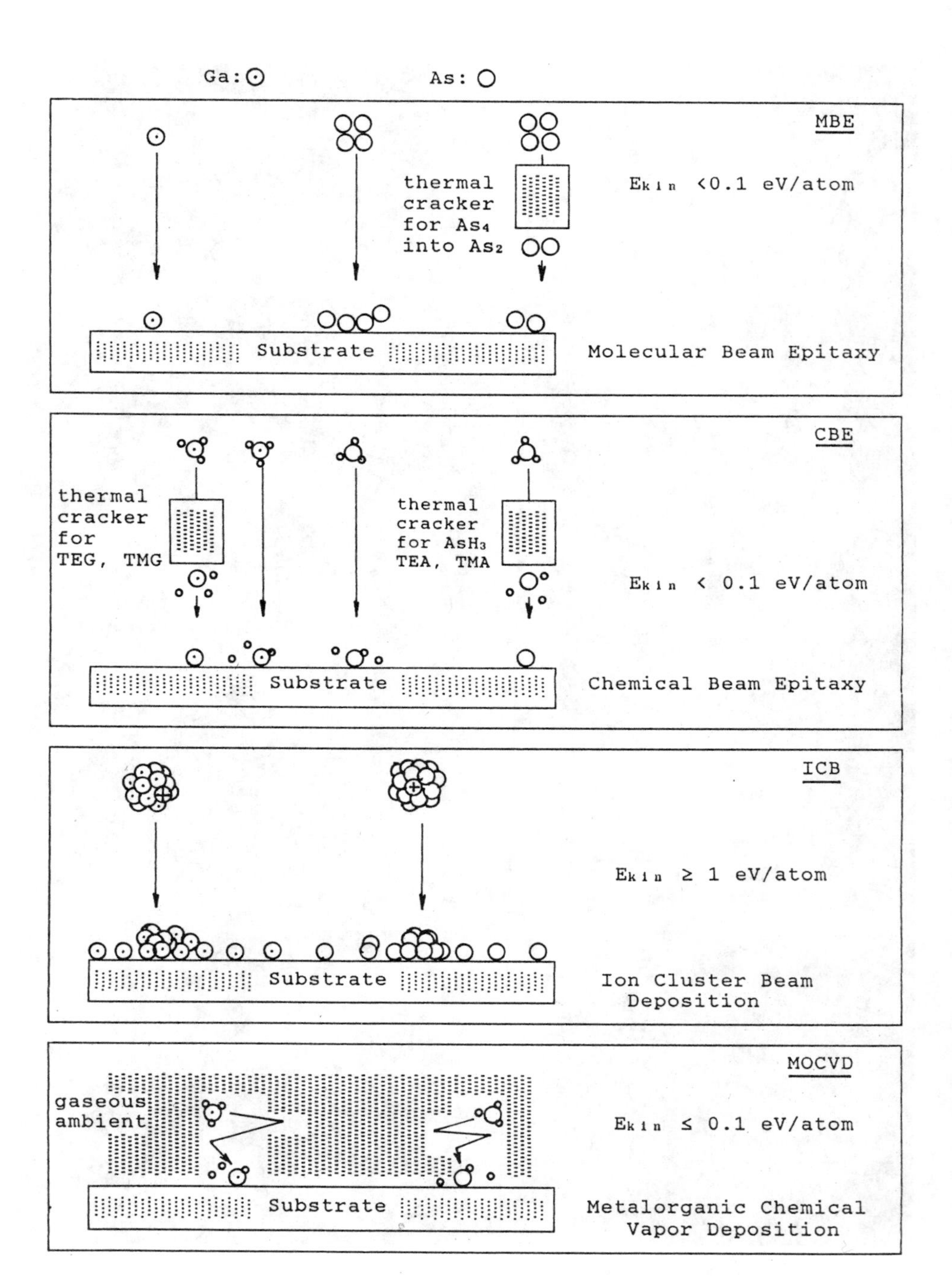

Fig 2. Schematic representation of elementary processes of different deposition techniques for GaAs. (Ga:⊙ ,As:○)

HIGH EFFICIENCY III-V SOLAR CELLS FOR USE WITH FLUORESCENT CONCENTRATORS

Author : KLAUS HEIDLER

Contract number : EN3S-0098-D (B)

Duration : 30 months 01.10.1986 - 31.03.1989

Total budget : CEC contribution: ECU 35 000

Head of project : Dr. Klaus Heidler, Fraunhofer-Institut für Solare Energiesysteme (ISE), Freiburg

Contractor : Fraunhofer-Gesellschaft e.V. (FhG)

Address : Fraunhofer-Gesellschaft e.V.
Leonrodstr. 54
D-8000 München 19

Summary

A FPC stack/GaAlAs solar cell system has been optimized using a Monte-Carlo computer model. The calculations show that the overall electrical efficiency can be increased and the thermal relaxation losses in the solar cells can be reduced significantly by matching each solar cell bandgap and dye emission spectrum individually. 7% overall electrical efficiency at an optical power concentration of 2.7 has been calculated in the best case for a three sheet stack, each sheet with a 400 · 400 · 2mm^3 geometry.

INTRODUCTION

FPC's combine the effect of a low level concentrator and a spectrum splitter (Goetzberger, 1977). Their unique feature is the ability to concentrate diffuse light too, thus eliminating the need for tracking and making low light applications highly attractive (Heidler and Kunzelmann, 1986).
FPC's reduce the solar cell area needed compared to directly exposed generators, but as their overall electrical efficiency is rather poor they need large areas.
One approach to increasing the electrical efficiency is the FPC stack. So far most of the work on the FPC was on single sheet FPC's, the stack being a complicated device with interdependences of many parameters. In recent publications (Heidler, 1982a, 1982b) we reported on a Monte-Carlo simulation programme which optimized the optical efficiency of a stack of up to three single sheet FPC's.
In this work, we report on the extension of this programme to optimize the

FPC : Fluorescent Planar Concentrator

electrical efficiency, which gives rise to some interesting results. We combined the optical data of a large variety of experimental FPC dyes with data for III-V solar cells recently published, and calculated the maximum electrical efficiency of FPC stacks with an optical power concentration of about 2.5.

THEORY

In this section we define and discuss important quantities involved in the optimization process. As most efficiencies are in terms of power, we omit the subscript "P" in order to improve readability. Quantum efficiencies are marked by a superscript "Q".
The optical power efficiency of a FPC stack is defined as

$$\eta_{opt} = \frac{\sum_{i=1}^{n} (\int P_{opt,i}(\lambda) \cdot d\lambda)}{\int P_{in}(\lambda) \cdot d\lambda} \qquad (1)$$

where $P_{opt,i}(\lambda)$ is the optical output power of the ith FPC,
P_{in} the total optical input power,
λ the wavelength, and
n the number of sheets in the stack.

If solar cells are optically coupled to the edges of the concentrator sheets, the electrical efficiency of the system can be defined analogously by

$$\eta_{el} = \frac{P_{el}}{P_{in}} = \frac{\sum_{i=1}^{n} (U_{oc,i} \cdot FF_i \cdot \int \phi_{p,i}(\lambda) \cdot \eta_Q(\lambda) \cdot e^- \cdot d\lambda)}{\int P_{in}(\lambda) \cdot d\lambda} \qquad (2)$$

where P_{el} is the total electrical output power,
$U_{oc,i}$ the open circuit voltage of the ith solar cell,
FF_i the fill factor of the ith solar cell,
$\phi_{p,i}(\lambda)$ the photon flux emitted from the edge of the ith concentrator,
$\eta_Q(\lambda)$ the quantum efficiency of the solar cell, and
e^- the electron charge.

A further physical quantity which has to be mentioned, when looking at the reduction of solar cell area by the FPC, is the optical power concentration, which is the ratio of output energy flux density to input energy flux density. As both the solar cell area needed and the system efficiency decrease with increasing optical power concentration, both efficiency and optical power concentration have to be specified, when the qualities of FPC's are described and compared.
An important loss mechanism introduced by going from η_{opt} to η_{el} is the difference between the energy of the emitted photons at the FPC edge and the energy bandgap of the solar cell. This energy difference is lost in the solar cell by thermal relaxation processes (Fig. 1). Thus, a significant increase of η_{el} can be reached, if solar cells of higher bandgaps (GaAs and GaAlAs) are used. Furthermore, the bandgaps of solar cells with a $Ga_{1-x}Al_xAs$ junction can be adapted to a specific emission spectrum by varying the aluminium content.
A useful quantity to describe the spectrum splitting feature is the solar cell efficiency with regard to the edge emission spectrum of the FPC dye, which is defined by the ratio of electrical to optical efficiency. It indicates the quality of the match between the solar cell and concentrator dye.

Similar to conventional concentrators, a solar cell area reduction factor f_r can now be defined. f_r is the factor by which the FPC reduces the solar cell area needed, while delivering the same output power as in direct exposure to the input spectrum.
Due to the spectrum splitting capability of the FPC, it is not only given by the optical concentration ratio, but by an additional factor, which is given by the ratio of the solar cell efficiency on the FPC to that for direct exposure, and is typically of the order of two.

PROGRAMME STRUCTURE AND OPTIMIZATION

In order to optimize the FPC stack/$Ga_{1-x}Al_xAs$ solar cell system, the calculation of the electrical efficiency has been included in the Monte-Carlo programme of Heidler (1982) as given in equation (2).
The most important optimization parameter is the dye concentration. Increasing it has two opposite effects on the efficiency; the resulting higher solar absorption acts to raise it, while the increasing reabsorption losses lower it. Thus an efficiency maximum exists for the best trade-off between the two contributing processes.
The dye concentration of one sheet at this maximum is not a free parameter of the overall electrical efficiency but is a function of the other dye concentrations in the stack. Thus, the dye concentrations have to be tuned successively several times to find the absolute efficiency maximum of one sequence. This procedure has to be repeated with all possible permutations of the stack. A flowchart of the programme is given in fig. 2.

DATA

As an input spectrum, the ASTM-IEC AM1.5 global reference spectrum has been used. Quantum efficiency, absorption and emission spectra of 22 dyes from different manufacturers have been measured. For the optimization, six dyes have been selected taking quantum efficiency and spectral match into account.
An extensive literature scan of the Interuniversity Micro-Electronics Center (IMEC) in Leuven (Belgium) has been evaluated to get the data of the best GaAlAs solar cells reported in the recent years. However, only cells for which the spectral response was given in the publication could be used for the calculation.
For U_{oc} and FF, the values for one sun intensity were taken. This underestimates P_{el} because in all calculated cases, the optical power concentration has been higher than 1, and the solar cell efficiency increases with irradiance in this concentration range due to an increase of U_{oc}.
As most response data were taken from graphics, the data sets have been tested by calculating the efficiency of the cells using standard input spectra (spectrum above and AM0 reference spectrum recommended by the World Radiation Laboratory) with the same AM value as given in the publication. In all cases, our calculated values were slightly lower than the published values. The calculated values were taken for optimization.

RESULTS

After careful pre-selection because of the huge number of possible stacks and the time needed for one optimization procedure, 15 different two sheet stacks and 4 different three sheet stacks have been optimized for a $400 \cdot 400\ mm^2$ surface.
The values of the best calculated two and three sheet stacks with an optical power concentration of more than 2.5 are reported in table 1, with dyes and solar cells as published by Iden (1984), Gale (1984), and Werthen (1985).

TABLE 1 Results

type	geometry	position	dye	dye concentration as given in (6)	solar cell	solar cell efficiency relative to the emission spectrum	overall electrical efficiency	optical power conc.	solar cell area reduction factor
two-sheet	400*400 *3 mm each	upper	BA339	2.25	GaAs homojunction	44.2 %	6.7 %	2.5	5.7
		lower	BA241	8.0	$Ga_{0.73}Al_{0.27}As$ homojunction	42.2 %			
three-sheet	400*400 *2 mm each	uppermost	BA241	1.5	$Ga_{0.73}Al_{0.27}As$ homojunction	42.7 %	7.0 %	2.7	6.4
		middle	BA339	2.0	GaAs homojunction	44.3 %			
		lowest	k 27	1.0	$Ga_{0.73}Al_{0.27}As$ homojunction	38.7 %			

Varying the sheet thickness and therefore the optical power concentration results only in a small change of the stack's efficiencies. The optimization of the two and three sheet stack with a sheet thickness of two and three mm, respectively, yielded an efficiency of 6.2 % at an optical power concentration of about 3.5 in the first case and 7.2 % at 1.8 in the second case. Thus, the efficiency is only a weak function of the dye concentration in the efficiency maximum.
Due to the better availability of GaAs homojunction cells, the two best stacks have been optimized with GaAs cells exclusively, where they yielded 6.6 % and 6.8 % with nearly the same optical power concentration as given in table 1. Thus, the use of GaAlAs homojunction cells results only in a small advantage, because their higher U_{OC} is partly offset by a quantum efficiency maximum typically 10 % lower than in the best reported GaAs cases. Thus, an increase of the quantum efficiency of these cells will result in a further improvement of the FPC device's efficiencies.

CONCLUSIONS

The relaxation losses of a FPC stack/GaAlAs solar cell system can be reduced significantly by using solar cells with higher bandgaps than Si and carefully matching spectral response and dye emission.
Three sheet stacks perform better than two sheet stacks.
Due to the rather low number of reported high efficiency GaAlAs homojunction cells, the dye/solar cell matching has not reached its theoretical optimum in most cases. From this point of view, further improvement of the system can be expected.
7 % overall electrical efficiency at an optical power concentration of 2.7 has been calculated. Experiments within the near future will show what efficiency is actually achievable with available solar cells.

ACKNOWLEDGEMENTS

We wish to thank the IMEC in Leuven (Belgium) for their detailed literature scan, A. Zastrow for stimulating discussions, H. Lautenschlager and F. Brucker for technical assistance and H. R. Wilson for both.

REFERENCES

Goetzberger, A., and Greubel, W. (1977). Solar energy conversion with fluorescent collectors. Applied Physics, 14, 123-139.
Heidler, K. and Kunzelmann, S. (1986). Solar wall clock with a fluorescent

planar concentrator (FPC). Proc. of the 7th EC PV Sol. En. Conf., Sevilla 1986, 201-206.
Werthen, J. G., and others (1985). Proc. of the 18th IEEE PV Spec. Conf., Las Vegas 1985, 300-303.
Gale, R. P., and others (1984). Proc. of the 17th IEEE PV Spec. Conf., Kissimmee 1984, 721-725.
Iden, R., and others (1984). BMFT Bericht, BMFT-FB-T84-164.
Heidler, K., and others (1982a). Fluorescent planar concentrator (FPC) Monte-Carlo computer model, limit efficiency and latest experimental results. Proc. of the 7th EC PV Sol. En. Conf., Stresa 1982, 682-686.
Heidler, K. (1982b). Doctoral thesis, Freiburg (FRG).

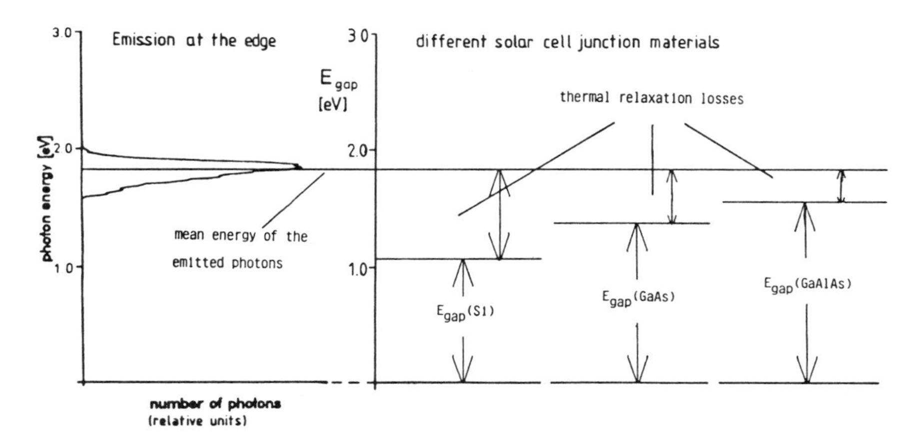

Fig. 1. Reduction of thermal relaxation losses in a FPC/solar cell system using solar cells with higher bandgaps. The losses of the dye BA339 with a mean emission photon energy of ≈ 1.8 eV, using Si, GaAs, and a GaAlAs homojunction cell with a bandgap of 1.6 eV are shown.

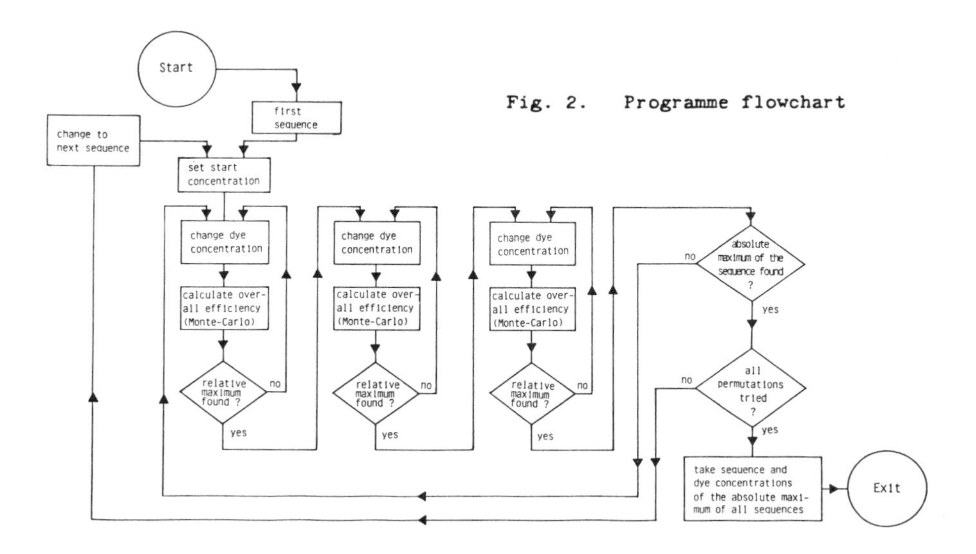

Fig. 2. Programme flowchart

HIGH EFFICIENCY III-V SOLAR CELLS FOR USE WITH FLUORESCENT CONCENTRATORS

Authors : G. Borghs, M. Mauk, P. De Meester, S. Xu, R. Baets

Contract number : EN3S-0097-B (GDF)

Duration : 12 months 1.1.1987 - 31.12.87

Total budget : 80 000 ECU CEC contribution : 40 000 ECU

Head of project : G. Borghs

Contractor : IMEC

Address : Kapeldreef, 75, 3030 Leuven, Belgium

Reporting period : 1.7.87 - 30.9.87

Summary

The present status of the AlGaAs/GaAs solar cell is briefly reviewed, with emphasis on MOCVD and MBE fabrication technology. Design and processing issues critical to the development of high efficiency devices are discussed. Methods of analysis and experimental results are presented and include solar performance data, spectral response and dark current-voltage measurements.

PROGRESS AND DEVELOPMENT OF MOCVD AND MBE AlGaAs/GaAs SOLAR CELLS

G. Borghs, R. Baets, M. Mauk, P. De Meester and S. Xu

1. Introduction. In comparison to all other photovoltaic materials, solar cells based on gallium arsenide have the highest conversion efficiencies. An efficiency of 23.7% (AM1.5 spectrum; 1 sun concentration) has recently been reported [1]. Furthermore, theoretical modeling indicates that a single junction GaAs solar cell with a one-sun conversion efficiency exceeding 25% is a realistic expectation. In this paper, we review progress in GaAs (and aluminum gallium arsenide– AlGaAs) solar cells and report on our group's activities in the development of MOCVD and MBE AlGaAs/GaAs solar cells.

The superior performance of gallium arsenide solar cells is attributed to: 1) the near optimal bandgap for conversion of the solar spectrum, 2) the strong absorption of light which in combination with the high mobility of electrons results in high quantum efficiencies, and 3) the ability to reduce surface recombination losses which in addition to further improving quantum efficiency also increases the open-circuit voltage. (The effect of surface recombination on open-circuit voltage is due to the dependence of the reverse-saturation ("dark") current of the junction on the front and back surface recombination velocity.)

Unlike silicon, where a native oxide effectively passivates the surface, devices based on GaAs ordinarily use heteroepitaxial layers for surface passivation. In the usual configuration, a thin "window" layer of $Al_xGa_{1-x}As$ (x in the range of 0.6 to 0.9) is epitaxially grown over the emitter. This lattice-matched layer reduces surface states yet is sufficiently transparent so that most photons are absorbed in the underlying GaAs layers and contribute to the photogenerated current. Similarly, though much less used at present, back surface recombination can be reduced by sandwiching a layer of AlGaAs between the substrate and the base. Here, passivation is due not only to a reduction in surface states but also to the minority carrier confinement created by the GaAs/AlGaAs heterojunction. This phenomenom is also, of course, present at the front window layer. These methods of surface passivation add a degree of complexity not seen in conventional silicon solar cells. Therefore, much of the effort in GaAs solar cell development concerns the improvement of heterostructures (e.g. AlGaAs/ GaAs junctions) to optimize device performance. Alternative methods of surface passivation are also of interest.

In practice, the advantages of GaAs and AlGaAs as photovoltaic materials have been effectively exploited due to the relatively advanced state of gallium arsenide technology. Although GaAs processing lags far behind that of silicon, it is nevertheless much more developed than processing techniques used for other promising photovoltaic materials. This is due to the wide use of GaAs in optoelectronic and microwave devices and high speed integrated circuits. Much of this technology is directly transferable to solar cell production. Three methods of crystal growth are widely used in the fabrication of GaAs and AlGaAs devices: Liquid Phase Epitaxy (LPE), Metal Organic Chemical Vapor Deposition (MOCVD) and Molecular Beam Epitaxy (MBE). Because MOCVD can be scaled-up for production of large area devices in large batch quantities, it is attractive as a cost-effective solar cell technology. MBE can produce devices with ultra-thin layers and excellent doping control and is therefore very useful in the development of novel device designs.

2. Solar Cell Structure and Processing. Figure 1a shows the basic heteroface AlGaAs/GaAs solar cell design [2] used for both MOCVD and MBE fabrication. Both *n on p* and *p on n* cells have been fabricated. An *n on p* cell is optimized with a thin ($\approx$ 200 nm), heavily-doped emitter so that most of the light absorption takes place in the base where the minority carrier (electron) diffusion length is long, thus assuring a high collection efficiency. The emitter is doped heavily to reduce series resistance. (Heavy doping of the window layer also reduces series resistance.) For *p on n* cells, the emitter is made much thicker (500 to 1200 nm) so that most of the carriers are photogenerated in the emitter where minority carrier diffusion length is long. The choice between *n on p* or *p on n* is often dictated by technological considerations. These aside, the optimum configuration (i.e. *n on p* or *p on n,* doping levels, and emitter and base thickness) depends on how effectively the surface recombination velocities can be reduced and what values of minority carrier diffusion lengths can be realized.

Figure 1b shows the shallow homojunction solar cell design [3]. This cell does not have an AlGaAs window layer and therefore its fabrication is considerably simpler than the heteroface cell. In this device, the effects of front surface recombination are reduced by using a shallow (50 to 150 nm) junction to minimize light absorbtion in the emitter. Again, the emitter is heavily doped (3 to 5 x 10^{18} cm^{-3}) to reduce series resistance. The heavy doping of the emitter also improves open-circuit voltage by reducing the emitter component of the dark current. This injected current must be minimized to reduce the deleterious effects of front surface recombination on the open-circuit voltage.

Ohmic contacts are applied either by evaporation or electroplating. For evaporation, a shadow mask or a *lift-off* technique is used to form the front contact grid. For ohmic contact to n-type GaAs, a Au:Ge alloy is evaporated; for contact to p-type GaAs, a Au:Zn alloy is evaporated. Both these alloys need to be annealed at approximately 350 °C. Alternatively, we have had success with electroplated gold contacts to both n and p-type GaAs. In this case, pure gold is plated through a photoresist pattern to form the top contact grid. This process is simpler than evaporation and requires no annealing. Its success is probably due to the heavy doping of the layers to which contact is made. Grid lines as narrow as 5 microns have been plated. Electroplating also is capable of depositing much thicker (up to ten microns) metallizations than evaporation, which is useful for reducing the series resistance of the cell. At any rate, evaporated contacts should also be thickened by electroplating to reduce series resistance lossses. We should mention that ohmic contact cannot be made directly to the window layer. A *contact* or *cap* layer of GaAs (300 nm thick) must be grown over the window layer to facilitate contact. This cap layer is selectively etched away (except under the metal contact grid) thus exposing the AlGaAs window layer. Finally, a film of silicon nitride is deposited by plasma-enhanced CVD on the front surface of the cell as a single-layer anti-reflection coating (ARC).

3. Surface Passivation Issues. As mentioned, surface passivation is effected in GaAs devices by the use of AlGaAs/GaAs heterostructures. The objective in designing the front window layer is to make it as transparent as possible while still providing sufficient surface passivation. To this end, the window layer should be thin to reduce unnecessary absorption. However, in the total optimization of the device, it should be recognized that the window layer also functions as a component of the anti-reflection coating [4, 5]. The window layer thickness is thus determined in part by including its effect in the optimization of the solar cell anti-reflection coating. We are using window layer with a thickness of about 50 nm. The aluminum fraction x of the $Al_xGa_{1-x}As$ window layer must also be optimized. Obviously, since the bandgap of AlGaAs increases with x, the window is more transparent for higher values of x. This suggests that an AlAs window be used. However, AlGaAs layers with a high aluminum content are very prone to degradation, primarily due to oxidation of the aluminum by ambient water vapor. In practice, a sufficient degree of encapsulation by the silicon nitride AR coating can protect the window layer. However, degradation problems can be minimized if the aluminum fraction is kept below about 0.9. The instability of the window layer is still a significant problem for AlGaAs/GaAs cells.

These difficulties have prompted interest in other methods of surface passivation which do not require an AlGaAs layer. We are investigating methods of chemical passivation where amorphous or polycrystaline films are applied to the solar cell to reduce surface states. As mentioned, surface passivation is less important in shallow homojunction solar cells. Significantly, efficiencies as high as 20% have been achieved in this type of cell [3]. We are fabricating shallow junction MBE cells (with no AlGaAs window layers) to determine the optimum junction depth and the effect of unconventional methods of surface passivation. In these solar cells, electroplating is particularly useful since alloying can be problematic with very thin emitters.

4. Device Analysis. Material and device analysis is indispensible for improving the solar cell design and processing. Useful material analysis includes SIMS, Auger and luminescence spectroscopy. In the present paper, however, we limit ourselves to a brief discussion of device analysis, and in particular, spectral response, and dark and illuminated current-voltage characteristics.

We have fabricated solar cells by both MOCVD and MBE. Table 1 summarizes solar cell performance data for a selected group of samples. All of the MOCVD cells shown had AlGaAs window layers and a single layer Si_3N_4 anti-reflection coating. The MBE cell was a shallow (200 nm) homojunction device with no window layer and no anti-reflection coating. For these reasons, its performance is somewhat less than the MOCVD cells.

Figure 2 shows the spectral response measurements of an MOCVD cell. This response was fairly typical. The spectral response and an independent measure of solar cell reflectivity was used to calculate the internal quantum efficiency of the device. Using a simulation program based on a simple model of an AlGaAs/GaAs solar cell, such as described in Hovel [6], it is possible to deduce the surface recombination velocity at the AlGaAs/GaAs interface. This determined value is based on the assumption that the emitter is transparent to minority carriers, i.e. bulk recombination is negligible. For the case shown, a recombination velocity of 5 x 10^3 cm/s is indicated.

Figure 3 shows a plot of open-circuit voltage versus short-circuit current for varying light intensities. Analysis of such plots has been described previously [7]. As is seen, both dark current mechanisms are present: space-charge region recombination (J_{o2} and where the slope m is approximately 2) and injection-diffusion (J_{o1} and where the slope m is 1) A value of the minority carrier lifetime τ_{SCR} in the space-charge region can be inferred from this data since

$$J_{o2} \approx \frac{q n_i W_{eff}}{\tau_{SCR}}$$

where q is the electronic charge, n_i is the intrinsic carrier concentration and W_{eff} is the effective width of the space-charge region, which can be estimated from the doping concentration of the emitter and base. In this particular case, a lifetime of 0.1 ns is indicated. This is about an order of magnitude less than "typical" values [8] and suggests room for improvement in the material quality of the junction region.

5. References.

1. S. P. Tobin, C. Bajgar, S. M. Vernon, L. M. Geoffroy, C. J. Keavney, M. M. Sanfacon, V. E. Haven and M. B. Spitzer, "A 23.7 Percent Efficient One-Sun GaAs Solar Cell" *Conf. Rec. 19th IEEE Photovoltaic Specialist Conf.* (1987) to be published.

2. J. M. Woodall and H. J. Hovel, "High Efficiency $Ga_{1-x}Al_xAs$-GaAs Solar Cells" *Applied Physics Letters* **21, 8** (15 Oct 1972) pp. 379-381.

3. J. C. C. Fan and Carl O. Bozler, "High Efficiency GaAs Shallow-Homojunction Solar Cells" *Conf. Rec. 13th IEEE Photovoltaic Specialists Conf.* (New York: IEEE Press) pp. 953-955.

4. A. Yoshikawa and H. Kasai, "Optimum Design for Window Layer Thickness of GaAlAs-GaAs Heteroface Solar Cell Regarding the Effect of Reflection Loss" *J. Applied Physics* **52(6)** (June 1981) pp. 4345-4347.

5. N. D. Nora and J. R. Hauser, "Anti-Reflection Layers for GaAs Solar Cells" *J. Applied Physics* **53(12)** (December 1982) pp. 8839-8846.

6. H. J. Hovel, Solar Cells (Academic Press, 1975).

7. P. D. De Moulin, C. S. Kyono, M. S. Lundstrom and M. R. Melloch, "Dark IV Characterization of GaAs P/N Heteroface Cells" *Conf. Rec. 19th IEEE Photovoltaic Specialists Conf.* to be published.

8. A. Fahrenbruch and R. Bube, *Fundamentals of Solar Cells* (New York: Academic Press, 1983) p. 305.

TABLE 1
Summary of Solar Cell Performance

sample number	cell type	config.	area (cm^2)	V_{oc} (mV)	J_{sc} (ma/cm^2)	FF –	eff. (%)
358	MOCVD	p/n	0.84	915	19.3	0.77	13.6
360	MOCVD	n/p	0.84	985	22.7	0.72	16.1
362	MOCVD	n/p	0.84	980	22.7	0.75	16.7
449	MOCVD	n/p	0.80	999	23.7	0.76	18.0
FT1	MOCVD	n/p	1.80	987	25.8	0.66	16.7
Q1	MBE	n/p	0.30	815	16.0	0.52	6.8

AlGaAs/ GaAs Heteroface Solar Cell

front metallization (1- 10 microns)
AR coating (50 - 100 nm)
GaAs contact (300 nm)
AlGaAs window (50 nm)
GaAs emitter (100 - 800 nm)
GaAs base (2-4 microns)
AlGaAs buffer (500 - 1000 nm)
GaAs substrate
back metallization

Figure 1a

Shallow GaAs Homojunction Solar Cell

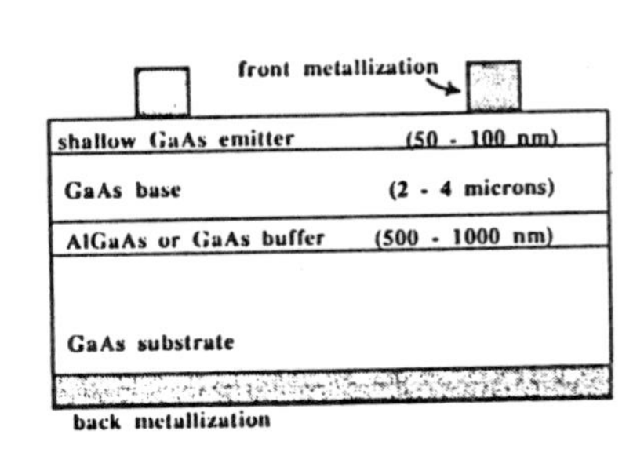

Figure 1b

EXTERNAL RELATIVE SPECTRAL RESPONSE

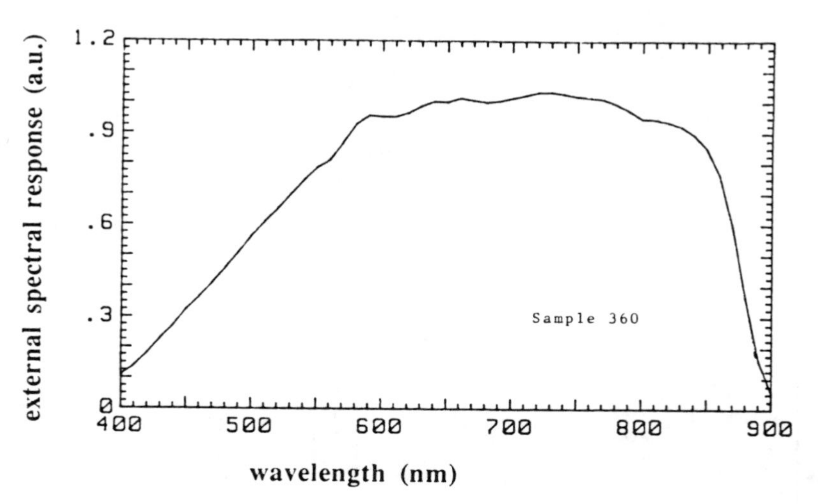

Figure 2

Short-Circuit Current vs. Open-Circuit Voltage for Varying Light Intensity

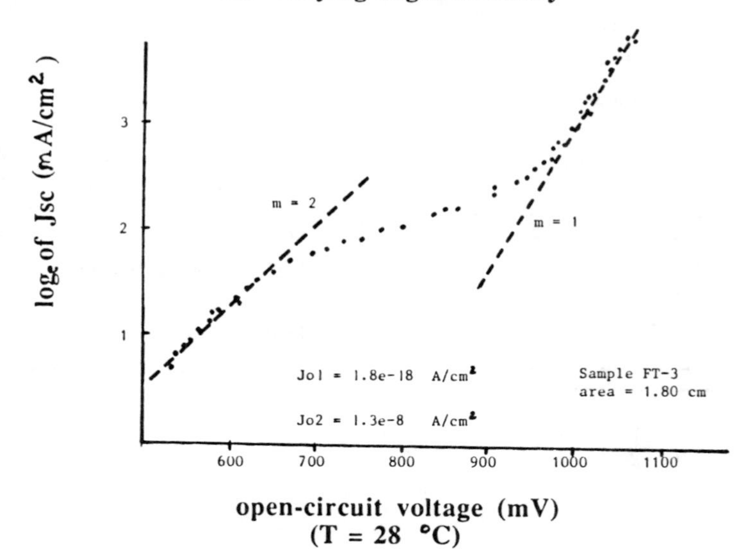

Figure 3

NON NUCLEAR R & D PROGRAMME
SUBPROGRAMME SOLAR ENERGY - PV

"HYDROGEN EFFECT ON THE ELECTRICAL PROPERTIES OF AMORPHOUS GALLIUM ARSENIDE".

Contract Number : EN-35-C2-141-F

Duration : 36 months 1 June 1986 - 31 May 1989

Financial CEC contribution : 100 % of the allowable marginal costs (526 000 FF)

Head of Project : Pr. H. CARCHANO

Contractor : Laboratoire des Matériaux Amorphes
Département de Photoélectricité

Address : Faculté des Sciences et Techniques de St.Jérôme
Avenue de l'Escadrille Normandie Niemen
13397 MARSEILLE Cedex 13 - FRANCE

Phone : 91.67.09.96

Summary

In the precedent report /6/ we have presented the results about the doping of a GaAs with molybdenium, sulfur and tellurium. With the aim of the amelioration of electronic properties we have studied the effect of hydrogenation on the electrical properties of R.F. sputtering a-GaAs. Our results show a considerable improvement of these properties upon hydrogenation. The electrical conductivity decreases by about four orders of magnitude while the activation energy increases by 0.31 eV. Doping of hydrogenated material by cosputtering with tellerium causes the electrical conductivity to increase by a factor of fifty and the Fermi level to shift by 0.18 eV with 1 % at of Te. The possibility of doping of this highly compensated a-GaAs is discussed.

1. EXPERIMENTAL TECHNIQUES

Films are deposited using a conventional diode R.F. Sputtering System in a total pressure of 50 mTorr. For the study of hydrogenation effect on the electrical properties of intrinsic material the hydrogen partial pressure ($PH2/P_T$) was varied between 0 and 20 %. The sputtering films are found to be amorphous up to substrat temperature < 50°C. Higher temperatures cause film crystallisation. This phenomenon is attribuated to the fact that the presence of hydrogen in the plasma reduces the growth rate to value which enhance crystallisation. For doping experiments, tellerium pieces of different sizes were placed on the GaAs target. The substrat temperature was maintained at 35°C and $PH2/P_T$ was kept constant at 20 %. Films composition was checked using EDAX.

2. RESULTS

The temperature variation of the d.c. conductivity is shown in figure 1. For a series of samples prepared under $PH2/P_T$ varying between 0 and 20 %. Upon hydrogenation, the room temperature electrical conductivity, σ_{RT}, is found to decrease by four orders of magnitude. This correspond to an increase from 0.39 to 0.7 eV in the activation energy, E_a, as $PH2/P_T$ varies from 0 to 20 %. The pre-exponential factor, σ_o, increase respectively from 10 to 1.1×10^3 $(\Omega.cm)^{-1}$. The variation of these three parameters as a function of $PH2/P_T$ is shown in figure 2.

For results concerning doping with tellerium of the hydrogenated material, we present in figure 3 the temperature variation of the d.c. electrical conductivity for a series of samples sputtered under $PH2/P_T = 20$ % and containing different amounts of Te. With 1 at % of dopant, σ_{RT} increases from 1.10^{-9} to 5×10^{-8} $(\Omega.cm)^{-1}$ and E_a decreases from 0.7 to 0.52 eV while σ_o drops from 1.4×10^3 to 38 $(\Omega.cm)^{-1}$. The variation of these parameters as a function of Te at. Concentration is summarised in figure 4.

3. DISCUSSION

3.1. Hydrogenation of intrinsic material

It is well known that hydrogen incorporation in amorphous silicon (a-Si) improve considerably its properties through dangling bonds compensation. It is possible, now, to produce a-Si films with only 10^{15} dangling bonds/cm^3 in the gap /1/. By analogy, we expect that hydrogenation of a-GaAs would yield the same effect. Our results given in figure 1 and 2 show the important role played by hydrogen to compensate material defects. According to the activation energy E_a and the pre-exponential factor σ_o values above room temperature, one can distinguish two cases :

- For $0 \leqslant PH2/P_T \leqslant 5$ %, the electrical conduction is carried out by hopping in the band tails localised states.
- For $10\ \% \leqslant PH2/P_T \leqslant 20$ %, σ_o and E_a values suggest a changement of the conduction mechanism to a thermally activated one in the band extended states.

This change may be explained by a narrowing of the band tails width following defect compensation by hydrogen. Our results agree well with those of Paul et al., Hargreaves et al. and Kussel /2,3,4/.

Variation of electrical conductivity at low temperature is presented in figure 5. With these results it is possible to study the effect of hydrogen on the density of midgap localised states $N(E_f)$ through MOTT relation /5/ concerning conduction by variable range hopping around Fermi level.

$N(E_f)$ is found to decrease from 2.7×10^{18} to 5.6×10^{17} $eV^{-1}cm^{-3}$ for $PH2/P_T$ equal to 0 and 15 % respectively.

3.2. Doping of hydrogenated material

After we have showed previously /6/, the faillure of doping of non hydrogenated a-GaAs, we discuss here results concerning doping of hydrogenated material. As it is indicated in figure 4, σ_{RT} increases by a factor of fifty and E_a decreases by 0.18 eV with 1 at % of Te. These results may not be explained by a widening of the band tails because photothermal deflection analyses show no increase of the band tails width. In the other side, the strong dependence of the Fermi level position for doped materials on the temperature would explain the important drop in the pre-exponential factor from 1.4×10^3 to 38 $(\Omega.cm)^{-1}$ for doped samples.

Robertson /7/ in his theoritical work showed that a-GaAs:H is highly compensated and no doping may be realized. This is caused essentially by the high energy difference between donor and dangling bond electronic levels in this material compared to a-Si:H. We think that the resulted Fermi level shift for tellerium doped a-GaAs:H samples is caused by the participation of midgap defect states what ever their type in the compensation process. However other works are necessary to confirm these primer results.

4. REFERENCES

/1/ J.A. REIMER, R.W. VAGHAN and J.C. KNIGHT, Phys. Rev. Lett. 44, 3, 193 (1980).

/2/ W. PAUL, T.D. MOUSTAKAS, D.A. ANDERSON and E. FREEMAN, in Proc. 7th Int. Conf. on "Amorphous and Liquid Semiconductors" ed. by W.E. SPEAR (Edinburgh, 1977) p.467.

/3/ M. HARGREAVES, M.J. THOMPSON and D. TURNER, J. Non Cryst. Solids 35-36, 403 (1980).

/4/ B. KUSEIL, Ph. D. MARBOURG, RFA (1984).

/5/ N.F. MOTT , J. Non Cryst. Solids 8-10, 1 (1972).

/6/ Report of this contract, june 1987.

/7/ J. ROBERTSON, Phys. Rev. B33, 6, 4399 (1986).

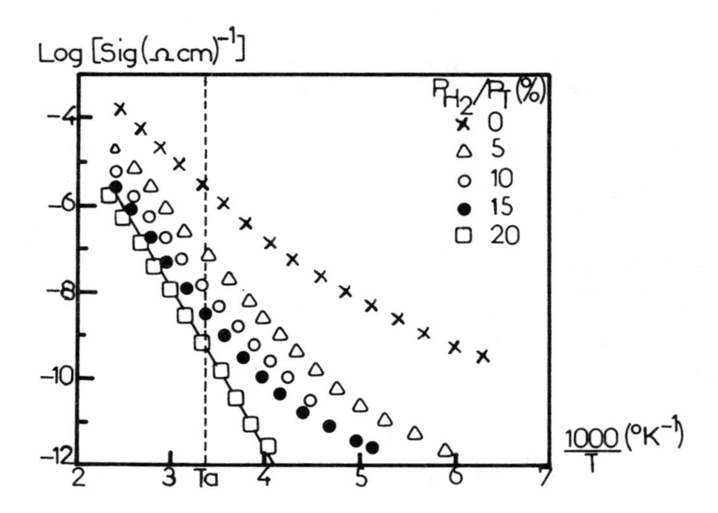

Fig.1. Temperature variation of the d.c. conductivity for a series of samples prepared with different $PH2/P_T$.

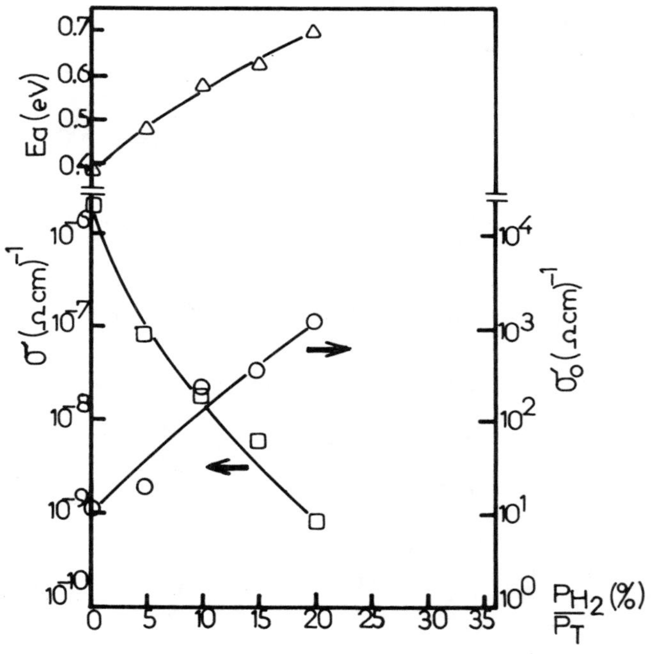

Fig.2. Electrical conductivity σ, activation energy E_a and the pre-exponential factor σ_o as a function of $PH2/P_T$.

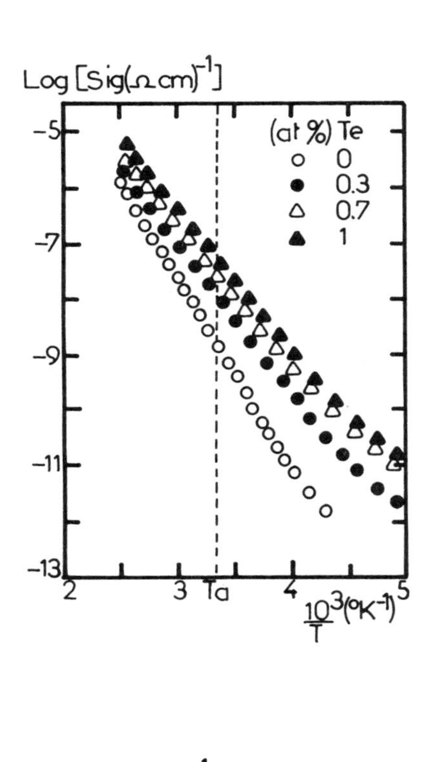

Fig.3. Temperature variation of the d.c. conductivity for a series of samples doped with different amounts of Te.

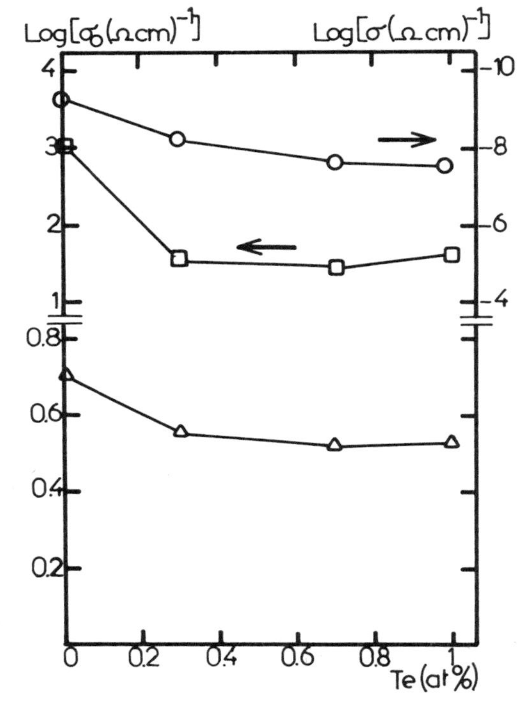

Fig.4. Electrical conductivity σ, activation energy E_a and pre-exponential factor σ_o as a function of Te at. concentration.

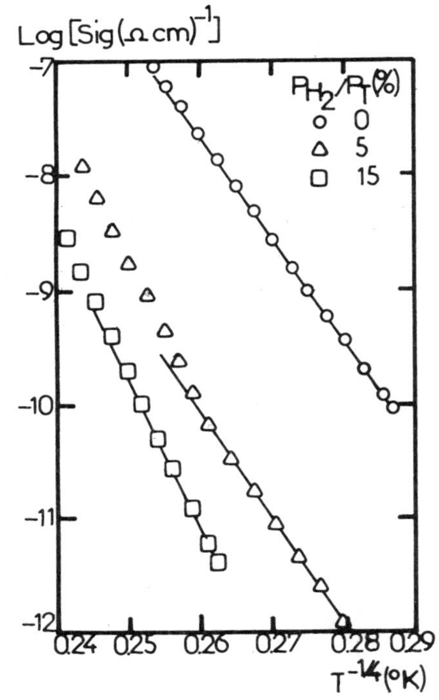

Fig.5. Electrical conductivity as a function of $T^{-1/4}$.

BAETS R.
IMEC
MAP
Kapeldreef 75
B - 3030 LEUVEN
Tel. 016/28.12.81

BAUER G.H.
Universität Stuttgart
Institut für Physikalische Elektronik
Keplerstr. 7
D - 7000 STUTTGART 1
Tel. 0711/685-7141
Telex 7-255 455 UNIV D

BEGHI M.
University of Milano
Department of Physical Chemistry
and Electrochemistry
Via Golgi 19
I - MILANO
Tel. 02/29.68.63

BOLLANI B.
CISE
Via Reggio Emilia 39
I - 20090 SEGRATE-MILANO
Tel. 02/21.67.23.72
Telex 311 643 I

BONNET D.
Battelle Institut e.V.
Physical Technologies and Devices
Am Römerhof 35
D - 6000 FRANKFURT/MAIN 90
Tel. 069/79.08-26.31
Telex 411 966

BORGHS G.
IMEC
MAP
Kapeldreef 75
B - 3030 LEUVEN
Tel. 016/28.12.81

BOUGNOT J.
Université des Sciences et Techniques
du Languedoc
Centre d'Electronique de Montpellier
Place Eugène Bataillon
F - 34060 MONTPELLIER CEDEX
Tel. 63.63.38.86
Telex USTMONT 490 944 F

BRONGERSMA H.H.
TH Eindhoven
Physics of Surfaces and Interfaces PSI
(F.O.G.)
Den Dolech 2, P.O. Box 513
NL - 5600 MB EINDHOVEN
Tel. 040/47.42.81
Telex 51163

BRUNO G.
University of Bari
Dept. of Chemistry
Via Amendola, 173
I - 70126 BARI
Tel. 080/58.07.07
Telex 812 274 CHIMIC I

BUCHER E.
Fachbereich Physik
Universität Konstanz
Postfach 5560
D - 7750 KONSTANZ
Tel. 07531/88-20.73

BUCKLEY R.W.
Twyford Church of England High School
Twyford Crescent
Acton
GB - LONDON W3 9PP
Tel. 0482/63.12.08 or 01/922.55.66

BULLOT J.
C.N.R.S.
Bât. 490
Université de Paris-Sud
F - 91405 ORSAY
Tel. 69.41.78.16

CAPEZZUTO P.
University of Bari
Dept. of Chemistry
Via Amendola, 173
I - 70126 BARI
Tel. 080/58.07.07
Telex 812 274 CHIMIC I

CARCHANO H.
Université de Droit, d'Economie
et des Sciences d'Aix-Marseille
Labo. des Matériaux Amorphes
13 rue Henri Poincaré
F - 13397 MARSEILLE CEDEX 13
Tel. 91/98.90.10
Telex FACSTJE 402 876 F

CAYMAX M.
IMEC
MAP
Kapeldreef 75
B - 3030 LEUVEN
Tel. 016/28.12.81
Telex 26 152

CHEN V.
Université des Sciences et Techniques
du Languedoc
Centre d'Electronique de Montpellier
Place Eugène Bataillon
F - 34060 MONTPELLIER CEDEX
Tel. 63.63.38.86
Telex USTMONT 490 944 F

CHRISTOU A.
Research Center of Crete
Inst. of Electronic Structure
and Laser
P.O. Box 1527
GR - HERAKLION, CRETE 711 10
Tel. 081/23.84.80-23.25.51
Telex 262 728 MPUC

CRAMAROSSA F.
University of Bari
Dept. of Chemistry
Via Amendola, 173
I - 70126 BARI
Tel. 080/58.07.07
Telex 812 274 CHIMIC I

DE CROON M.H.J.M.
Catholic University of Nijmegen
Faculty of Science
Toernooiveld
NL - 6523 ED NIJMEGEN
Tel. 08/55.88.33
Telex 48 228

DE MEESTER P.
IMEC
MAP
Kapeldreef 75
B - 3030 LEUVEN
Tel. 016/28.12.81

de ROSNY G.
L.P.I.C.M.. - Ecole Polytechnique
F - 91128 PALAISEAU
Tel. 1/69.41.82.00

FRIGERI C.
Instituto MASPEC
C.N.R.
Via Chiavari 18/A
I - 43100 PARMA
Tel. 0521/968.41
Telex 531 639 MASPEC I

DELGADO L.
Instituto de Energias Renovables
C.I.E.M.A.T. - J.E.N.
Energia Solar Fotovoltaica
Avda. Complutense 22
E - 28040 MADRID
Tel. 4.49.64.00 ext. 1672
Telex 23555 JUVIG E

FRIGERIO G.
Instituto MASPEC
C.N.R.
Via Chiavari 18/A
I - 43100 PARMA
Tel. 0521/968.41
Telex 531 639 MASPEC I

DUCHEMIN S.
Université des Sciences et Techniques du Languedoc
Centre d'Electronique de Montpellier
Place Eugène Bataillon
F - 34060 MONTPELLIER CEDEX
Tel. 63.63.38.86
Telex USTMONT 490 944 F

GAROZZO M.
E.N.E.A.
Department FARE, Division ELE-ISE
Casaccia
Via Anguillarese 301
I - 00060 ROMA
Tel. 06/69.48.40.89
Telex 613 296

FLORES C.
CISE
Via Reggio Emilia 39
I - 20090 SEGRATE-MILANO
Tel. 02/21.67.23.72
Telex 311 643 I

GILING L.J.
Catholic University of Nijmegen
Faculty of Science
Toernooiveld
NL - 6523 ED NIJMEGEN
Tel. 08/55.88.33
Telex 48 228

FOGARASSY E.
Centre de Recherches Nucléaires de Strasbourg, Laboratoire PHASE
23 rue du Loess
F - 67037 STRASBOURG CEDEX
Tel. 88/28.66.57
Telex 890 032 CNRS CRO

GINATTA F.
Elettrochimica Marco Ginatta
Via Brofferio 1
I - 10121 TORINO
Tel. 11/51.26.66
Telex 220 542 ELCHEM I

GISLON R.
E.N.E.A.
FARE/ELE/ISF
Via Anguillarese 301
I - 00060 S.M. GALERIA - ROMA
Tel. 06/69.48.33.04
Telex 613 296 ENEACA

KAVADIAS S.
Institute of Chemical
Engineering and High Temperature
Chemical Processes
P.O. Box 1239
GR - 26110 PATRAS
Tel. 061/99.30.35

GUIMARAES L.
Universidade Nova de Lisboa
Faculdaded de Ciencas e Tecnologia
Quinta da Torre
P - 2825 MONTE DE CAPARICA
Tel. 2954386

KLEEFSTRA M.
Delft University of Technology
Fac. of Electrical Engineering
Laboratory of Electrical Materials
Mekelweg 4
NL - 2628 CD DELFT
Tel. 015/78.35.96
Telex 38 151 BHTHD

GUTIERREZ M.T.
Instituto de Energias Renovables
C.I.E.M.A.T. - J.E.N.
Energia Solar Fotovoltaica
Avda. Complutense 22
E - 28040 MADRID
Tel. 1/449.64.00
Telex 23555 JUVIG E

LUQUE A.
Universidad Politecnica de Madrid
E.T.S.I. Telecomunicacion
Ciudad Universitaria
E - 28040 MADRID
Tel. 341/244.10.60

HEIDLER K.
Fraunhofer-Gesellschaft
Institut für Solare Energiesysteme
Oltmannstrasse 22
D - 7800 FREIBURG
Tel. 0761/401.64-36
Telex 07 72 510

MARGADONNA D.
Italsolar S.p.A.
Via A. d'Andrea n.6
I - 00048 NETTUNO (ROMA)
Tel. 06/980.26.44

HILL R.
Newcastle upon Tyne Polytechnic
School of Physics
Ellison Place
GB - NEWCASTLE UPON TYNE NE1 8ST
Tel. 091/232.60.02
Telex 53 519 NEWPOL G

MARIS O.
Saint-Gobain Recherche
39 quai Lucien Lefranc
F - 93304 AUBERVILLIERS CEDEX
Tel. 1/48.34.92.19
Telex 213350 F

MATARAS D.
Institute of Chemical
Engineering and High Temperature
Chemical Processes
P.O. Box 1239
GR - 26110 PATRAS
Tel. 061/99.30.35

MAUK M.
IMEC
MAP
Kapeldreef 75
B - 3030 LEUVEN
Tel. 016/28.12.81

METSELAAR J.W.
Delft University of Technology
Fac. of Electrical Engineering
Laboratory of Electrical Materials
Mekelweg 4
NL - 2628 CD DELFT
Tel. 015/78.35.96
Telex 38 151 BHTHD

MEURIS M.
IMEC
MAP
Kapeldreef 75
B - 3030 LEUVEN
Tel. 016/28.12.81
Telex 26 152

MEURIS M.
IMEC
Kapeldreef 75
B - 3030 LEUVEN
Tel. 016/28.12.81

MORENZA J.L.
Universidad de Barcelona
Dept. de Fisica Aplicada i Electronica
Av. Diagonal 645
E - 08007 BARCELONA
Tel. 318.42.60
Telex 54 549

NAGELS P.
S.C.K./C.E.N.
Materials Physics Department
Boeretang 200
B - 2400 MOL
Tel. 014/31.18.01
Telex 31922

NARDUCCI D.
University of Milano
Department of Physical Chemistry
and Electrochemistry
Via Golgi 19
I - MILANO
Tel. 02/29.68.63

PASSONI D.
CISE
Via Reggio Emilia 39
I - 20090 SEGRATE-MILANO
Tel. 02/21.67.23.72
Telex 311 643 I

PERUZZI R.
Italsolar SpA
Via A. d'Andrea 6
I - 00048 NETTUNO (ROMA)
Tel.06/980.26.44

PIZZINI S.
University of Milano
Department of Physical Chemistry
and Electrochemistry
Via Golgi 19
I - MILANO
Tel. 02/29.68.63

RAPAKOULIAS D.
Institute of Chemical
Engineering and High Temperature
Chemical Processes
P.O. Box 1239
GR - 26110 PATRAS
Tel. 061/99.30.35

REVEL G.
CNRS/LPS
1 place Aristide Briand
F - 92195 MEUDON
Tel. 1/45.34.45.50
Telex 204 135

RICHTER H.
Battelle Institut e.V.
Physical Technologies and Devices
Am Römerhof 35
D - 6000 FRANKFURT 90
Tel. 069/79.08.24.88
Telex 411 966

ROCH C.
Universidad de Barcelona
Dept. de Fisica Aplicada i Electronica
Av. Diagonal 645
E - 08007 BARCELONA
Tel. 318.42.60
Telex 54 549

RODOT M.
CNRS/LPS
1 place Aristide Briand
F - 92195 MEUDON
Tel. 1/45.34.45.50
Telex 204 135

ROMAN P.
Instituto de Energias Renovables
C.I.E.M.A.T. - J.E.N.
Energia Solar Fotovoltaica
Avda. Complutense, 22
E - 28040 MADRID
Tel. 1/449.64.00
Telex 23555 JUVIG E

ROMEO N.
Università di Parma
Dipartimento di Fisica
Via M. d'Azeglio 85
I - 43100 PARMA
Tel. 05521/463.46
Telex 530 327 UNIV PR I

SALA G.
Universidad Politecnica de Madrid
E.T.S.I. Telecomunicacion
Ciudad Universitaria
E - 28040 MADRID
Tel. 341/244.10.60

SANDRINELLI A.
University of Milano
Department of Physical Chemistry
and Electrochemistry
Via Golgi 19
I - MILANO
Tel. 02/29.68.63

SARDIN G.
Universidad de Barcelona
Dept. di Fisica Aplicada i Electronica
Av. Diagonal 645
E - 08007 BARCELONA
Tel. 318.42.60
Telex 54 549

SCHMORANZER H.
FB Physik
Universität Kaiserslautern
Postfach 3049
D-6750 KAISERSLAUTERN
Tel. 0631/205-2329
Telex 04-5627 UNIKL D

SAUSSEY H.
Photowatt International S.A.
131, route de l'Empereur
F - 92500 Rueil-Malmaison
Tel. 1/47.08.05.05
Telex 202 084 F

SCHOCK H.W.
Universität Stuttgart
Institut für Physikalische Elektronik
Keplerstr. 7
D - 7000 STUTTGART 1
Tel. 0711/685-71.41
Telex 7 255 455 UNIV D

SAVELLI M.
Université des Sciences et Techniques du Languedoc
Centre d'Electronique de Montpellier
Place Eugène Bataillon
F - 34060 MONTPELLIER CEDEX
Tel. 63.63.38.86
Telex USTMONT 490 944 F

SHAMSI T.
TEAM S.r.l.
Dipartimento di Ricerca e Sviluppo
Via G. Marconi 46/20
I - 21027 ISPRA (VARESE)
Tel. 0332/78.17.77
Telex 380 042 EUR I

SCHMIDBAUR H.
Anorganisch-chemisches Institut
Technische Universität München
Lichtenbergstrasse 4
D - 8046 GARCHING
Tel. 089/32.09-30.00

SIFFERT P.
Centre de Recherches Nucléaires de Strasbourg, Laboratoire PHASE
23 rue du Loess
F - 67037 STRASBOURG CEDEX
Tel. 88.28.66.57
Telex 890 032 CNRS CRO

SCHMITT J.P.M.
Solems S.A.
3, rue Léon Blum
Z.I. Les Glaises
F - 91120 PALAISEAU
Tel. 6/013.34.40
Telex 690549 F

SPOSITO R.
Italsolar SpA
Via A. d'Andrea 6
I - 00048 NETTUNO (ROMA)
Tel. 06/980.26.44

VAN DAAL H.J.
TH Eindhoven
Physics of Surfaces and Interfaces PSI
(F.O.G.)
Den Dolech 2, P.O. Box 513
NL - 5600 MB EINDHOVEN
Tel. 040/47.42.81
Telex 51163

VANDERVORST W.
IMEC
MAP
Kapeldreef 75
B - 3030 LEUVEN
016/28.12.81
Telex 26 152

VANDERVORST W.
IMEC
Kapeldreef 75
B - 3030 LEUVEN
Tel. 016/28.12.81

VEDEL J.
Ecole Nationale Supérieure de
Chimie de Paris
Laboratoire d'Electrochimie
Analytique et Appliquée
11, rue Pierre et Marie Curie
F - 75231 PARIS CEDEX 05
Tel. 1/354.53.95

VERIE C.H.
C.N.R.S.
Laboratoire de Physique du Solide
et Energie Solaire
Parc de Sophia Antipolis
Rue Bernard Grégory
F - 06560 VALBONNE
Tel. 93.95.42.11
Telex 970 006 F

VERLINDE A.S.
TH Eindhoven
Physics of Surfaces and Interfaces PSI
(F.O.G.)
Den Dolech 2, P.O. Box 513
NL - 5600 MB EINDHOVEN
Tel. 040/47.42.81
Telex 51163

VILATO P.
Saint-Gobain Recherche
39 Quai Lucien Lefranc
F - 93304 AUBERVILLIERS CEDEX
Tel; 1/48.34.92.19
Telex 213350 F

WILLEKE G.
IMEC
MAP
Kapeldreef 75
B - 3030 LEUVEN
Tel. 016/28.12.81
Telex 26152

WINTERLING G.
MBB GmbH
Energy and Process Tech.
ZI 12-MBB
P.O. Box 801109
D - 8000 MUNCHEN 80
Tel. 089/460.05-312
Telex 5287 470

XU S.
IMEC
MAP
Kapeldreef 75
B - 3030 LEUVEN
Tel. 016/28.12.81

YOYOTTE J.C.
Université des Sciences et Techniques
du Languedoc
Centre d'Electronique de Montpellier
Place Eugène Bataillon
F - 34060 MONTPELLIER CEDEX
Tel. 63.63.38.86
Telex USTMONT 490 944 F

ZANOTTI L.
Instituto MASPEC
C.N.R.
Via Chiavari 18/A
I - 43100 PARMA
Tel. 0521/968.41
Telex 531 639 MASPEC I